Materials for Energy Harvesting and Storage

Materials for Energy Harvesting and Storage

Editors

Xia Lu
Xueyi Lu

Basel • Beijing • Wuhan • Barcelona • Belgrade • Novi Sad • Cluj • Manchester

Editors
Xia Lu
Sun Yat-sen University
Shenzhen
China

Xueyi Lu
Sun Yat-sen University
Shenzhen
China

Editorial Office
MDPI
St. Alban-Anlage 66
4052 Basel, Switzerland

This is a reprint of articles from the Topic published online in the open access journals *Batteries* (ISSN 2313-0105), *Designs* (ISSN 2411-9660), *Energies* (ISSN 1996-1073), and *Materials* (ISSN 1996-1944) (available at: https://www.mdpi.com/topics/materials_energy_harvesting_storage).

For citation purposes, cite each article independently as indicated on the article page online and as indicated below:

Lastname, A.A.; Lastname, B.B. Article Title. *Journal Name* **Year**, *Volume Number*, Page Range.

ISBN 978-3-7258-1185-4 (Hbk)
ISBN 978-3-7258-1186-1 (PDF)
doi.org/10.3390/books978-3-7258-1186-1

© 2024 by the authors. Articles in this book are Open Access and distributed under the Creative Commons Attribution (CC BY) license. The book as a whole is distributed by MDPI under the terms and conditions of the Creative Commons Attribution-NonCommercial-NoDerivs (CC BY-NC-ND) license.

Contents

Kaito Fukushima, So Yoon Lee, Kenichi Tanaka, Kodai Sasaki and Takahiro Ishizaki
Effect of Surface Modification for Carbon Cathode Materials on Charge–Discharge Performance of Li-Air Batteries
Reprinted from: *Materials* 2022, 15, 3270, doi:10.3390/ma15093270 1

Lyubomir Soserov, Delyana Marinova, Violeta Koleva, Antonia Stoyanova and Radostina Stoyanova
Comparison of the Properties of Ni–Mn Hydroxides/Oxides with Ni–Mn Phosphates for the Purpose of Hybrid Supercapacitors
Reprinted from: *Batteries* 2022, 8, 51, doi:10.3390/batteries8060051 12

Xinli Li, Ben Su, Wendong Xue and Junnan Zhang
Synthesis and Electrochemical Characterization of $LiNi_{0.5}Co_{0.2}Mn_{0.3}O_2$ Cathode Material by Solid-Phase Reaction
Reprinted from: *Materials* 2022, 15, 3931, doi:10.3390/ma15113931 25

Wenshan Gou, Zhao Xu, Xueyu Lin, Yifei Sun, Xuguang Han, Mengmeng Liu and Yan Zhang
Boosting Lithium Storage of a Metal-Organic Framework via Zinc Doping
Reprinted from: *Materials* 2022, 15, 4186, doi:10.3390/ma15124186 35

Xinli Li, Ben Su, Wendong Xue and Junnan Zhang
The Effects of Ru^{4+} Doping on $LiNi_{0.5}Mn_{1.5}O_4$ with Two Crystal Structures
Reprinted from: *Materials* 2022, 15, 4273, doi:10.3390/ma15124273 46

Jun Liu, Yuan Liu, Jiaqi Wang, Xiaohu Wang, Xuelei Li, Jingshun Liu, et al.
Hierarchical and Heterogeneous Porosity Construction and Nitrogen Doping Enabling Flexible Carbon Nanofiber Anodes with High Performance for Lithium-Ion Batteries
Reprinted from: *Materials* 2022, 15, 4387, doi:10.3390/ma15134387 56

Ádám Révész, Marcell Gajdics, Miratul Alifah, Viktória Kovács Kis, Erhard Schafler, Lajos Károly Varga, et al.
Thermal, Microstructural and Electrochemical Hydriding Performance of a $Mg_{65}Ni_{20}Cu_5Y_{10}$ Metallic Glass Catalyzed by CNT and Processed by High-Pressure Torsion
Reprinted from: *Energies* 2022, 15, 5710, doi:10.3390/en15155710 68

Wanwan Zhang, Meiyan Cui, Jindou Tian, Pengfeng Jiang, Guoyu Qian and Xia Lu
Two Magnetic Orderings and a Spin–Flop Transition in Mixed Valence Compound $Mn_3O(SeO_3)_3$
Reprinted from: *Materials* 2022, 15, 5773, doi:10.3390/ma15165773 83

Wendy Pantoja, Jaime Andres Perez-Taborda and Alba Avila
Tug-of-War in the Selection of Materials for Battery Technologies
Reprinted from: *Batteries* 2022, 8, 105, doi:10.3390/batteries8090105 92

Zainab Haider Abdulrahman, Dhafer Manea Hachim, Ahmed Salim Naser Al-murshedi, Furkan Kamil, Ahmed Al-Manea and Talal Yusaf
Comparative Performances of Natural Dyes Extracted from Mentha Leaves, Helianthus Annuus Leaves, and Fragaria Fruit for Dye-Sensitized Solar Cells
Reprinted from: *Designs* 2022, 6, 100, doi:10.3390/designs6060100 162

Kamil Roman, Witold Rzodkiewicz and Marek Hryniewicz
Analysis of Forest Biomass Wood Briquette Structure According to Different Tests of Density
Reprinted from: *Energies* 2023, 16, 2850, doi:10.3390/en16062850 175

Tomasz Suponik, Krzysztof Labus and Rafał Morga
Assessment of the Suitability of Coke Material for Proppants in the Hydraulic Fracturing of Coals
Reprinted from: *Materials* **2023**, *16*, 4083, doi:10.3390/ma16114083 . . . . **189**

Yutaka Moritomo, Masato Sarukura, Hiroki Iwaizumi and Ichiro Nagai
Partial Oxidation Synthesis of Prussian Blue Analogues for Thermo-Rechargeable Battery
Reprinted from: *Batteries* **2023**, *9*, 393, doi:10.3390/batteries9080393 . . . . **202**

Article

Effect of Surface Modification for Carbon Cathode Materials on Charge–Discharge Performance of Li-Air Batteries

Kaito Fukushima [1], So Yoon Lee [2], Kenichi Tanaka [1], Kodai Sasaki [1] and Takahiro Ishizaki [2,*]

[1] Materials Science and Engineering, Graduate School of Engineering and Science, Shibaura Institute of Technology, 3-7-5 Toyosu, Koto-ku, Tokyo 135-8548, Japan; mb21045@shibaura-it.ac.jp (K.F.); mb21035@shibaura-it.ac.jp (K.T.); mb22019@shibaura-it.ac.jp (K.S.)

[2] Department of Materials Science and Engineering, College of Engineering, Shibaura Institute of Technology, 3-7-5 Toyosu, Koto-ku, Tokyo 135-8548, Japan; soyoon@shibaura-it.ac.jp

* Correspondence: ishizaki@shibaura-it.ac.jp; Tel.: +81-3-5859-8115

Abstract: Li-air batteries have attracted considerable attention as rechargeable secondary batteries with a high theoretical energy density of 11,400 kWh/g. However, the commercial application of Li-air batteries is hindered by issues such as low energy efficiency and a short lifetime (cycle numbers). To overcome these issues, it is important to select appropriate cathode materials that facilitate high battery performance. Carbon materials are expected to be ideal materials for cathodes due to their high electrical conductivity and porosity. The physicochemical properties of carbon materials are known to affect the performance of Li-air batteries because the redox reaction of oxygen, which is an important reaction for determining the performance of Li-air batteries, occurs on the carbon materials. In this study, we evaluated the effect of the surface modification of carbon cathode materials on the charge–discharge performance of Li-air batteries using commercial Ketjenblack (KB) and KB subjected to vacuum ultraviolet (VUV) irradiation as cathodes. The surface wettability of KB changed from hydrophobic to hydrophilic as a result of the VUV irradiation. The ratio of COOH and OH groups on the KB surface increased after VUV irradiation. Raman spectra demonstrated that no structural change in the KB before and after VUV irradiation was observed. The charge and discharge capacities of a Li-air battery using VUV-irradiated KB as the cathode decreased compared to original KB, whereas the cycling performance of the Li-air battery improved considerably. The sizes and shapes of the discharge products formed on the cathodes changed considerably due to the VUV irradiation. The difference in the cycling performance of the Li-air battery was discussed from the viewpoint of the chemical properties of KB and VUV-irradiated KB.

Keywords: carbon; surface modification; Li-air battery; charge–discharge performance

1. Introduction

Li-air batteries have attracted considerable attention as rechargeable secondary batteries due to their high theoretical energy density of 11,400 kWh/g [1]. Particularly, Li-air batteries are expected to replace Li-ion batteries for electric vehicles that require a high energy density. However, the commercial use of Li-air batteries is hindered by issues such as low energy efficiency and a short lifetime (cycle numbers) [2]. The significant issues include a high overpotential difference between charging and discharging, types of products generated from the discharging reaction, and low efficiency of the formation and decomposition of such products [2]. To clarify these issues, it is necessary to understand the role of the materials used for electrodes, the electrolytes in the batteries, and other factors [2]. The cathode material is an important part of a cell because it hosts the charging and discharging reactions, which have a significant effect on the performance of Li-air batteries. Therefore, it is important to select a cathode material that facilitates excellent battery performance.

Carbon materials have good electrical conductivity and porosity as well as oxygen-reducing abilities for the discharging reaction; therefore, they are considered as promising candidate materials for cathode materials in Li-air batteries [3–6]. Among the various properties of carbon materials, it has been reported that specific surface area, porous structure, and crystallinity are important factors affecting the performance of Li-air batteries [7–11]. From the viewpoint of the specific surface area of the cathode material, it is considered that the surface condition of carbon affects the redox reaction of oxygen because an oxygen reduction reaction occurs during discharging and an oxygen evolution reaction occurs during charging [12]. Therefore, it has been reported that there is a relationship between the functional groups on the surface, which affects the condition of the carbon surface and, subsequently, the performance of Li-air batteries [10,13,14]. However, these reports have discussed the issues combined with (1) the change in the functional groups on the surface of carbon materials (chemical change) and (2) the change in the crystallinity and morphology of carbon materials (physical change). Therefore, it was found that only the combined effect of the carbon material (physicochemical effect) considerably affected the battery performance. Consequently, the effect of functional groups on the performance of Li-air batteries has not yet been clarified.

In this study, we selected commercial Ketjenblack (KB) as a representative carbon material for the cathode material and modified the surface functional group to investigate the effect of surface modification on the charge–discharge performance of Li-air batteries. In particular, we investigated the effect of the surface condition of KB on the performance of Li-air batteries.

2. Materials and Methods

2.1. Preparation of Cathode Materials

A dispersion medium was prepared by 1-methyl-2-pyrrolidone (NMP) containing 1 wt.% polyvinylidene fluoride (PVDF) by stirring for 24 h at room temperature. The dispersion medium was controlled at a weight ratio of commercial Ketjenblack (KB: EC600JD) and PVDF of 1:9. After this step, the carbon slurry was prepared by adding NMP to this medium, and ultrasonication was performed for 30 min to obtain a homogeneous carbon slurry. The obtained carbon slurry was dropped to carbon paper (TORAY Co., TGP-H-060) and dried at 120 °C for over 12 h using a vacuum drying oven to remove any residual solvent. This cathode material is denoted as KB in this study.

2.2. Surface Modification of the Cathode Material

Surface modification of the commercial Ketjenblack was carried out by irradiating a vacuum ultraviolet (VUV) light with a wavelength of 172 nm for 30 min. After that, the cathode materials using KB after VUV irradiation were obtained through the above-mentioned procedures. Hereafter, the cathode materials using KB after VUV irradiation are denoted as KB + VUV in this study.

2.3. Evaluation of the Charge–Discharge Performance

Li-air cells were assembled for the charge–discharge tests. KB and KB + VUV were used as the cathode materials. Tetraethylene glycol dimethyl ether (TEGDME) was dried for several days over freshly activated molecular sieves (type 4 Å) before use. An ether-based electrolyte was prepared in an Ar-filled glovebox (H_2O and O_2 levels < 1 ppm) by dissolving 1 M lithium bis(trifluoromethanesulfonyl)imide in TEGDME. Swagelok-type cells (MTI, Tokyo, Japan: EQ-STC-LI-AIR) were assembled in an Ar-filled glovebox (Miwa Seisakusho, Ama, Japan) (H_2O and O_2 levels < 1 ppm) using lithium metal foil as the anode, a glass-fiber membrane (Toray Co., Kyoto, Japan: Whatman GF/A) immersed in the electrolyte as the separator, and Ni foam as the gas diffusion layer.

The cell was connected to a charging–discharging device (Hokuto Denko, Tokyo, Japan: HJ1005SD8, HJ1001SD8C), and the performance of the Li-air battery was evaluated. To evaluate the battery performance, a charge–discharge test (cutoff voltage 2.0–4.5 V, current value 0.2 A/g) and a cycling performance (cutoff voltage 2.0–5.0 V, current value 0.2 A/g, cutoff capacity 1000 mAh/g) were performed under an oxygen gas flow rate of 50 mL/min.

For cathode characterization, the cells were first transferred to the Ar-filled glove box and disassembled inside it to extract the cathodes. The cathodes were then rinsed with DME and dried for further characterization.

2.4. Characterization of Cathode Materials

To evaluate the changes in the surface conditions before and after VUV irradiation of KB, contact angle measurement (Kyowa interface: DM-501, Saitama, Japan), X-ray photoelectron spectroscopy (XPS; JEOL JPS-9010MC, Tokyo, Japan), and Raman spectroscopy (JEOL, Tokyo, Japan spectroscope: NRS-5100) were performed. For contact angle measurement, 10 µL of the ultra-pure water with a resistance of 18.2 MΩ was used. The MgKα line was used as the radiation source for XPS, and the measurements were performed under conditions of 10 kV and 25 mA; the C 1s spectrum was used for the charge-up correction of the obtained spectra. An excitation laser with a wavelength of 532.1 nm was used for Raman spectroscopy.

To analyze the products generated on the cathode surface after the charge–discharge test, the cells after the discharge tests were first transferred to the Ar-filled glove box and disassembled inside it to extract the cathodes, and the cathode material was rinsed with dehydrated dimethoxyethane (DME). After drying, the cathode material was introduced into a general-purpose atmosphere separator (Rigaku, Tokyo, Japan) in a glove box and sealed to prevent contact with the atmosphere. Under this condition, X-ray diffraction (XRD; Rigaku: Smart Lab, Tokyo, Japan) analysis was conducted without interacting with the atmosphere. CuKα radiation was used as the radiation source for XRD analysis; the measurements were performed at 40 kV and 30 mA. Additionally, Fourier transform infrared (FT-IR) spectroscopy (Shimadzu: IRTracer-100, Kyoto, Japan) and a field emission scanning electron microscopy (FE-SEM; JEOL, JSM-7610F) were conducted to evaluate the products formed on the cathode surfaces. FT-IR spectroscopy was conducted from 400 to 1600 cm^{-1}; the resolution was 4 cm^{-1}. FE-SEM was carried out at an acceleration voltage of 5.0 kV.

3. Results and Discussion

3.1. Surface Wettability

To evaluate the change in the surface condition of KB before and after VUV irradiation, static contact angles of KB and KB + VUV were evaluated. Figure 1 shows the behavior of the water droplets on the cathode materials. As shown in Figure 1a, the static contact angle (θ) of KB was approximately 126°, indicating high hydrophobicity. However, when the water droplets were dropped on the KB + VUV surface, the water droplets immediately wet the surface and spread, indicating that the KB + VUV surface exhibited high hydrophilicity (Figure 1b). Moreover, it can be inferred that a polar functional group having high affinity toward water was introduced onto the carbon surface in the cathode material after VUV irradiation.

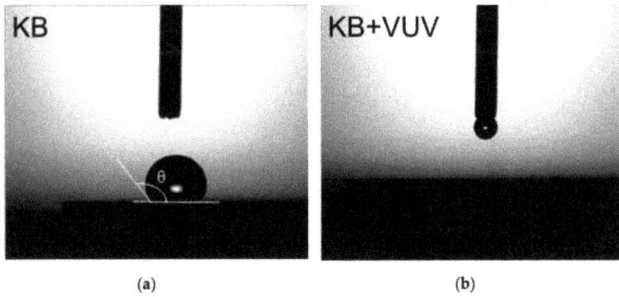

Figure 1. Behavior of a water droplet on (**a**) KB and (**b**) KB + VUV surfaces.

3.2. XPS Results

The chemical bonding states on the cathode material surfaces before and after VUV irradiation were evaluated by XPS. Figure 2 shows the O 1 s spectra of the cathode materials before and after surface modification. The atomic concentrations of O on the KB and KB + VUV surfaces were estimated to be 15.8 and 19.1%, respectively. This indicates that the atomic O contents increased by the VUV irradiation. One peak was observed in the O1s spectra before and after the VUV irradiation, and the peak was deconvoluted into four components corresponding to C=O, –OH, O–C=O, and COOH bonds at around 531.4, 532.2, 533.4, and 534.5 eV, respectively [15,16]. These bond species are present on the surfaces of both cathode materials; however, the peak area assigned to COOH bonds for KB was very small. Figure 3 shows the ratio of each binding species obtained from the results of the peak separation of the O 1s spectrum. The relative component ratio of COOH bonds for KB + VUV increased to be about 40%. The results of the atomic concentration and the relative component ratio indicate that the existence ratio of the COOH bonds on KB samples increased considerably via the VUV light irradiation. The ratio of the C=O bond decreased on the KB + VUV surface, and the ratio of the functional group derived from COOH increased. However, a slight change in the ratio of other bond species, i.e., -OH and O-C=O bonds, was observed. From these results, it is assumed that COOH was introduced by VUV irradiation; therefore, the KB + VUV surface was hydrophilized.

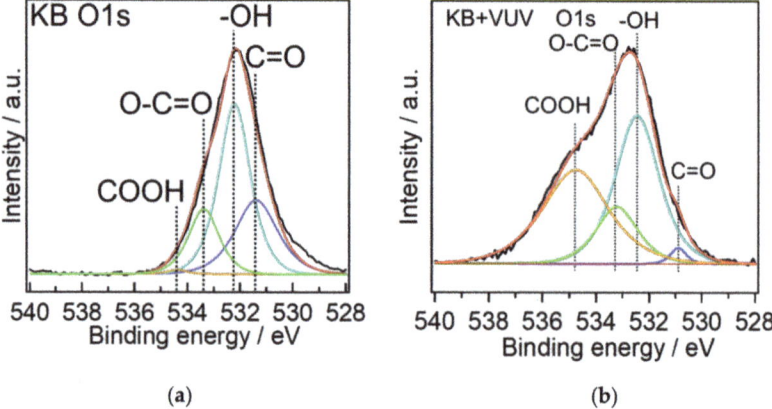

Figure 2. XPS O 1s spectra of (**a**) KB and (**b**) KB + VUV samples. Light green, light blue, blue, orange, and red lines show O-C=O, -OH, C=O, -COOH bonds, and overall fitting, respectively.

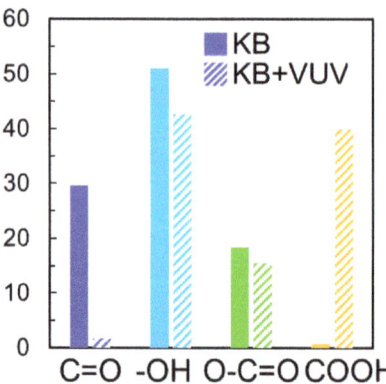

Figure 3. Relative component ratios of chemical bonds for KB and KB + VUV.

3.3. Raman Spectroscopy

The change in the crystallinity of the KB and KB + VUV was evaluated through Raman spectroscopy. The obtained Raman spectra are shown in Figure 4. The spectra in Figure 4 show two peaks at around 1350 and 1590 cm^{-1} before and after surface modification, respectively. These peaks indicate that the D band at around 1350 cm^{-1} was caused by the defect structure in the crystal plane of carbon and the G band at around 1590 cm^{-1} was due to the graphite structure. The peak intensity ratio (I_D/I_G) of the G and D bands is used as an index of the crystallinity of the carbon material [17]; a larger value of this ratio indicates that there are more defects in the material [18]. The I_D/I_G ratios of KB and KB + VUV were almost the same. Therefore, it is considered that the surface modification by VUV irritation had almost little effect on the crystallinity of the carbon material; however, it could change only the functional groups existing on the surface of the cathode material, as shown in Figures 2 and 3.

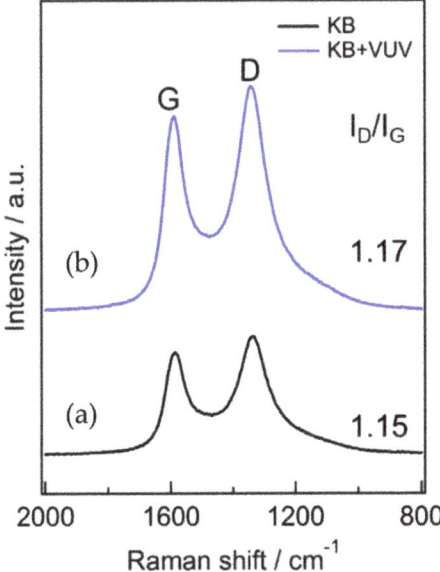

Figure 4. Raman spectra of (a) KB and (b) KB + VUV.

3.4. Charge–Discharge Performances

The full charge–discharge curves are shown in Figure 5. The discharge capacity was estimated to be ~9500 mAh/g for KB and ~8850 mAh/g for KB + VUV, indicating that KB had a slightly larger discharge capacity than KB + VUV. The discharge voltage was ~2.7 V for both samples, and no change was observed before and after surface modification. The charging overpotential was decreased by the surface modification. In addition, the results of the cycling performance are shown in Figures 6 and 7. The discharge capacity of KB began to decrease from the 14th cycle and decreased abruptly after the 15th cycle. After surface modification, the discharge capacity of KB + VUV decreased from the 18th cycle; however, the capacity was maintained until the 37th cycle. This difference in the cycling performance can be attributed to the introduction of hydrophilic functional groups on the carbon surface after surface modification, which has an effect on the charging–discharging reaction in the Li-air battery. Therefore, the change in the surface states of both cathode materials before and after charge–discharge tests were investigated in detail.

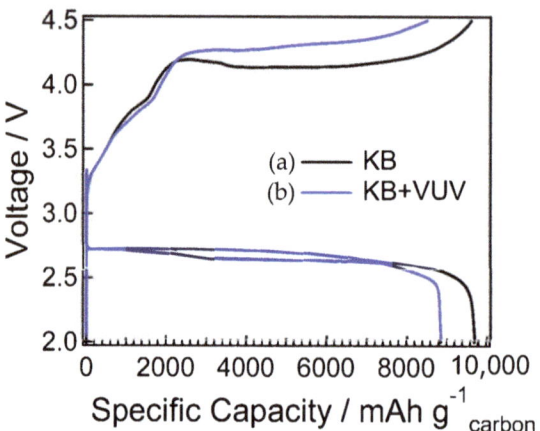

Figure 5. Full discharge and charge curves of Li-air batteries using (a) KB and (b) KB + VUV as cathodes.

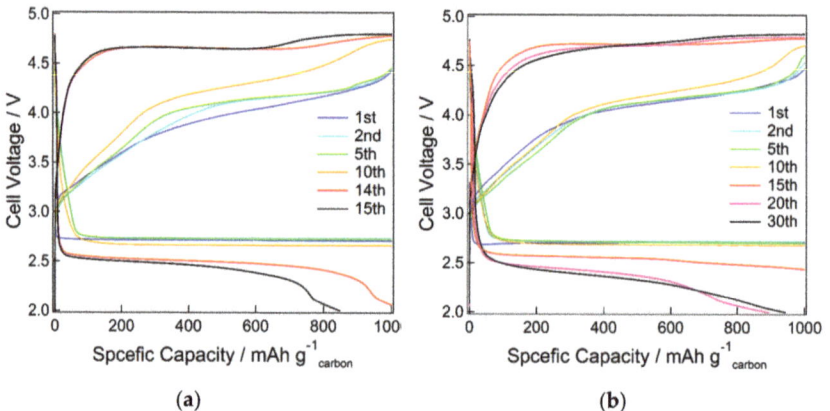

Figure 6. Cycling performance of Li-air batteries using (a) KB and (b) KB + VUV as cathodes at a limited capacity of 1000 mAh g^{-1}.

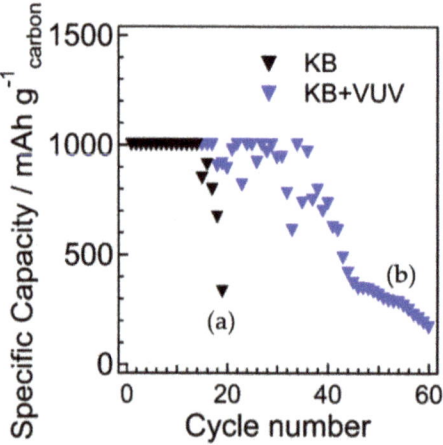

Figure 7. Cycling performance as a function of cycle numbers for Li-air batteries using (a) KB and (b) KB + VUV as cathodes at a limited capacity of 1000 mAh g^{-1}.

3.5. Analysis of the Cathode Material Surfaces before and after Charging and Discharging

XRD analysis was performed to examine the discharge products generated on the cathode surface before and after surface modification. Figure 8 shows the XRD patterns of the samples. It has been reported that the crystallinity of lithium peroxide, which is a discharge product, changed depending on the surface states of carbon [10]. Therefore, we unified the discharge time; the discharge capacity was measured at 5.0 mAh to observe the surface states of the cathode materials. On both XRD patterns, three clear peaks attributable to carbon paper were observed at around $2\theta = 26°$, $43.5°$, and $54°$. In addition to these peaks, on the XRD patterns of both cathode material surfaces, a few peaks assigned to lithium peroxide were observed at around $2\theta = 33°$, $35°$, and $59°$, indicating that the generated discharge product was mainly lithium peroxide. Additionally, in the XRD pattern of KB + VUV, a peak assigned to lithium carbonate was also observed at around $2\theta = 50°$. This means that the lithium carbonate was also considered to be formed on the cathode sample for KB + VUV. It is considered that the presence of the lithium carbonate formed on the cathode sample for KB + VUV during discharge induced the increase in the overpotential for the charging process of KB + VUV.

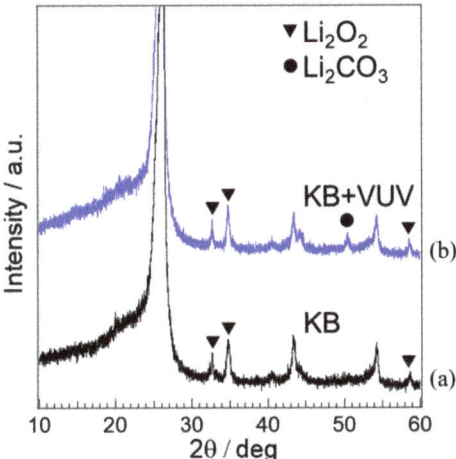

Figure 8. XRD patterns of (a) KB and (b) KB + VUV surfaces after the 1st discharge test.

The surface states of the cathode materials after charging and discharging were observed using FE-SEM. The SEM images are shown in Figure 9. In the case of KB, thin-film substances were formed on the surface after the 1st discharge cycle (Figure 9a); however, these products were not present on the surface of the cathode material after charging (Figure 9d). Therefore, the discharge products were formed during the 1st discharge cycle, and were decomposed upon further charging. The shape of the products generated on KB during discharge tended to change after the 5th (Figure 9b) and 10th cycles (Figure 9c); large and assembled particles were observed on the cathode material at the 10th discharge cycle. After the 5th charge cycle, the discharge products were not significantly decomposed, and existed on the thin films as flakes. Needle-shaped or flat plate-shaped substances were produced as the discharge products delivered from the 1st discharge cycle of KB + VUV (Figure 9g); the shape of this substance was different from that of the substance formed from KB. After charging, the generated products were rarely observed on the surface of the cathode material, and it was confirmed that the discharge products were decomposed by the charging reaction (Figure 9j); the same decomposition trend was observed for KB and KB + VUV. After the 5th discharge cycle, the discharge products (Figure 9h) were a granular substance (Figure 9b) of the same size as those obtained from KB; however, after the 5th charge cycle, the granular substance was decomposed. Its shape was significantly different from that of KB, indicating that many pores could be confirmed on the surface of the cathode material. The shape of the discharge product generated on the KB + VUV changed after the 10th cycle (Figure 9i); it was identical to that of the thin-film substance after the 1st discharge cycle of KB. After the 10th charge cycle, the thin-film substance was decomposed, and the existence of the pores was confirmed. However, it can be observed that the residual pores were fewer after the 10th charge cycle than those on the surface after the 5th charge cycle.

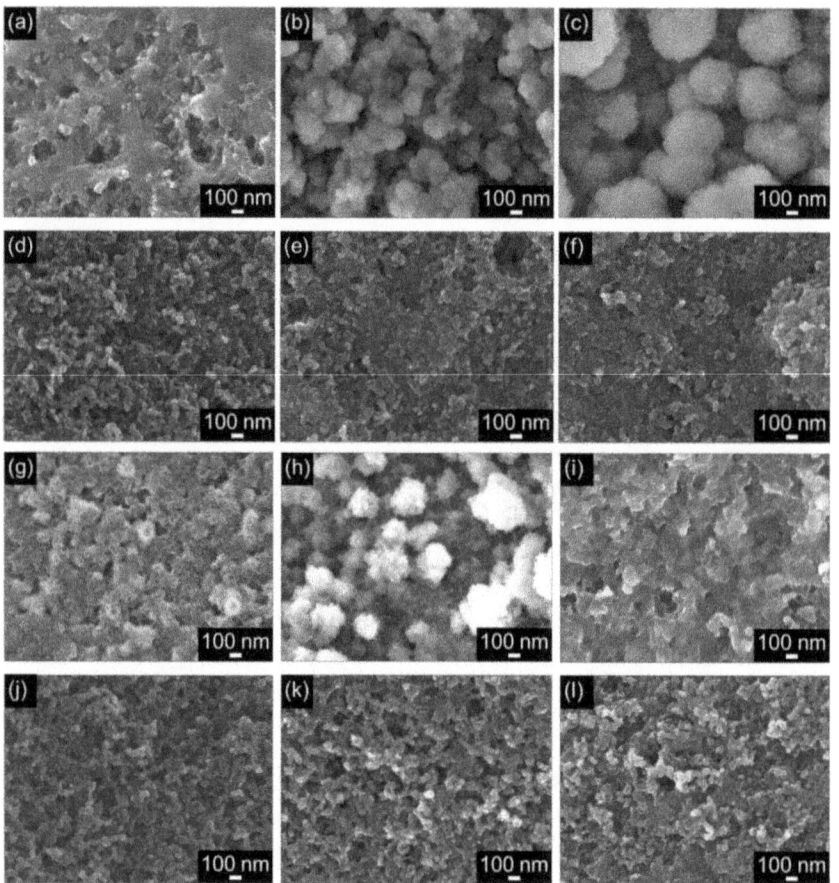

Figure 9. SEM images of (**a–f**) KB and (**g–l**) KB + VUV surfaces after (**a**) 1st discharge, (**b**) 5th discharge, (**c**) 10th discharge, (**d**) 1st charge, (**e**) 5th charge, (**f**) 10th charge, (**g**) 1st discharge, (**h**) 5th discharge, (**i**) 10th discharge, (**j**) 1st charge, (**k**) 5th charge, and (**l**) 10th charge cycles.

Comparing the surfaces of KB with KB + VUV after the 10th cycle charge–discharge tests, the number of pores on the surface was significantly different. Because it has been reported that pores served as a diffusion path for O_2 and Li ions, it is considered that the number of pores has a significant influence on the charge–discharge characteristics of Li-air batteries.

The chemical bonds of the substances formed on the cathode materials after the charge–discharge tests were analyzed by FT-IR. The FT-IR spectra are shown in Figure 10. A peak derived from lithium peroxide (Li_2O_2) as the discharge product was observed at near 550 cm^{-2}, and peaks associated with lithium carbonate as the discharge product were present at near 1500, 1450, and 870 cm^{-1} [19,20]. In both KB and KB + VUV, the peaks derived from lithium peroxide (Li_2O_2) and lithium carbonate (Li_2CO_3) were observed after the 1st discharge cycle; however, the peak intensity derived from Li_2CO_3 was lower for

KB + VUV than that of KB. Additionally, no peaks derived from Li_2O_2 and Li_2CO_3 were observed after charging, indicating that the generated product decomposed after charging. Comparing the states after the 10th cycle charge–discharge tests, the peaks derived from Li_2O_2 and Li_2CO_3 in both KB and KB + VUV after discharging were detected; however, the FT-IR spectra after the 10th charge cycles were different. In the case of KB, a peak corresponding to Li_2CO_3 was observed after the 10th charge cycle. However, in the case of KB + VUV, no peak associated with Li_2CO_3 was observed after the 10th charge cycle. This indicates that the accumulation of the Li_2CO_3-derived products on the KB surface became higher than that of KB + VUV surface with an increase in the cycle numbers of the charging–discharging reaction. Itkis et al. revealed that a superoxide radical (O_2^-) were produced on carbon materials during the discharging reaction, and epoxy groups and carbonates were generated by the reaction of the radical with the carbon materials, which deaccelerated the charging reaction [21]. Therefore, it is considered that a large amount of Li_2CO_3 was generated on the KB surface, which blocked the diffusion path of Li ions and O_2, resulting in the shortening of the cycling performance. However, in this study, a functional group such as a carboxyl group was introduced by the surface modification on the cathode material; thus, it can be considered that the reaction between the superoxide radical and carbon was inhibited, leading to the suppression of the formation of epoxy groups and carbonates. Furthermore, the enlargement of the discharge product was suppressed during repeated charge–discharge cycles; therefore, the space for generating the discharge product was maintained, and the diffusion path of oxygen and lithium ions required for the discharging reaction was maintained. It is presumed that it could be maintained over a long cycle. Based on this consideration, this phenomenon plays a significant role in improving the cycling performance of KB + VUV.

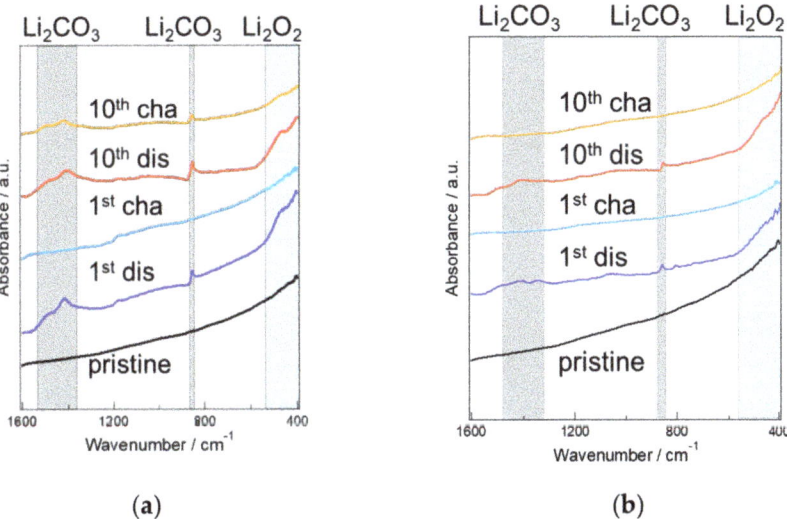

Figure 10. FT-IR spectra of (**a**) KB and (**b**) KB + VUV before and after cycling tests.

4. Conclusions

This study clarified that the size and shape of the generated discharge product could be changed, and the cycling performance of the Li-air battery could be improved by the surface modification of KB through VUV irradiation, leading to the introduction of the COOH group. This result revealed that the functional group on the carbon surface was one of the important factors that improved the cycling performance. In the future, the cycling performance of Li-air batteries can be further improved by clarifying how the type of the functional group affects the shape of the generated discharge product and the charging–discharging reaction.

Author Contributions: Conceptualization and methodology, T.I.; experimental and data analysis, K.F., K.T. and K.S.; writing—original draft preparation, S.Y.L., K.F. and T.I.; writing—review and editing, S.Y.L. and T.I.; supervision, T.I.; project administration and funding acquisition, T.I. All authors have read and agreed to the published version of the manuscript.

Funding: This research was funded by Strategic International Collaborative Research Program (SICORP) grant number JPMJSC18H1, Japan from the Japan Science and Technology Agency (JST), and by Grant-in-Aid for Challenging Research Exploratory (No. 21K18835) from the Japan Society for the Promotion of Science (JSPS).

Institutional Review Board Statement: Not applicable.

Informed Consent Statement: Not applicable.

Data Availability Statement: Not applicable.

Acknowledgments: Shibaura Institute of Technology assisted in meeting the publication costs of this article.

Conflicts of Interest: The authors declare no conflict of interest.

References

1. Girishkumar, G.; McCloskey, B.; Luntz, A.; Swanson, S.; Wilcke, W. Lithium−Air Battery: Promise and Challenges. *J. Phys. Chem. Lett.* **2010**, *1*, 2193–2203. [CrossRef]
2. Jung, K.-N.; Kim, J.; Yamauchi, Y.; Park, M.-S.; Lee, J.-W.; Kim, J. Rechargeable lithium–air batteries: A perspective on the development of oxygen electrodes. *J. Mater. Chem. A* **2016**, *4*, 14050–14068. [CrossRef]
3. Chen, Y.; Li, F.; Tang, D.-M.; Jian, Z.; Liu, C.; Golberg, D.; Yamada, A.; Zhou, H. Multi-walled carbon nanotube papers as binder-free cathodes for large capacity and reversible non-aqueous Li-O_2 batteries. *J. Mater. Chem. A* **2013**, *1*, 13076–13081. [CrossRef]
4. Lin, H.; Liua, Z.; Mao, Y.; Liu, X.; Fang, Y.; Liu, Y.; Wang, D.; Xie, J. Effect of nitrogen-doped carbon/Ketjenblack composite on the morphology of Li_2O_2 for high-energy-density Li–air batteries. *Carbon* **2016**, *96*, 965–971. [CrossRef]
5. Xie, J.; Yao, X.; Cheng, Q.; Madden, I.P.; Dornath, P.; Chang, C.-C.; Fan, W.; Wang, D. Three Dimensionally Ordered Mesoporous Carbon as a Stable, High-Performance Li–O_2 Battery Cathode. *Angew. Chem. Int. Ed.* **2015**, *54*, 4299–4303. [CrossRef]
6. Xiao, J.; Mei, D.; Li, X.; Xu, W.; Wang, D.; Graff, G.L.; Bennett, W.D.; Nie, Z.; Saraf, L.V.; Aksay, I.A.; et al. Hierarchically Porous Graphene as a Lithium–Air Battery Electrode. *Nano Lett.* **2011**, *11*, 5071–5078. [CrossRef]
7. Xiao, J.; Wang, D.; Xu, W.; Wang, D.; Williford, R.E.; Liu, J.; Zhang, J.-G. Optimization of Air Electrode for Li/Air Batteries. *J. Electrochem. Soc.* **2010**, *157*, A487–A492. [CrossRef]
8. Tran, C.; Yang, X.-Q.; Qu, D. Investigation of the gas-diffusion-electrode used as lithium/air cathode in non-aqueous electrolyte and the importance of carbon material porosity. *J. Power Sour.* **2010**, *195*, 2057–2063. [CrossRef]
9. Ding, N.; Chien, S.W.; Hor, T.S.A.; Lum, R.; Zong, Y.; Liu, Z. Influence of carbon pore size on the discharge capacity of Li–O_2 batteries Influence of carbon pore size on the discharge capacity of Li–O_2 batteries. *J. Mater. Chem. A* **2014**, *2*, 12433–12441. [CrossRef]
10. Wong, R.; Dutta, A.; Yang, C.; Yamanaka, K.; Ohta, T.; Nakao, A.; Waki, K.; Byon, H.R. Structurally Tuning Li_2O_2 by Controlling the Surface Properties of Carbon Electrodes: Implications for Li–O_2 Batteries. *Chem. Mater.* **2016**, *28*, 8006–8015. [CrossRef]
11. Belova, A.; Kwabi, D.; Yashina, L.; Shao-Horn, Y.; Itkis, D. On the Mechanism of Oxygen Reduction in Aprotic Li-Air Batteries: The Role of Carbon Electrode Surface Structure. *J. Phys. Chem. C* **2017**, *121*, 1569–1577. [CrossRef]
12. Shu, C.; Wang, J.; Long, J.; Liu, H.; Dou, S. Understanding the Reaction Chemistry during Charging in Aprotic Lithium–Oxygen Batteries: Existing Problems and Solutions. *Adv. Mater.* **2019**, *31*, 1804587. [CrossRef] [PubMed]
13. Xia, G.; Shen, S.; Zhu, F.; Xie, J.; Hu, Y.; Zhu, K.; Zhang, J. Effect of oxygen-containing functional groups of carbon materials on the performance of Li–O_2 batteries. *Electrochem. Commun.* **2015**, *60*, 26–29. [CrossRef]
14. Qian, Z.; Sun, B.; Du, L.; Lou, S.; Du, C.; Zuo, P.; Ma, Y.; Cheng, X.; Gao, Y.Z.; Yin, G. Insights into the role of oxygen functional groups and defects in the rechargeable nonaqueous Li–O_2 batteries. *Electrochim. Acta* **2018**, *292*, 838–845. [CrossRef]
15. Yang, D.; Velamakanni, A.; Bozoklu, G.; Park, S.; Stoller, M.; Piner, R.D.; Stankovich, S.; Jung, I.; Field, D.A.; Ventrice, C.A., Jr.; et al. Chemical analysis of graphene oxide films after heat and chemical treatments by X-ray photoelectron and Micro-Raman spectroscopy. *Carbon* **2009**, *47*, 145–152. [CrossRef]
16. Johnson, P.S.; Cook, P.L.; Liu, X.; Yang, W.; Bai, Y.; Abbott, N.L.; Himpsel, F.J. Universal mechanism for breaking amide bonds by ionizing radiation. *J. Chem. Phys.* **2011**, *135*, 044702. [CrossRef] [PubMed]
17. Wu, Z.-S.; Ren, W.; Gao, L.; Zhao, J.; Chen, Z.; Liu, B.; Tang, D.; Yu, B.; Jiang, C.; Cheng, H.-M. Synthesis of Graphene Sheets with High Electrical Conductivity and Good Thermal Stability by Hydrogen Arc Discharge Exfoliation. *ACS Nano* **2009**, *3*, 411–417. [CrossRef]
18. Soin, N.; Sinha Roy, S.; Roy, S.; Hazra, K.S.; Misra, D.S.; Lim, T.H.; Hetherington, C.J.; McLaughlin, J.A. Enhanced and Stable Field Emission from in Situ Nitrogen-Doped Few-Layered Graphene Nanoflakes. *J. Phys. Chem. C* **2011**, *115*, 5366–5372. [CrossRef]

19. Mizuno, F.; Nakanishi, S.; Kotani, Y.; Yokoishi, S. Rechargeable Li-Air Batteries Carbonate-Based Liquid Electrolytes. *Electrochemistry* **2010**, *78*, 403–405. [CrossRef]
20. Freunberger Sutefan, A.; Chen, Y.; Peng, Z.; Griffin John, M.; Hardwick Laurence, J.; Bardé, F.; Novák, P.; Bruce Peter, G. Reactions in the Rechargeble Lithium-O_2 Battery with Alkyl Carbonate Electrolytes. *J. Am. Chem. Soc.* **2011**, *133*, 8040–8047. [CrossRef]
21. Itkis, D.M.; Semenenko, D.A.; Kataev, E.Y.; Belova, A.I.; Neudachina, V.S.; Sirotina, A.P.; Hävecker, M.; Teschner, D.; Gericke, A.K.; Dudin, P.; et al. Reactivity of Carbon in Lithium−Oxygen Battery Positive Electrodes. *Nano Lett.* **2013**, *13*, 4697–4701. [CrossRef] [PubMed]

Article

Comparison of the Properties of Ni–Mn Hydroxides/Oxides with Ni–Mn Phosphates for the Purpose of Hybrid Supercapacitors

Lyubomir Soserov [1], Delyana Marinova [2], Violeta Koleva [2], Antonia Stoyanova [1] and Radostina Stoyanova [2,*]

[1] Institute of Electrochemistry and Energy Systems, Bulgarian Academy of Sciences, Acad. G. Bonchev Str., Bldg. 10, 1113 Sofia, Bulgaria; l_stefanov@iees.bas.bg (L.S.); antonia.stoyanova@iees.bas.bg (A.S.)
[2] Institute of General and Inorganic Chemistry, Bulgarian Academy of Sciences, Acad. G. Bonchev Str., Bldg. 11, 1113 Sofia, Bulgaria; manasieva@svr.igic.bas.bg (D.M.); vkoleva@svr.igic.bas.bg (V.K.)
* Correspondence: radstoy@svr.igic.bas.bg; Tel.: +35-9979-3915

Abstract: This study aims to quantify the synergistic effect of Ni^{2+} and Mn^{2+} ions on the capacitive performance of oxide, hydroxide and phosphate electrodes in alkaline electrolytes. Three types of phases containing both nickel and manganese in a ratio of one-to-one were selected due to their stability in alkaline media: oxides with ilmenite and spinel structures ($NiMnO_3$ and $Ni_{1.5}Mn_{1.5}O_4$); hydroxides with layered structures (β-$Ni_{1/2}Mn_{1/2}(OH)_2$); and phosphates with olivine and maricite structures ($LiNi_{1/2}Mn_{1/2}PO_4$ and $NaNi_{1/2}Mn_{1/2}PO_4$). In the mixed hydroxides and phosphates, Ni^{2+} and Mn^{2+} ions randomly occupied one crystallographic site, whereas in the ilmenite oxide, a common face was shared by the Ni^{2+} and Mn^{4+} ions. The electrochemical parameters of the Ni–Mn compositions were evaluated in asymmetric hybrid supercapacitor cells working with alkaline electrolytes and activated carbon as a negative electrode. A comparative analysis of oxides, hydroxides and phosphates enabled us to differentiate the effects of nickel and manganese ions, structures and morphologies on their capacitive performance. Thus, the best performed electrode was predicted. The electrode composition should simultaneously contain Ni and Mn ions, and their morphologies should comprise spherical aggregates. This was an ilmenite $NiMnO_3$, which delivers high energy and power density (i.e., 65 W h kg^{-1} at 3200 W kg^{-1}) and exhibits a good cycling stability (i.e., around 96% after 5000 cycles at a current load of 240 mA g^{-1}).

Keywords: hybrid supercapacitors; Ni/Mn oxides; hydroxides and phosphates; synergetic effect; capacitance performance; alkaline electrolyte

Citation: Soserov, L.; Marinova, D.; Koleva, V.; Stoyanova, A.; Stoyanova, R. Comparison of the Properties of Ni–Mn Hydroxides/Oxides with Ni–Mn Phosphates for the Purpose of Hybrid Supercapacitors. *Batteries* **2022**, *8*, 51. https://doi.org/10.3390/batteries8060051

Academic Editors: Xia Lu and Xueyi Lu

Received: 29 April 2022
Accepted: 24 May 2022
Published: 30 May 2022

Publisher's Note: MDPI stays neutral with regard to jurisdictional claims in published maps and institutional affiliations.

Copyright: © 2022 by the authors. Licensee MDPI, Basel, Switzerland. This article is an open access article distributed under the terms and conditions of the Creative Commons Attribution (CC BY) license (https://creativecommons.org/licenses/by/4.0/).

1. Introduction

The elaboration of hybrid supercapacitors with improved energy density and cycling stability is a current challenge that requires identification of the most suitable electrode materials [1–4]. In this context, transitional metal oxides or hydroxides are considered as attractive electrode materials due to their capability to store energy by different mechanisms [1,5–10]. For example, MnO_2 displays a classical pseudocapacitive mechanism based on fast surface redox reactions with the participation of Mn^{3+} and Mn^{4+} ions [5,6], whereas $Ni(OH)_2$ is characterized by reversible electrochemical redox reactions with Ni^{2+}/Ni^{3+} pair which is concomitant with ion/molecule intercalation [7–10]. The electrochemical storage mechanisms become more diverse when ternary metal oxides/hydroxides having multiple oxidation states are used. According to A/B/O notation, ternary metal compounds can be categorized mainly into three groups: AB_2O_4, $ABO_{2/3/4}$ and $A_3B_2O_8$ [11,12]. In comparison with MnO_2, mixed nickel manganese oxide ($NiMn_2O_4$) stored in aqueous electrolytes (e.g., 1 M Na_2SO_4), the vast majority of capacitance (91%) is by intercalation, and only 9% is by a capacitive mechanism [13]. The performance of oxides depends also on the type of the crystal structure; it has been found that $NiMnO_3$ with an ilmenite structure outperforms $NiMn_2O_4$ with a spinel structure [14]. The ratio between Ni and Mn is also of importance [15,16]; the best capacitive properties have been established for Ni–Mn oxide with Ni:Mn = 1:3. The replacement of oxides with hydroxides has a positive impact on the electrochemical performance of mixed nickel manganese compounds [17,18]. Layered double hydroxides (Ni–Mn LDH), as well as Ni–Mn LDH deposited on reduced graphene oxide, exhibit high faradaic pseudocapacitance, which makes them attractive electrodes for hybrid supercapacitors [12,17,18].

The improved storage performance of mixed Ni–Mn oxides/hydroxides is directly related with the synergistic effect of Ni and Mn ions [19]. The next question is whether the Ni/Mn effect is specific for oxides/hydroxides. In this context, phosphate compounds represent an alternative towards oxides/hydroxides due to their stability in aqueous and carbonate-based electrolytes [20]. Irrespective of this, intensive studies on phosphate-based supercapacitors started in 2012 with $NH_4CoPO_4 \cdot H_2O$ [21,22]. Regarding alkaline transition metal phosphates, the first report appeared in 2015, with the electrode being lithium manganese phosphate, $LiMnPO_4$, with an olivine-type structure [23,24]. It has been found that nano-crystalline $LiMnPO_4$ coated with a thick carbon layer delivers high capacitance when lithium aqueous electrolytes (such as LiOH and Li_2SO_4) are used. This phospho-olivine shows non-faradic behavior in neutral aqueous electrolytes, whereas in alkaline electrolytes, the faradic kind of the capacitive profiles is more pronounced. Furthermore, the capacitive performance of the phospho-olivine is amplified when the composite between $LiMnPO_4$ and reduced graphene oxide aerogel is formed [25]. Like $LiMnPO_4$, nickel analogue $LiNiPO_4$ stores electrochemical energy by faradaic and non-faradaic mechanisms [26]. Recently, it has been reported that sodium manganese and sodium nickel phosphates ($NaMnPO_4$ and $NaNiPO_4$) have maricite structures that operate through the same mechanisms in NaOH electrolytes; battery-like reversible redox processes are owed to Mn^{2+}/Mn^{3+} and Ni^{2+}/Ni^{3+} redox pairs concomitant with adsorption/desorption reactions at the electrode/electrolyte interface [27–29]. In neutral electrolytes, such as NaCl and Na_2SO_4, however, the non-faradaic mechanism prevails [27]. The nickel compound $NaNiPO_4$ displays higher specific capacitance than the manganese one, $NaMnPO_4$, with a voltametric specific capacitance of 390 F g^{-1} vs. 219 F g^{-1} at a scan rate of 2 mV s^{-1} [27]. Mixed sodium manganese-nickel-cobalt phosphate (i.e., $NaMn_{1/3}Ni_{1/3}Co_{1/3}PO_4$) demonstrates stable capacitive performance in hybrid supercapacitors irrespective of the kind used electrolytes (i.e., 2 M NaOH solution and 1 M $NaPF_6$ in EC/DEC/DMC) [30,31]. Regardless of these few reports, the capacitive performance of phosphates is still far from that which is desired. That is why the challenge is how to improve the capacitive performance of phosphates.

This study aims to quantify the synergistic effect of nickel and manganese ions on the capacitive performance of oxide, hydroxide and phosphate electrodes in alkaline electrolytes. For all electrodes, the ratio of Ni-to-Mn was selected to be one-to-one. Two types of phosphate phases stable in alkaline media were selected: $LiNi_{1/2}Mn_{1/2}PO_4$ with an olivine structure and $NaNi_{1/2}Mn_{1/2}PO_4$ with a maricite structure. As a measure of the phosphate performance, mixed Ni–Mn oxides with ilmenite and spinel structures, as well as mixed Ni–Mn hydroxides with layered structures, were used as references. For the phosphates and hydroxides, Ni^{2+} and Mn^{2+} ions randomly occupied one crystallographic site, whereas in the ilmenite oxide, a common face was shared by Ni^{2+} and Mn^{4+} ions (Figure 1).

The electrochemical parameters were evaluated in hybrid supercapacitor cells working with alkaline electrolytes and activated carbon as a negative electrode [10]. The comparative analysis of the phosphates and oxides/hydroxides is of significance to obtain insight into the synergistic effect of nickel and manganese on the capacitive performance of electrodes.

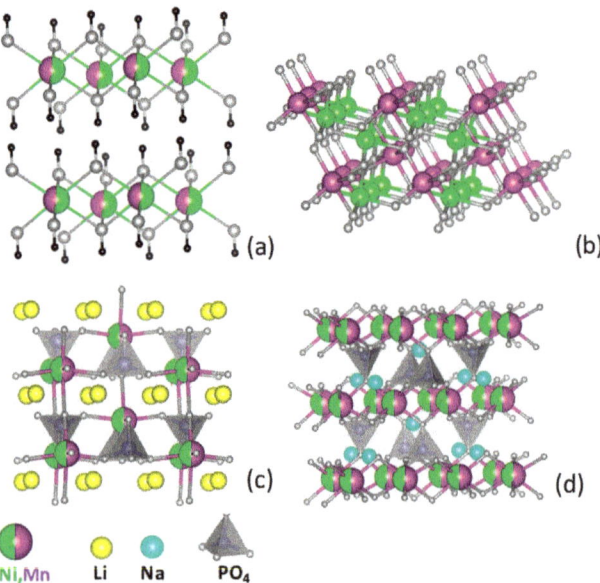

Figure 1. Schematic view of the crystal structures of mixed Ni–Mn phases: (**a**) β-Ni$_{1/2}$Mn$_{1/2}$(OH)$_2$; (**b**) NiMnO$_3$; (**c**) LiNi$_{1/2}$Mn$_{1/2}$PO$_4$; and (**d**) NaNi$_{1/2}$Mn$_{1/2}$PO$_4$. The structures are adopted from: β-Ni(OH)$_2$ (COD 9011314), ilmenite (ICSD 31853), phospho-olivine (COD 4002560) and maricite (COD 1530437).

2. Results and Discussion

For the preparation of single phases containing both nickel and manganese, specific synthetic procedures were adopted. Table 1 summarizes all the experimental conditions necessary for the synthesis of the given electrode, as well as its notation. Mixed nickel-manganese hydroxide, Ni$_{1/2}$Mn$_{1/2}$(OH)$_2$, is prepared by the co-precipitation of a nickel-manganese aqueous solution with KOH. Interestingly, the kind of the used Ni and Mn salts affects the composition of the precipitated hydroxides (Figure S1); single Ni$_{1/2}$Mn$_{1/2}$(OH)$_2$ phase, which is isostructural to the well-known β-Ni(OH)$_2$ (Figure 1a, ICSD 28101), is obtained only in the case when nickel and manganese nitrate salts are used. The nickel and manganese sulfate salts yield a phase mixture between α- and β-type Ni$_x$Mn$_{1-x}$(OH)$_2$, as well as individual Mn(OH)$_2$ and β-Ni(OH)$_2$ (Figure S1). In this study, single β-Ni$_{1/2}$Mn$_{1/2}$(OH)$_2$ phase was only tested as an electrode in a supercapacitor cell. Contrary to the hydroxides, precipitation in the presence of CO$_2$ leads to the formation of a single carbonate phase, Ni$_{1/2}$Mn$_{1/2}$CO$_3$, irrespective of the kind of the used nickel and manganese salts. The common features of β-Ni$_{1/2}$Mn$_{1/2}$(OH)$_2$ and Ni$_{1/2}$Mn$_{1/2}$CO$_3$ are the random distributions of Ni^{2+} and Mn^{2+} ions in hydroxide and carbonate crystal structures (Figure S1).

Table 1. Experimental conditions, preparation methods, phase compositions and labeling of the studied hydroxides, oxides and phosphates.

Sample	Description	Preparation Method	T, °C	Annealing Time, hs	Phase Composition	Labeling
1	Ni–Mn hydroxide	Co-precipitation from nitrate salts	25	-	β-type Ni$_{0.5}$Mn$_{0.5}$(OH)$_2$	N-OH
2	Ni–Mn oxide	Thermal decomposition of hydroxides prepared from nitrates	400	3	Mixture of ilmenite NiMnO$_3$ and spinel Ni$_{1.5}$Mn$_{1.5}$O$_4$	IS-O
3	Ni–Mn oxide	Thermal decomposition of Ni$_{1/2}$Mn$_{1/2}$CO$_3$	400	3	Single ilmenite NiMnO$_3$ phase	I-O
4	Li–Ni–Mn phosphate	Li–Ni–Mn phosphate-formate precursor	500	10	Single olivine phase LiNi$_{1/2}$Mn$_{1/2}$PO$_4$	LP
5	Na–Ni–Mn phosphate	Na–Ni–Mn phosphate-formate precursor	700	10	Single maricite phase NaNi$_{1/2}$Mn$_{1/2}$PO$_4$	NP

Oxide electrodes are prepared by the thermal decomposition of the corresponding carbonate and hydroxide phases (Figure S1, Table 1). Although thermal decomposition of the carbonate phase, $Ni_{1/2}Mn_{1/2}CO_3$, yields single $NiMnO_3$ phase with an ilmenite type of structure (Figure 1b, ICSD 31853), a mixture between oxide phases with ilmenite and spinel structures (i.e., $NiMnO_3$ and $Ni_{1.5}Mn_{1.5}O_4$) is formed after the decomposition of a single β-$Ni_{1/2}Mn_{1/2}(OH)_2$ phase. As electrodes in a supercapacitor cell, $NiMnO_3$ ilmenite and an oxide mixture, "$NiMnO_3$+$Ni_{1.5}Mn_{1.5}O_4$", are used.

The phosphate phases $LiNi_{1/2}Mn_{1/2}PO_4$ and $NaNi_{1/2}Mn_{1/2}PO_4$ are obtained from the lithium and sodium phosphate-formate precursors. XRD patterns evidence that lithium and sodium compounds crystallize in different types of structures, although the same synthetic method (Figure S1): $LiNi_{1/2}Mn_{1/2}PO_4$ adopts an olivine-type structure (Figure 1c), whereas $NaNi_{1/2}Mn_{1/2}PO_4$ crystallizes in a maricite-type structure (Figure 1d). The olivine- and maricite-type structures are closely related to each other; they have the same PO_4 framework but with a reverse distribution of M^+ and M^{2+} ions over the two octahedral sites (4a and 4c) [32,33]. It is of importance that Ni^{2+} and Mn^{2+} ions are randomly distributed on the given octahedral positions in the two structures, as discussed in the supporting information.

Specific surface area is another important factor contributing to the electrochemical performance of materials [34,35]. Figure 2 compares the specific surface area for all the samples. Hydroxides, oxides and phosphates are typical mesoporous materials (Figure 2). For β-$Ni_{0.5}Mn_{0.5}(OH)_2$ and its oxide-derived product (IS-O), the isotherms show characteristic H1-type hysteresis loops associated with the narrow distribution of relatively uniform cylindrical-like pores (Figure 2a,c) [36]. The calculated specific surface areas, total pore volumes and pore size distributions are collected in Table 2. The hydroxides and oxides have close porous characteristics with high specific surface areas (varying between 106 and 128 m^2 g^{-1}) and total pore volumes (i.e., varying between 0.25 and 0.35 cm^3 g^{-1}). However, close inspection of the pore size distribution curves (Figure 2 and Table 2) reveals that the mean pore size was slightly shifted after the thermal decomposition of the hydroxide to an oxide (i.e., from 8 to 11 nm). This means that the H_2O evolution from a hydroxide caused an opening of pores in the mesopore range. The phosphates exhibited low specific surface areas: 7 m^2 g^{-1} for $LiNi_{1/2}Mn_{1/2}PO_4$ and about 1 m^2 g^{-1} for $N37Ni_{1/2}Mn_{1/2}PO_4$ (Table 2). (For the sake of convenience, the isotherm of $NaNi_{1/2}Mn_{1/2}PO_4$ is not shown due to its lowest specific surface area.) For lithium compounds, the pores were distributed in a broad range from 5 to 100 nm, with mesopores with sizes between 10 and 50 nm being predominant. The different porosity of hydroxides/oxides and phosphates is related with the preparation conditions; phosphates are prepared at higher temperatures using longer heating times in comparison with oxides (Table 1).

Table 2. Specific surface area S_{BET}, total pore volume V_t and pore size distribution for the studied compounds.

Samples	Detailed Description	S_{BET}, m^2 g^{-1}	V_t, cm^3 g^{-1}	Pore size Distribution, nm
N-OH	β-$Ni_{1/2}Mn_{1/2}(OH)_2$	117	0.25	Uniform narrow pore size distribution between 3 and 12 nm; mean pore size of 8 nm
I-O	$NiMnO_3$	128	0.35	Narrow pore size distribution between 3 and 12 nm; mean pore size of 10 nm
IS-O	$NiMnO_3$ + $Ni_{1.5}Mn_{1.5}O_4$	106	0.30	Uniform narrow pore size distribution between 3 and 18 nm; mean pore size of 11 nm
LP	$LiNi_{1/2}Mn_{1/2}PO_4$	7	0.04	Broad pore size distribution between 5 and 100 nm, with mesopores between 10 and 50 nm being predominant
NP	$NaNi_{1/2}Mn_{1/2}PO_4$	$\approx$1	-	-

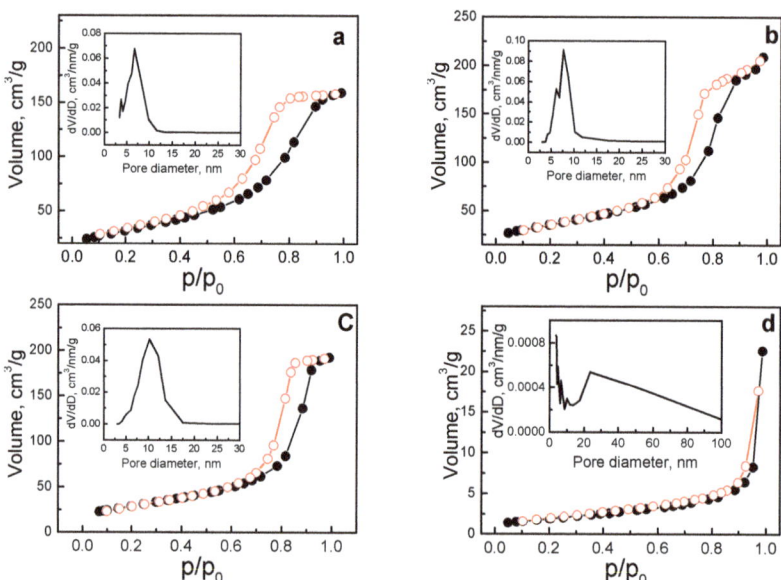

Figure 2. Adsorption–desorption isotherms of: (**a**) β-Ni$_{1/2}$Mn$_{1/2}$(OH)$_2$; (**b**) ilmenite NiMnO$_3$; (**c**) mixture of NiMnO$_3$ and Ni$_{1.5}$Mn$_{1.5}$O$_4$; and (**d**) LiNi$_{1/2}$Mn$_{1/2}$PO$_4$. Filled and open symbols denote adsorption and desorption curves, respectively. The insets show the pore size distribution curves.

In synchrony with the porosity, the morphology of the samples is also specific. The morphology consisted of micrometric aggregates with various shapes (Figure 3). For NiMnO$_3$, spherical aggregates with sizes of around 1–5 µm dominated, whereas for β-Ni$_{1/2}$Mn$_{1/2}$(OH)$_2$ and its oxide-derivative, unshaped aggregates with sizes larger than 10 µm appeared. For NaNi$_{1/2}$Mn$_{1/2}$PO$_4$, having a lower specific surface area, primary well-shaped particles inside aggregates could be distinguished, with the particle dimensions being of 0.4–0.8 µm. For LiNi$_{1/2}$Mn$_{1/2}$PO$_4$, spherical aggregates were mostly observed.

Figure 3. SEM images: (**a**) β-Ni$_{1/2}$Mn$_{1/2}$(OH)$_2$; (**b**) ilmenite NiMnO$_3$; (**c**) mixture of NiMnO$_3$ and Ni$_{1.5}$Mn$_{1.5}$O$_4$; (**d**) olivine LiNi$_{1/2}$Mn$_{1/2}$PO$_4$; and (**e**) maricite NaNi$_{1/2}$Mn$_{1/2}$PO$_4$.

All the samples used as positive electrodes in asymmetric electrochemical cells displayed charge/discharge curves whose shapes are typical for supercapacitor behavior [5–9] (Figure 4). The comparison shows that hydroxides, oxides and phosphates delivered different capacitances, with the highest being for ilmenite I-O and the lowest being for olivine LP. This reflects a current–resistance iR drop calculated from the discharge curve. I-O exhibited the lowest iR drop (i.e., drop in voltage ΔV of 0.029 ± 0.002 V at 240 mA g^{-1}), and the highest value was calculated for LP (i.e., drop in voltage ΔV of 0.045 ± 0.002 V at 240 mA g^{-1}).

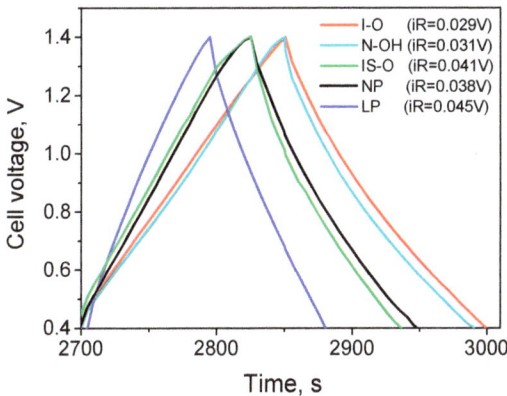

Figure 4. Galvanostatic charge–discharge curves of supercapacitor cells with different composite electrodes at a current load of 240 mA g^{-1}. The drop in voltage ΔV (i.e., iR) is shown.

To more precisely compare the electrochemical performances of all the samples, two experimental protocols are used. Firstly, the electrochemical cells are cycled at a constant current load that increases stepwise from 30 to 900 mA g^{-1} for 25 cycles per each step (Figure 5). Secondly, the electrochemical cells are tested at a current load of 240 mA g^{-1} for 5000 cycles (Figure 6). According to the first protocol, the highest capacitance is delivered by the ilmenite oxide I-O regardless of the current load, with 175 F g^{-1} and 115 F g^{-1} at a current load of 30 and 900 mA h g^{-1}, respectively (Figure 5). The hydroxide N-OH displays lower capacitance than that of the ilmenite oxide I-O at a low current load, whereas at high current loads, the capacitances of hydroxides and oxides become comparable. This reveals a better rate performance of the hydroxides in comparison with that of the oxides. To understand the effect of the mixing of Ni and Mn, Figure 5 gives the capacitance behavior of α,β-Ni(OH)$_2$ reference, for which it has been shown to display the best performance among nickel hydroxide modifications [9]. This comparison indicates high capacitance values for mixed β-Ni$_{1/2}$Mn$_{1/2}$(OH)$_2$ hydroxide at lower current loads, whereas at high current loads, the capacitances of β-Ni$_{1/2}$Mn$_{1/2}$(OH)$_2$ and α,β-Ni(OH)$_2$ tend towards each other. When a spinel oxide Ni$_{1.5}$Mn$_{1.5}$O$_4$ is mixed with ilmenite oxide NiMnO$_3$ (IS-O), there is a decrease in the capacitance, thus indicating worse supercapacitor behavior of the spinel. It is worth mentioning that the better performance of ilmenite NiMnO$_3$ compared to that of the spinel with a composition of NiMn$_2$O$_4$ has been previously established [14]. Given that both oxides have close specific surface areas, the different capacitances disclose the effect of crystal structures on the supercapacitor behavior of oxides. This is more evidence for the complex features of energy storage for oxides, including capacitive and faradaic mechanisms. That is why the energy storage of oxides is calculated and presented in Figure 5 with units of F g^{-1} and mA h g^{-1}.

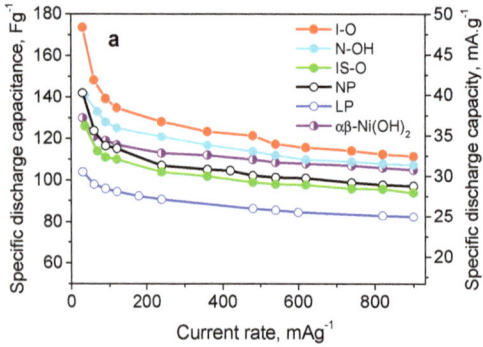
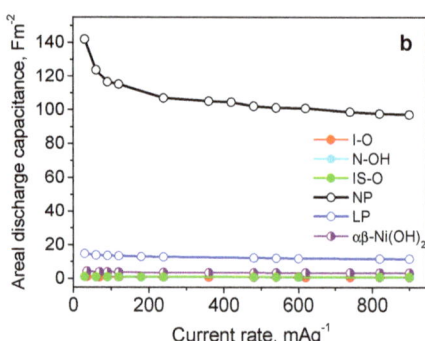

Figure 5. Specific discharge capacitance (**a**) and areal discharge capacitance (**b**) as a function of the current load of supercapacitor cells with different composite electrodes. For the sake of comparison, the discharge capacitance is also calculated in mA h g^{-1}.

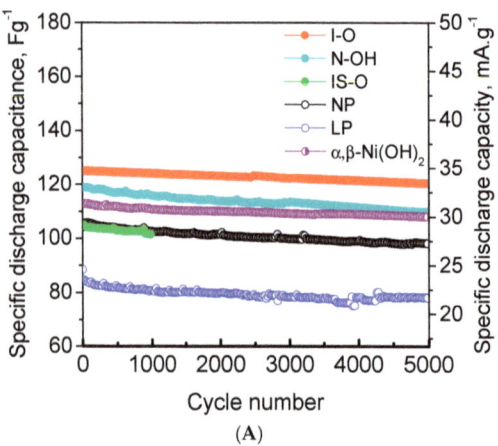
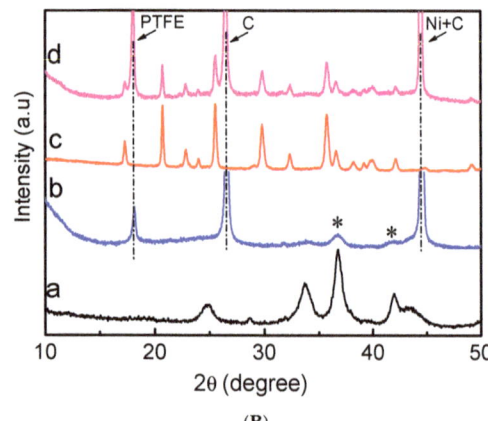

Figure 6. (**A**) Specific discharge capacitance (capacity) as a function of the cycle number at 240 mA g^{-1} of supercapacitor cells with different composite electrodes. (**B**) Ex situ XRD patterns of ilmenite (b) and phospho-olivine (d) electrodes after 5000 cycles. For the sake of comparison, the pristine ilmenite (a) and phospho-olivine (c) are also given. Symbols (*), (PTFE), (C) and (Ni) denote the peaks due to the ilmenite phase, PTFE, graphite and Ni foam, respectively.

The lowest capacitance was observed for the olivine LP, with 105 F g^{-1} and 90 F g^{-1} at a current load of 30 and 900 mA h g^{-1}, respectively. However, the olivine LP outperformed the oxides and hydroxides with respect to the rate capability. Going from 30 to 900 mA g^{-1}, the capacitance loss was around 50 F g^{-1} for I-O and only 15 F g^{-1} for LP. Irrespective of the lowest specific surface area, the maricite NP demonstrated a capacitance that was close to that of the mixed oxide "ilmenite-spinel". To outline the performance of maricite NP, the discharge capacitance was calculated per specific surface area (Figure 5). According to this scheme, the maricite NP delivered the highest areal capacitance, followed by the phospho-olivine (Figure 5). On one hand, this illustrates the impact of the crystal structure on the supercapacitor behavior of phosphates, which mimics that of the oxides. On the other hand, the maricite NP may be of interest as an electrode material in hybrid supercapacitors if its specific surface area is increased drastically.

The second protocol of electrochemical testing is based on an extended number of cycles, and it enables the further differentiation of samples as electrode materials (Figure 6A). After 5000 cycles, the ilmenite I-O delivered the highest capacitance (i.e., of around

125 F g^{-1}) and better cycling stability (i.e., around of 96%). The oxide mixture between ilmenite and spinel underperforms the single ilmenite phase, thus supporting once again the better capacitive performance of the ilmenite phase. The N-OH hydroxide is characterized by a slightly lower capacitance than that of ilmenite I-O (i.e., around of 115 F g^{-1}), but the cycling stability was worse (around 92%). It is noticeable that α,β-Ni(OH)$_2$ reference, having a lower capacitance than that of N-OH (113 versus 119 F g^{-1}), displayed better cycling stability (i.e., around 95%). In comparison with oxides and hydroxides, the performance of the phosphate electrodes was worse; the capacitance was lower than 100 F g^{-1}, and the cycling stability tended to 92% regardless of the crystal structure (i.e., maricite or olivine). It is of importance that the capacitances of mixed Ni–Mn phosphates are among the highest values reported in the literature, in which single Mn and Ni phosphates are mainly examined [23–29]. Moreover, single maricite phosphate NP and mixed ilmenite and spinel oxides IS-O had comparable capacitances regardless of their different specific surface areas (Table 2). The above data allow the outlining of two important features; the capacitance depends mainly on whether the electrode simultaneously contains nickel and manganese and, to a lesser extent, on the type of anionic constituents. On the other hand, the cycling stability is a function of the morphology; it appears that better cycling stability is achieved at spherical aggregates.

Cycling stability is directly associated with the chemical stability of electrodes in alkaline electrolytes. For that reason, Figure 6B gives the ex situ XRD patterns of ilmenite and phospho-olivine after 5000 cycles at 240 mA g^{-1}. These electrodes were selected since they exhibit the best and worst performances. As one can see, both XRD patterns remained unchanged, thus indicating the chemical stability of the ilmenite and phospho-olivine phases during cycling in alkaline electrolytes.

To rationalize the electrochemical performance of oxides, hydroxides and phosphates, Figure 7 compares the relationship between energy density and power density. For the sake of comparison, the data available in the literature are also given. At lower power densities, the specific energy density decreases in following the order: I-O > N-OH~NP > α,β-Ni(OH)$_2$ > LP, whereas at a high power density, the specific energy of I-O, N-OH and >α,β-Ni(OH)$_2$ becomes close and higher than that of the phosphates. It is of importance that the ilmenite NiMnO$_3$ still displays a high energy density at the highest power density (i.e., 65 W h kg^{-1} at 3200 W kg^{-1}).

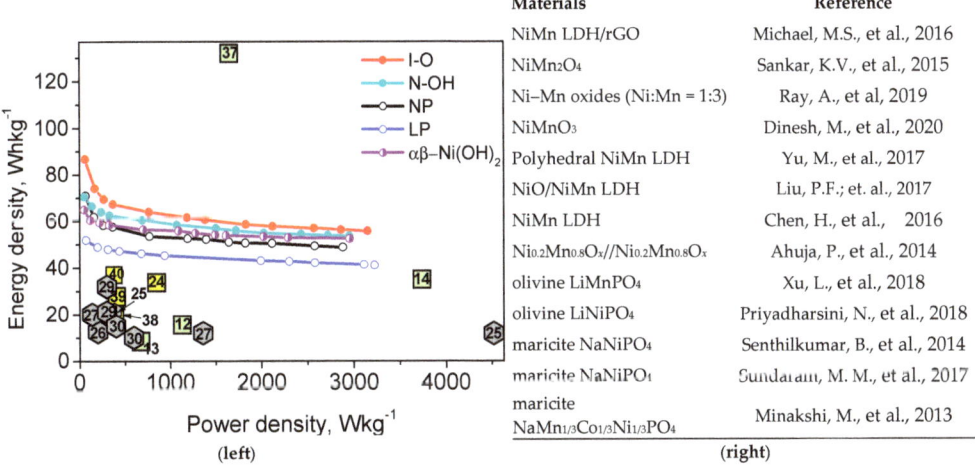

Figure 7. (**left**) Energy density versus power density (Ragone plot) for supercapacitors with different composite electrodes. The literature data for oxides, hydroxides and phosphates are indicated as ⬢, ⬛ and ⬤, respectively (**right**). The used references are given in the figure: [12–14,24–27,29,30,37–40].

This can be related with the synergistic effect of Ni and Mn elements, as well as with the specific morphology and texture of I-O. The capacitance performance of ilmenite I-O is one of the good performances reported in literature (Figure 7). In comparison with ilmenite I-O with a Ni-to-Mn ratio of one-to-one, composite $Ni_{0.2}Mn_{0.8}O_x//Ni_{0.2}Mn_{0.8}O_x$ oxides with a Ni-to-Mn ratio of one-to-four exhibit an energy density of around 38 W h kg^{-1} at 3800 W kg^{-1} power density, whereas the energy and the power density of Ni–Mn oxide with a one-to-three ratio reach around 130 W h kg^{-1} at 1700 W kg^{-1} [23,26]. The ilmenite $NiMnO_3$ prepared by the hydrothermal method displays significantly lower energy density (i.e., around 10 W h kg^{-1} at 700 W kg^{-1} [22]) than that of ilmenite $NiMnO_3$ prepared by us using the co-precipitation method. Interestingly, the hydrothermal-derived $NiMnO_3$, having worse performance, is characterized by a low specific surface area (i.e., around 21 m^2 g^{-1}) and irregularly shaped particles with a size of 100–200 nm, whereas co-precipitate-derived $NiMnO_3$, having the best performance, possesses a high specific surface area (around 125–130 m^2 g^{-1}) and spherical morphology. It is well recognized that spherical morphology is an important factor contributing to the improved performance of battery-like materials due to a higher volume-to-surface ratio of the electrode and its better wetting with electrolytes [41]. Given that the energy storage mechanism of Ni–Mn oxides in supercapacitors is more complex than the single capacitive and faradaic one, it is not surprising that spherical morphology has a favorable effect in asymmetric supercapacitors too.

Both LP and NP phosphates deliver lower specific energy densities than those of oxides and hydroxides (Figure 7), with NP being slightly better than LP. Keeping a Ni-to-Mn ratio of one-to -one, this implies, at first glance, that the capacitive performance of an electrode depends further on the anionic constituents and on the type of structure. Considering that the morphology of phosphates is not optimized, it appears that the phosphates are also suitable for supercapacitor applications. Moreover, the energies and power densities of the phosphates prepared by us, which simultaneously contained Ni and Mn, were more than two times higher than those of phosphates containing one element, such as Ni or Mn (Figure 7). This supports once again the leading effect of Ni and Mn elements on the performance of electrodes in hybrid supercapacitors.

3. Conclusions

Three types of electrodes were evaluated in hybrid supercapacitor cells with alkaline electrolytes: oxides with ilmenite and spinel structures, hydroxides isostructural to β-Ni(OH)$_2$ and phosphates with olivine and maricite structures. The common feature between them is that their crystal structures are able to simultaneously accommodate nickel and manganese ions. The capacitance performances of oxides, hydroxides and phosphates depend mainly on whether the electrode contains simultaneously contains nickel and manganese in a ratio of one-to-one and, to a lesser extent, on the type of anionic constituents. Cycling stability becomes better when the morphology consists of spherical aggregates. Based on these findings, one can predict the electrode with the best capacitance performance; the electrode composition should simultaneously contain Ni and Mn ions, and the morphology should comprise spherical aggregates. A proof-of-concept is demonstrated by $NiMnO_3$ with an ilmenite structure and optimized morphology; it delivers high energy and power density (i.e., 65 W h kg^{-1} at 3200 W kg^{-1}) and exhibits good cycling stability (i.e., around 96% after 5000 cycles at a current load of 240 mA g^{-1}). The capacitive performance of olivine and maricite phosphates outperforms the previously reported phosphates by more than two times due to the synergistic effect of Ni^{2+} and Mn^{2+} ions. Further optimization of phosphate morphology is needed in order to reach the capacitance performance of $NiMnO_3$ ilmenite.

4. Material and Methods

4.1. Synthesis

The mixed Ni–Mn hydroxides were prepared via the classical co-precipitation method from nickel and manganese salts (1:1 mole ratio) and KOH as a precipitant. Two kinds of salts were used: nitrates and sulfates. The mixed Ni–Mn oxides were obtained by the thermal decomposition of the following at 400 °C: (i) corresponding hydroxides; (ii) calcite-type $Ni_{1/2}Mn_{1/2}CO_3$ prepared by a co-precipitation from the nitrate salts with $NaHCO_3$ in a flow of CO_2. For the synthesis of $LiNi_{1/2}Mn_{1/2}PO_4$ and $NaNi_{1/2}Mn_{1/2}PO_4$, we adopted the phosphate-formate precursor method developed previously for the preparation of electrochemically active $LiMPO_4$ (M = Fe, Mn, Co, Ni) and $NaMPO_4$ [32,33]. The synthetic procedure consists of mixing aqueous solutions of $Ni(HCOO)_2 \cdot 2H_2O$, $Mn(HCOO)_2 \cdot 2H_2O$ and LiH_2PO_4, accordingly NaH_2PO_4, taken in a mole ratio of 1:1:2. The metal-phosphate-formate solutions were frozen instantly with liquid nitrogen and subjected to freeze-drying for about 18 h in vacuum (20–30 mbar) using an Alpha-Christ Freeze Dryer. Thus, the obtained solid phosphate-formate precursors were pre-decomposed in an air atmosphere at 350 °C for 4 h. The solid products were further annealed at temperatures between 500 and 700 °C.

4.2. Characterization Methods

The XRD patterns of the oxides/hydroxides and phosphates were recorded on a Bruker D8 Advance diffractometer using CuKα radiation (LynxEye detector). The Ni and Mn contents in the mixed composition were determined by inductively coupled plasma atomic emission spectrometry (ICP-AES). The morphology of the electrode materials was examined by JEOL JSM-5510 SEM. The porous texture of the samples was studied by low-temperature (77.4 K) nitrogen adsorption using the Quantachrome (Boynton Beach, FL, USA) NOVA 1200e instrument. The specific surface area was evaluated by the BET method at a relative pressure p/p_o in a range of 0.10–0.30. The total pore volume was calculated according to Gurwitsch's rule at $p/p_o = 0.99$. The pore size distribution was estimated by using the Barett–Joyner–Halenda method.

4.3. Electrochemical Characterization

Two-electrode cells were used to monitor the electrochemical performances of the electrodes. The cell was constructed from a positive electrode (consisting of a mixture between activated carbon (AC) and oxides/hydroxides in a content of 25 wt.%) and a negative electrode (containing only AC) with the mass ratio between them being 1:1. As a binder and a conductive additive, we used polytetraflourethylene (PTFE) (10 wt.%) and graphite ABG 1005 EG1 (10 wt.%). The electrolyte contained 7 M KOH solutions with additives of 35 g l-1 LiOH. The charge–discharge curves were recorded on the Arbin Instrument System BT-2000. The capacitance (F g-1) was calculated from the charge–discharge curves using the following equations [12,13]:

$$C = (I \times \Delta t)/(m \times \Delta V) \quad (1)$$

where I (A), Δt (s), m (g) and ΔV (V) are the discharge current, discharge time, mass of the active material and voltage window, respectively. Based on the capacitance, the energy densities (E, W h kg^{-1}) and power densities (P, W kg^{-1})) were calculated as [44]:

$$E = C \Delta V2/2 \quad (2)$$

$$P = E \times 3600/t \quad (3)$$

Supplementary Materials: The following supporting information can be downloaded at: https://www.mdpi.com/article/10.3390/batteries8060051/s1. Figure S1: XRD patterns of Ni–Mn-based mixed hydroxides, oxides and phosphates. References [27,32,33,45–47] are cited in the supplementary materials.

Author Contributions: Conceptualization, R.S.; methodology, V.K., A.S. and R.S.; validation, L.S. and D.M.; investigation, L.S., D.M. and V.K.; data curation, L.S. and D.M.; writing—original draft preparation, V.K., A.S. and R.S.; writing—review and editing, R.S., V.K. and A.S.; visualization, V.K. and L.S.; project administration, A.S. and R.S.; funding acquisition, A.S. All authors have read and agreed to the published version of the manuscript.

Funding: This research was funded by the Bulgarian National Science Fund, grant number KP-06-OPR 04/5 "Innovative hybrid supercapacitors as a challenge for efficient, safe and environmental energy storage" and by the Operational Programme "Science and Education for Smart Growth" 2014-2020, co-funded by the EU from the European Regional Development Fund, grant number No BG05M2OP001-1.001-0008.

Institutional Review Board Statement: Not applicable.

Informed Consent Statement: Not applicable.

Data Availability Statement: Not applicable.

Conflicts of Interest: The authors declare no conflict of interest. The funders had no role in the design of the study; in the collection, analyses or interpretation of the data; in the writing of the manuscript; or in the decision to publish the results.

References

1. Choi, C.; Ashby, D.S.; Butts, D.M.; DeBlock, R.H.; Wei, G.; Lau, J.; Dunn, B. Achieving high energy density and high power density with pseudocapacitive materials. *Nat. Rev. Mater.* **2020**, *5*, 5–19. [CrossRef]
2. Chatterjee, D.; Nandi, A. A review on the recent advances in hybrid supercapacitors. *J. Mater. Chem. A* **2021**, *9*, 15880–15918. [CrossRef]
3. Xu, C.; Yang, H.; Li, Y.; Wang, J.; Lu, X. Surface engineering for advanced aqueous supercapacitors: A Review. *ChemElectroChem* **2020**, *7*, 586–593. [CrossRef]
4. Zang, X.; Shen, C.; Sanghadasa, M.; Lin, L. High-voltage supercapacitors based on aqueous electrolytes. *ChemElectroChem* **2019**, *6*, 976–988. [CrossRef]
5. Majumdar, D. Review on current progress of MnO_2-based ternary nanocomposites for supercapacitor applications. *ChemElectroChem* **2021**, *8*, 291–336. [CrossRef]
6. Hsieh, Y.-C.; Lee, K.-T.; Lin, Y.-P.; Wu, N.-L.; Donne, S.W. Investigation on capacity fading of aqueous $MnO_2 \cdot nH_2O$ electrochemical capacitor. *J. Power Sources* **2008**, *177*, 660–664. [CrossRef]
7. Guo, J.; Zhao, Y.; Jiang, N.; Liu, A.; Gao, L.; Li, Y.H.; Wang, H.; Ma, T. In-situ grown $Ni(OH)_2$ nanosheets on Ni foam for hybrid supercapacitors with high electrochemical performance. *J. Electrochem. Soc.* **2018**, *165*, A882. [CrossRef]
8. Mozaffari, S.A.; Najafi, S.H.M.; Norouzi, Z. Hierarchical NiO@$Ni(OH)_2$ nanoarrays as high-performance supercapacitor electrode material. *Electrochim. Acta* **2021**, *368*, 137633. [CrossRef]
9. Ramesh, S.; Karuppasamy, K.; Yadav, M.H.; Lee, J.-J.; Kim, H.-S.; Kim, H.-S.; Kim, J.-H. $Ni(OH)_2$-decorated nitrogen doped MWCNT nanosheets as an efficient electrode for high performance supercapacitors. *Sci. Rep.* **2019**, *9*, 6034. [CrossRef]
10. Soserov, L.; Stoyanova, A.; Boyadzhieva, T.; Koleva, V.; Kalapsazova, M. Nickel-manganese structured and multiphase composites as electrodes for hybrid supercapacitors. *Electrochim. Acta* **2018**, *283*, 1063–1071. [CrossRef]
11. Chen, D.; Wang, Q.; Wang, R.; Shen, G. Ternary oxide nanostructured materials for supercapacitors. *J. Mater. Chem. A* **2015**, *3*, 10158–10173. [CrossRef]
12. Sankar, K.V.; Surendran, S.; Pandi, S.; Allin, M.; Nithya, V.D.; Lee, Y.S.; Selvan, R.K. Studies on the electrochemical intercalation/de-intercalation mechanism of $NiMn_2O_4$ for high stable pseudocapacitor electrodes. *RSC Adv.* **2015**, *5*, 27649–27656. [CrossRef]
13. Dinesh, M.; Haldorai, Y.; Thangavelu, R.; Kumar, R. Mn–Ni binary metal oxide for high-performance supercapacitor and electro-catalyst for oxygen evolution reaction. *Ceram. Int.* **2020**, *46*, 28006–28012. [CrossRef]
14. Ahuja, P.; Ujjain, S.K.; Sharma, S.R.; Singh, G. Enhanced supercapacitor performance by incorporating nickel in manganese oxide. *RSC Adv.* **2014**, *4*, 57192–57199. [CrossRef]
15. Li, M.; Cheng, J.P.; Wang, J.; Liu, F.; Zhang, X.B. The growth of nickel-manganese and cobalt-manganese layered double hydroxides on reduced graphene oxide for supercapacitor. *Electrochim. Acta* **2016**, *206*, 108–115. [CrossRef]
16. Wang, R.; Wu, J. Structure and basic properties of ternary metal oxides and their prospects for application in supercapacitors. In *Metal Oxides in Supercapacitors*; Elsevier: Amsterdam, The Netherlands, 2017; pp. 99–132.
17. Singh, A.K.; Sarkar, D.; Khan, G.G.; Mandal, K. Hydrogenated NiO nanoblock architecture for high performance pseudocapacitor. *ACS Appl. Mater. Interfaces* **2014**, *6*, 4684–4692. [CrossRef]

18. Lu, Q.; Lattanzi, M.W.; Chen, Y.P.; Kou, X.M.; Li, W.F.; Fan, X.; Unruh, K.M.; Chen, J.G.; Xiao, J.Q. Supercapacitor electrodes with high-energy and power densities prepared from monolithic NiO/Ni nanocomposites. *Angew. Chem. Int. Ed.* **2011**, *50*, 6847–6850. [CrossRef]
19. Li, X.; Xin, M.; Guo, S.; Cai, T.; Du, D.; Xing, W.; Zhao, L.; Guo, W.; Xue, Q.; Yan, Z. Insight of synergistic effect of different active metal ions in layered double hydroxides on their electrochemical behaviors. *Electrochim. Acta* **2017**, *253*, 302–310. [CrossRef]
20. Kim, H.; Hong, J.; Park, K.-Y.; Kim, H.; Kim, S.-W.; Kang, K. Aqueous rechargeable Li and Na ion batteries. *Chem. Rev.* **2014**, *114*, 11788–11827. [CrossRef]
21. Pang, H.; Yan, Z.; Wang, W.; Chen, J.; Zhang, J.; Zheng, H. Facile fabrication of $NH_4CoPO_4 \cdot H_2O$ nano/microstructures and their primarily application as electrochemical supercapacitor. *Nanoscale* **2012**, *4*, 5946–5953. [CrossRef]
22. Li, X.; Xiao, X.; Li, Q.; Wei, J.; Xue, H.; Pang, H. Metal (M = Co, Ni) phosphate based materials for high-performance supercapacitors. *Inorg. Chem. Front.* **2018**, *5*, 11–28. [CrossRef]
23. Prabaharan, S.R.S.; Anslin Star, R.; Kulkarni, A.R.; Michael, M.S. Nano-composite $LiMnPO_4$ as new insertion electrode for electrochemical supercapacitors. *Curr. Appl. Phys.* **2015**, *15*, 1624–1633. [CrossRef]
24. Michael, M.S.; Kulkarni, A.R.; Prabaharan, S.R.S. Design of monolayer porous carbon-embedded hybrid-$LiMnPO_4$ for high energy density Li-ion capacitors. *J. Nanosci. Nanotechnol.* **2016**, *16*, 7314–7324. [CrossRef]
25. Xu, L.; Wang, S.; Zhang, X.; He, T.; Lu, F.; Li, H.; Ye, J. A facile method of preparing $LiMnPO_4$/reduced graphene oxide aerogel as cathodic material for aqueous lithium-ion hybrid supercapacitors. *Appl. Surf. Sci.* **2018**, *428*, 977–985. [CrossRef]
26. Priyadharsini, N.; Rupa Kasturi, P.; Shanmugavani, A.; Surendran, S.; Shanmugapriya, S.; Kalai Selvan, R. Effect of chelating agent on the sol-gel thermolysis synthesis of $LiNiPO_4$ and its electrochemical properties for hybrid capacitors. *J. Phys. Chem. Solids* **2018**, *119*, 183–192. [CrossRef]
27. Senthilkumar, B.; Sankar, K.V.; Vasylechko, L.; Lee, Y.-S.; Selvan, R.K. Synthesis and electrochemical performances of maricite-$NaMPO_4$ (M = Ni, Co, Mn) electrodes for hybrid supercapacitors. *RSC Adv.* **2014**, *4*, 53192–53200. [CrossRef]
28. Minakshi, M.; Mitchell, D.; Jones, R.; Alenazey, F.; Watcharatharapong, T.; Chakrabortyf, S.; Ahuja, R. Synthesis, structural and electrochemical properties of sodium nickel phosphate for energy storage devices. *Nanoscale* **2016**, *8*, 11291–11305. [CrossRef]
29. Sundaram, M.M.; Mitchell, D.R.G. Dispersion of Ni^{2+} ions via acetate precursor in the preparation of $NaNiPO_4$ nanoparticles: Effect of acetate vs. nitrate on the capacitive energy storage properties. *Dalton Trans.* **2017**, *46*, 13704–13713. [CrossRef]
30. Minakshi, M.; Meyrick, D.; Appadoo, D. Maricite ($NaMn_{1/3}Ni_{1/3}Co_{1/3}PO_4$)/Activated carbon: Hybrid capacitor. *Energy Fuels* **2013**, *27*, 3516–3522. [CrossRef]
31. Sundaram, M.M.; Watcharatharapong, T.; Chakraborty, S.; Ahuja, R.; Duraisamy, S.; Rao, T.; Munichandraiah, N. Synthesis, and crystal and electronic structure of sodium metal phosphate for use as a hybrid capacitor in non-aqueous electrolyte. *Dalton Trans.* **2015**, *44*, 20108–20120. [CrossRef]
32. Koleva, V.; Stoyanova, R.; Zhecheva, E. Nano-crystalline $LiMnPO_4$ prepared by a new phosphate–formate precursor method. *Mater. Chem. Phys.* **2010**, *121*, 370–377. [CrossRef]
33. Koleva, V.; Boyadzhieva, T.; Zhecheva, E.; Nihtianova, D.; Simova, S.; Tyuliev, G.; Stoyanova, R. Precursor-based methods for low-temperature synthesis of defectless $NaMnPO_4$ with an olivine- and maricite-type structure. *Cryst. Eng. Comm.* **2013**, *15*, 9080–9089. [CrossRef]
34. Roberts, A.J.; Slade, R.C.T. Effect of specific surface area on capacitance in asymmetric carbon/α-MnO_2 supercapacitors. *Electrochim. Acta* **2010**, *55*, 7460–7469. [CrossRef]
35. Zheng, S.; Zhang, J.; Deng, H.; Du, Y.; Shi, X. Chitin derived nitrogen-doped porous carbons with ultrahigh specific surface area and tailored hierarchical porosity for high performance supercapacitors. *J. Bioresour. Bioprod.* **2021**, *6*, 142–151. [CrossRef]
36. Sing, K.S.W.; Williams, R.T. Physisorption hysteresis and the characterization of nanoporous materials. *Adsorpt. Sci. Technol.* **2004**, *22*, 773–782. [CrossRef]
37. Ray, A.; Roy, A.; Saha, S.; Ghosh, M.; Chowdhury, S.R.; Maiyalagan, T.; Bhattacharya, S.K.; Das, S. Das, Electrochemical energy storage properties of Ni Mn oxide electrodes for advance asymmetric supercapacitor application. *Langmuir* **2019**, *35*, 8257–8267.
38. Liu, P.F.; Zhou, J.J.; Li, G.C.; Wu, M.K.; Tao, K.; Yi, F.Y.; Zhao, W.N.; Han, L. A hierarchical NiO/NiMn-layered double hydroxide nanosheet array on Ni foam for high performance supercapacitors. *Dalton Trans.* **2017**, *46*, 7388–7391. [CrossRef]
39. Chen, H.; Ai, Y.; Liu, F.; Chang, X.; Xue, Y.; Huang, Q.; Wang, C.; Lin, H.; Han, S. Carbon-coated hierarchical Ni–Mn layered double hydroxide nanoarrays on Ni foam for flexible high-capacitance supercapacitors. *Electrochim. Acta* **2016**, *213*, 55–65. [CrossRef]
40. Bucher, N.; Hartung, S.; Nagasubramanian, A.; Cheah, Y.L.; Hoster, H.E.; Madhavi, S. Layered Na_xMnO_{2+z} in sodium ion batteries–influence of morphology on cycle performance. *ACS Appl. Mater. Interfaces* **2014**, *6*, 8059–8065. [CrossRef]
41. Yu, M.; Liu, R.; Liu, J.; Li, S.; Ma, Y. Polyhedral-like NiMn-layered double hydroxide/porous carbon as electrode for enhanced electrochemical performance supercapacitors. *Small* **2017**, *13*, 1702616. [CrossRef]
42. Wang, T.; Zhang, S.; Yan, X.; Lyu, M.; Wang, L.; Bell, J.; Wang, H. 2-Methylimidazole-derived Ni–Co layered double hydroxide nanosheets as high rate capability and high energy ensity storage material in hybrid supercapacitors. *ACS Appl. Mat. Interfaces* **2017**, *9*, 15510–15524. [CrossRef] [PubMed]
43. Khan, Y.; Hussain, S.; Söderlind, F.; Käll, P.-O.; Abbasi, M.A.; Durrani, S.K. Honeycomb β-$Ni(OH)_2$ films grown on 3D nickel foam substrates at low temperature. *Mater. Lett.* **2012**, *69*, 37–40. [CrossRef]

44. Huang, J.; Xu, P.; Cao, D.; Zhou, X.; Wang, G. Asymmetric supercapacitors based on β-Ni(OH)$_2$ nanosheets and activated carbon with high energy density. *J. Power Sources* **2014**, *246*, 371–376. [CrossRef]
45. Pertlik, F. Structures of hydrothermally synthesized cobalt(II) carbonate and nickel(II) carbonate. *Acta Crystallogr. Sect. C Cryst. Struct. Commun.* **1986**, *42*, 4. [CrossRef]
46. Maslen, E.N.; Strel'tsov, V.A.; Strel'tsova, N.R.; Ishizawa, N. Electron density and optical anisotropy in rhombohedral carbonates.III.Synchrotron X-ray studies of CaCO$_3$, MgCO$_3$ and MnCO$_3$. *Acta Crystallogr. B* **1995**, *51*, 929–939. [CrossRef]
47. Koleva, V.; Zhecheva, E.; Stoyanova, R. Ordered Olivine-Type Lithium–Cobalt and Lithium–Nickel Phosphates Prepared by a New Precursor Method. *Eur. J. Inorg. Chem.* **2010**, *2010*, 4091–4099. [CrossRef]

Article

Synthesis and Electrochemical Characterization of LiNi$_{0.5}$Co$_{0.2}$Mn$_{0.3}$O$_2$ Cathode Material by Solid-Phase Reaction

Xinli Li [1], Ben Su [1], Wendong Xue [1,*] and Junnan Zhang [2]

[1] School of Materials Science and Engineering, University of Science and Technology Beijing, Beijing 100083, China; xinbattery@163.com (X.L.); suben111@163.com (B.S.)
[2] Shandong Wina Green Power Technology Co., Ltd., Weifang 261000, China; jnzhang@winabattery.com
* Correspondence: xuewendong@ustb.edu.cn; Tel.: +86-185-0134-1077

Abstract: In this paper, using four carbonates as raw materials, the cathode material LiNi$_{0.5}$Co$_{0.2}$Mn$_{0.3}$O$_2$ was prepared with the "ball milling-calcining" solid-phase synthesis method. The specific reaction process, which consists of the decomposition of the raw materials and the generation of target products, was investigated thoroughly using the TG-DSC technique. XRD, SEM and charge/discharge test methods were utilized to explore the influence of different sintering temperatures on the structure, morphology and electrochemical performance of the LiNi$_{0.5}$Co$_{0.2}$Mn$_{0.3}$O$_2$ cathode. The results show that 900~1000 °C is the appropriate synthesis temperature range. LiNi$_{0.5}$Co$_{0.2}$Mn$_{0.3}$O$_2$ synthesized at 1000 °C delivers optimal cycling stability at 0.5 C. Meanwhile, its initial discharge specific capacity and coulomb efficiency reached 167.2 mAh g^{-1} and 97.89%, respectively. In addition, the high-rate performance of the cathode sample prepared at 900 °C is particularly noteworthy. Cycling at 0.5 C, 1 C, 1.5 C and 2 C, the corresponding discharge specific capacity of the sample exhibited 148.1 mAh g^{-1}, 143.1 mAh g^{-1}, 140 mAh g^{-1} and 138.9 mAh g^{-1}, respectively.

Keywords: solid-phase synthesis; carbonate; temperature; LiNi$_{0.5}$Co$_{0.2}$Mn$_{0.3}$O$_2$

1. Introduction

With the decline in global fossil fuel reserves and the enhancement of environmental awareness, the social acceptance and demand for new energy vehicles represented by electric vehicles have continued to rise [1,2]. As one of the indispensable components of lithium-ion batteries, cathode materials have a decisive influence on the performance of lithium-ion batteries. Due to its excellent electrochemical performance, LiNi$_x$Co$_y$Mn$_z$O$_2$ with its layered structure is regarded as the most promising cathode material to replace LiCoO$_2$ [3–5].

Currently, the main synthesis process of LiNi$_x$Co$_y$Mn$_z$O$_2$ cathode materials is the preparation of precursors using co-precipitation, which are then mixed and sintered with a lithium source to obtain the target product [6–8]. In this way, the mixing degree of the raw materials reaches molecular level, and the modifications of the target product in the next step are easy to achieve [9–11]. However, this method of co-precipitation has a high production cost and strict requirements regarding synthesis conditions, such as the temperature, atmosphere, and concentrations. Meanwhile, the reaction process inevitably generates toxic waste liquid. In addition, some researchers have prepared a LiNi$_x$Co$_y$Mn$_z$O$_2$ cathode material using sol–gel and hydrothermal methods [12,13]. These synthesis methods also have limitations, such as complicated processes and high costs. By contrast, solid-phase synthesis is a simple and effective method for preparing cathode materials. The solid-phase synthesis method consists of only two steps: mechanical mixing and high-temperature calcination. As noted in past studies, the LiNi$_{1/3}$Co$_{1/3}$Mn$_{1/3}$O$_2$ cathode prepared using the low-temperature solid-state reaction was compared with samples synthesized with coprecipitation. Interestingly, the former was observed to deliver a great electrochemical performance, just like that of the latter [14]. Optimizing the raw materials by using α-MnO_2 nanorods, the $LiNi_{1/3}Co_{1/3}Mn_{1/3}O_2$ prepared with a low-temperature solid-state reaction at 900 °C shows the best electrochemical properties [15]. It is generally accepted that the microscopic morphology of $Li[Ni_xCo_yMn_z]O_2$ cathode materials is closely related to the stoichiometric ratio of transition metal elements. The size of the primary particles decreases sharply with the increase in Nickel content, which leads to the expansion of the contact area between the electrodes and the electrolytes, and, in turn, it causes deterioration in the cycling stability of the cathode material [16]. A relatively large number of particles can be accessed by the "ball milling–calcining" solid-phase synthesis method, which may be beneficial to the energy density of batteries. In addition, the $LiNi_{0.5}Co_{0.2}Mn_{0.3}O_2$ prepared with solid-state synthesis not only exhibits good electrochemical performance but is also suitable for the next steps in the modification of cathode materials, such as coating and doping [17]. However, the traditional solid-phase synthesis method also has limitations, mainly manifested in the uneven element distribution of the target product caused by poor mixing uniformity of the raw materials. Optimizing the ball milling process can reduce inhomogeneity and mitigate its adverse effects on the electrochemical performance of the materials. However, in previous solid-phase synthesis studies, most researchers selected acetate/oxalate as the raw materials to obtain $LiNi_xCo_yMn_zO_2$ [18], but rarely used carbonate. In terms of cost, acetate and oxalate are more expensive than carbonates.

In this paper, on the basis of optimizing the ball milling method and using carbonates as the raw materials, the $LiNi_{0.5}Co_{0.2}Mn_{0.3}O_2$ ternary cathode material was synthesized by using the "ball milling–calcining" pure solid-phase synthesis method. Thermogravimetric–differential scanning calorimetry (TG-DSC) was conducted to gain an insight into the reaction process, which includes the decomposition of MCO_3 (M = Li, Ni, Co, Mn) and the generation of target products. Furthermore, using a variety of electrochemical testing methods, XRD, SEM, etc., the effects of calcination temperatures on the characteristics of the as-prepared $LiNi_{0.5}Co_{0.2}Mn_{0.3}O_2$ cathode materials were comprehensively studied.

2. Materials and Methods

2.1. Preparation of Cathode

Stoichiometric amounts of $NiCO_3$ (98%, Macklin, Shanghai, China), $CoCO_3 \cdot xH_2O$ (CP, Macklin, Shanghai, China) and $MnCO_3$ (99.95%, Macklin, Shanghai, China), and 5% excess of Li_2CO_3 were mixed in a ball grinding jar. Using alcohol as a grinding aid and adding large, medium and small zirconia balls, the ball grinding jar was put into a planetary ball mill for mechanical mixing. The ball-to-material ratio was 2:1 and the weight ratio of large/medium/small zirconia balls was 3:4:3. The ball mill was milled in two directions at a speed of 500 rad/min for 2 h. Finally, to synthesize the $LiNi_{0.5}Co_{0.2}Mn_{0.3}O_2$ cathode material, the mixture was heated in an oxygen atmosphere for 7 h at various temperatures, namely, 800 °C, 900 °C, 1000 °C and 1100 °C.

2.2. Ex Site Characterizations

The thermal decomposition behaviors of the mixture in the range of room temperature to 1100 °C were investigated using TG-DSC analysis (NETZSCH STA 449 F5 Jupiter) at the heating rate of 10 K min^{-1}. The crystalline phase of the as-synthesized materials was identified using powder X-ray diffraction (XRD, Ultima IV, Tokyo, Japan) with Cu Kα radiation (2θ = 10~90°). The morphology of synthesized $LiNi_{0.5}Co_{0.2}Mn_{0.3}O_2$ was observed using field emission scanning electron microscopy (FESEM Quanta TEG 450, Hillsboro, OH, USA) in a high vacuum environment.

2.3. Electrochemical Measurements

An amount of 80 wt% active material (as-synthesized powder), 10 wt% acetylene black (conducting additive), and 10 wt% polyvinylidene fluoride (PVDF, binder) were mixed in N-methyl-2-pyrrolidone (NMP) to prepare the slurry, which then was coated on an aluminum foil. Subsequently, the prepared electrode film was dried at 90 °C for 12 h in a vacuum environment. All the electrochemical tests were conducted using the CR2025-type coin cell (HF-Kejing, Hefei, China) in this paper. Using the as-prepared sample as the cathode, the lithium foil as the anode, 1 M $LiPF_6$ dissolved in EC/DMC/DEC (1:1:1 by mass) as the electrolyte and Celgard 2400 membrane as the separator, the $LiNi_{0.5}Co_{0.2}Mn_{0.3}O_2$/Li half cells were assembled in an argon-filled glovebox. The cycling performance and high-rate charge/discharge capacity of the $LiNi_{0.5}Co_{0.2}Mn_{0.3}O_2$/Li half cells were measured using a LAND CT2001A tester (Wuhan, China) in a voltage range of 2.8~4.3 V at room temperature.

3. Results and Discussion

In the process of synthesizing cathode materials via the traditional solid-phase synthesis method, the mixing uniformity of the raw materials is also a key factor affecting the performance of the as-synthesized materials. Under the ideal conditions of complete mixing of the raw materials, the decomposition products produced by the various carbonates decomposed at different temperatures gradually diffuse and form the target product. In fact, the preparation of $LiNi_{0.5}Co_{0.2}Mn_{0.3}O_2$ cathode materials using solid-phase synthesis involves four different transition metal salts, which are very difficult to mix completely and uniformly. In most cases, the distribution of different carbonates in the mixed raw materials may be as shown in Figure 1b.

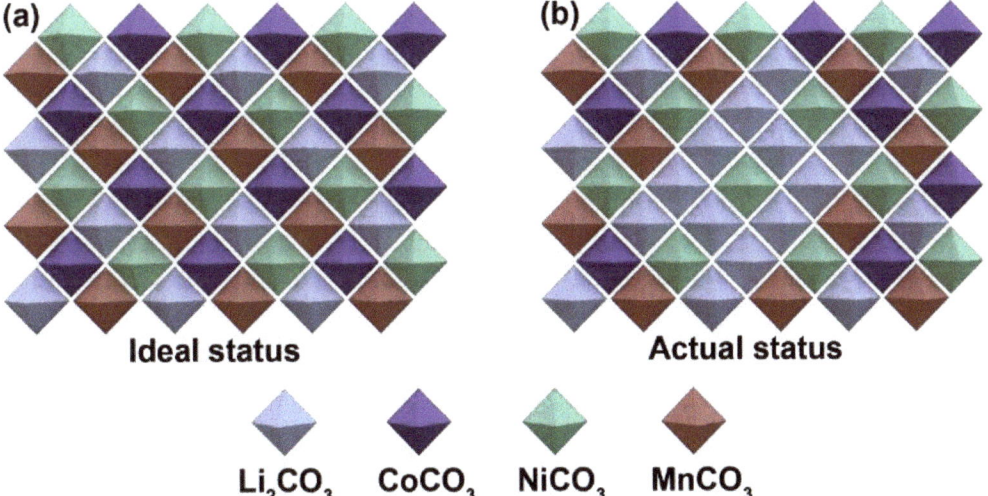

Figure 1. Estimation of the distribution of raw materials: (a) ideal status and (b) actual status.

In this situation, it is likely that one or more carbonates with a high decomposition temperature partially/completely encapsulate the carbonates with a low decomposition temperature, meaning that the wrapped material cannot gradually decompose with the increase in temperature and finally melt at a high temperature. This leads to the generation of heterogeneous phases and the loss of electrochemical performance. In this paper, we use alcohol as a grinding aid and adjust the ball milling process to enhance the mixing degree of the raw materials.

Under the condition that the raw material is a mixture rather than a pure substance, the temperature at which the decomposition reaction begins and completes may change. Especially for Li_2CO_3, transition metal salts may play a catalytic role in the decomposition of lithium carbonate. For the carbonate raw materials involved in this study ($MnCO_3$, $CoCO_3$, $NiCO_3$ and Li_2CO_3), the weight loss of the raw materials in the temperature range of 30~1100 °C is mainly related to the desorption of bound water and the gas generated by the decomposition reaction. In the differential scanning calorimetry (DSC) curve, it is generally accepted that the endothermic peak corresponds to the reaction of water loss, gas removal, the sublimation process, the decomposition process, etc., while the exothermic peak is related to the reaction of oxidation, crystallization, chemisorption, etc. In order to explore the reaction process of the carbonate raw materials at a high temperature, the compound was analyzed using thermogravimetric–differential scanning calorimetry (TG-DSC), as shown in Figure 2.

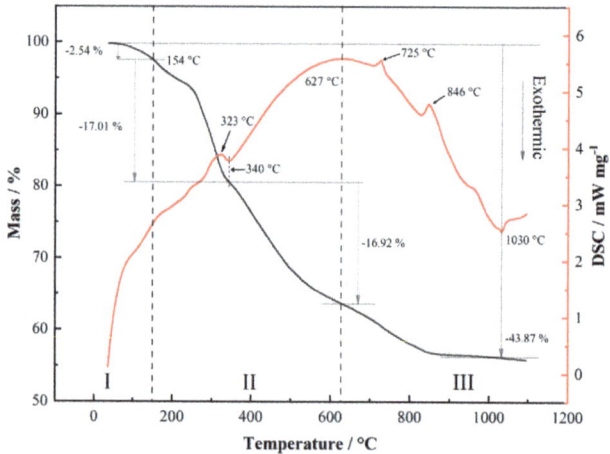

Figure 2. TG-DSC curves of raw materials at a heating rate of 10 K/min in air atmosphere.

Combined with the basic principle of high-temperature solid-state reactions and the property of the materials, it is considered that the sintering process of the raw materials is mainly divided into three steps, namely, (I) dehydration, (II) the decomposition of dehydrated intermediate and (III) the generation of the target product. Step (I) occurs in the range of room temperature to 154 °C. The first mass loss of 2.54% can be attributed to the removal of adsorbed water from the carbonate mixture and the dehydration of crystallized water in $CoCO_3 \cdot xH_2O$. Step (II) occurs in the temperature range of 154~627 °C, in which the decomposition reactions of $NiCO_3$, $MnCO_3$ and $CoCO_3$ are completed and the mass loss is caused by the decomposition of dehydrated intermediate. The endothermic response near 323 °C represents a one-time melt decomposition of $NiCO_3$ that has not been decomposed [19–22]. According to the TG curve, the actual mass loss (17.01%) is higher than the theoretical calculated value of 16.77% in the temperature range of 154~340 °C due to the decomposition of other carbonates, in addition to the decomposition of $NiCO_3$. The mass loss between 340 °C and 627 °C is mainly caused by the decomposition of $MnCO_3$ and $CoCO_3$. However, the phenomenon of mass loss still exists after 627 °C. Step (III), which occurs at 627~1030 °C, is the target product generation step. After the temperature reaches 627 °C, the DSC curve mainly shows an exothermic reaction. At this step, although the decomposition reaction of carbonates still occurs, the endothermic heat of the reaction is already much smaller than the exothermic heat of the oxidation reaction and crystallization of the target $LiNi_{0.5}Co_{0.2}Mn_{0.3}O_2$. This infers that there is still a small amount of mixing inhomogeneity, and some of the $MnCO_3$ and $CoCO_3$ particles melt at 725 °C at one time, thereby generating an endothermic peak. Furthermore, transition metal oxides produced by the decomposition of transition metal carbonates have a catalytic effect on the decomposition of Li_2CO_3, and the one-time melting decomposition of the last raw material Li_2CO_3 generates a final endothermic peak near 846 °C. The mixture at temperatures above 846 °C remains basically unchanged, while the DSC curve shows that there remains an exothermic process. This proves that there would be no new decomposition reactions above this temperature.

According to the TG curve, the total mass loss of the mixture from room temperature to 1030 °C is calculated to be 43.87%. After deducting the 2.54% mass loss caused by the dehydration process before 154 °C, the actual reaction mass loss is about 41.33%, which is similar to the theoretically calculated value of 40.71%. After 1030 °C, the reactions once again change from an endothermic process to an exothermic process, while the TG curve remains basically unchanged. It is believed that the high-temperature oxygen loss and sintering process of the product mainly occur above this temperature.

X-ray diffraction (XRD) patterns of $LiNi_{0.5}Co_{0.2}Mn_{0.3}O_2$ synthesized at various temperatures (800 °C, 900 °C, 1000 °C and 1100 °C) are shown in Figure 3.

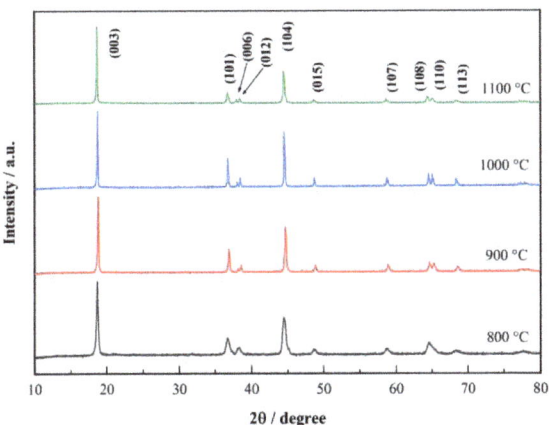

Figure 3. XRD spectra of $LiNi_{0.5}Co_{0.2}Mn_{0.3}O_2$ synthesized at 800 °C, 900 °C, 1000 °C and 1100 °C.

The diffraction patterns of all samples exhibit the typical characteristics of a layered α-NaFeO$_2$ structure with a group space of R$\bar{3}$m [23,24]. In addition, no peak of the impurity phase is observed in the patterns of samples synthesized at different temperatures. The split peaks of (006)/(102) and (108)/(110) are regarded as the characteristics of the layered structures [25,26]. Moreover, the ratio of $I_{(003)}/I_{(104)}$ reflects the degree of cation mixing between Ni^{2+} and Li^+ located at the 3a and 3b sites in an ideal layered structure [27,28]. For the $LiNi_{0.5}Co_{0.2}Mn_{0.3}O_2$ prepared at 800 °C, the $I_{(003)}/I_{(104)}$ ratio is significantly less than 1.2, and the indistinguishable (108)/(110) peaks indicate that the sample is contaminated by the rock salt, which has an adverse effect on the specific capacity and stability of the material. The two split peaks of (006)/(102) and (108)/(110) on the patterns reveal the well-defined layered structure of the samples synthesized at temperatures of 900 °C and above [29]. Furthermore, the $I_{(003)}/I_{(104)}$ ratio of the samples synthesized at 900 °C and 1000 °C is around 1.2, which can be expected to show good electrochemical properties in the above samples. Combined with the analysis of the TG-DSC curve, it can be inferred that the raw materials cannot be completely converted to the target $LiNi_{0.5}Co_{0.2}Mn_{0.3}O_2$ cathode material at the calcination temperature of 800 °C. Additionally, there are some impurities generated with the emergence of the target product at 800 °C.

The particle size has a direct impact on the ease of Li$^+$ deintercalation. An excessively large particle size means a longer path for Li$^+$ to migrate from the interior to the surface, thereby affecting the kinetic rate of the Li$^+$ deintercalation reaction and resulting in a lower rate performance [30]. However, the too small size of primary particles also increases the contact area between the electrodes and the electrolytes, which further aggravates the complex side reactions at the electrode/electrolyte interface. In other words, the too large/small size of particles damages the electrochemical performance of the materials, such as their cycling stability and high-rate charge/discharge capacity [31].

SEM images of $LiNi_{0.5}Co_{0.2}Mn_{0.3}O_2$ synthesized at different temperatures are given in Figure 4.

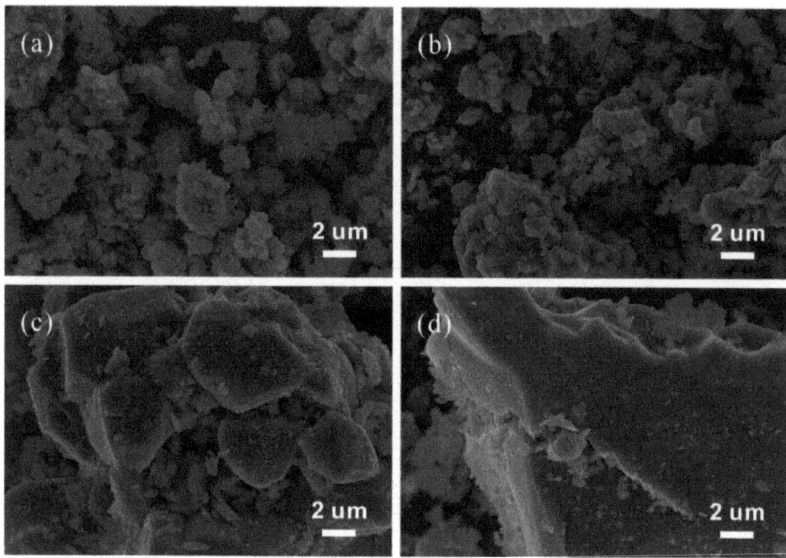

Figure 4. SEM images of LiNi$_{0.5}$Co$_{0.2}$Mn$_{0.3}$O$_2$ synthesized at different temperatures: (**a**) 800 °C, (**b**) 900 °C, (**c**) 1000 °C and (**d**) 1100 °C.

The particles of the samples are secondary particles composed of irregular primary particles with the obvious phenomenon of agglomeration, and they are different from the spherical LiNi$_{0.5}$Co$_{0.2}$Mn$_{0.3}$O$_2$ particles prepared using the co-precipitation method. Samples synthesized at the same temperature have primary particles with roughly the same size, indicating that the raw materials are uniformly mixed. In addition, it can be clearly seen from the SEM images that the size of primary particles increases with the increase in the sintering temperature (from 800 °C to 1100 °C), which is due to the rise in the calcination temperature promoting the growth of crystal grains.

All the LiNi$_{0.5}$Co$_{0.2}$Mn$_{0.3}$O$_2$/Li half cells were precycled at 0.1 C in a voltage range of 2.8~4.3 V at room temperature for five cycles to activate the cells. The first charge/discharge curves of the four samples are shown in Figure 5.

With the sintering temperature increasing from 800 °C to 900 °C and 1000 °C, the charge/discharge specific capacity of the corresponding LiNi$_{0.5}$Co$_{0.2}$Mn$_{0.3}$O$_2$/Li half cells rises from 148/134 mAh g^{-1} to 164.8/161 mAh g^{-1} and 170.8/167.2 mAh g^{-1}. However, the charge/discharge specific capacity of the cathode material synthesized at 1100 °C decreases sharply to 98.5/89.1 mAh g^{-1}. In addition, the coulomb efficiency of the half cells with LiNi$_{0.5}$Co$_{0.2}$Mn$_{0.3}$O$_2$ synthesized at 900 °C and 1000 °C reaches 97.69% and 97.89%, respectively. This is also consistent with the results of XRD and SEM. Furthermore, the specific capacity of LiNi$_{0.5}$Co$_{0.2}$Mn$_{0.3}$O$_2$ prepared at 800 °C, 900 °C and 1000 °C is higher than that of the cathode materials synthesized using the co-precipitation method (159 mAh g^{-1}).

The cycling performance of LiNi$_{0.5}$Co$_{0.2}$Mn$_{0.3}$O$_2$/Li half cells with cathode materials prepared at different temperatures was measured at a rate of 0.5 C at room temperature (Figure 6). Upon completing cycling 50 cycles, the half cells with LiNi$_{0.5}$Co$_{0.2}$Mn$_{0.3}$O$_2$ synthesized at 800 °C, 900 °C, 1000 °C and 1100 °C were observed to deliver capacity retention of 80.52% (at 50th ≈ 107.9 mAh g^{-1}), 78.76% (at 50th ≈ 126.8 mAh g^{-1}), 81.94% (at 50th ≈ 137 mAh g^{-1}) and 47.59% (at 50th ≈ 42.4 mAh g^{-1}), respectively. In terms of charge/discharge specific capacity and cycling stability, LiNi$_{0.5}$Co$_{0.2}$Mn$_{0.3}$O$_2$ prepared at 1000 °C shows an optimal property. In addition, the poor performance of LiNi$_{0.5}$Co$_{0.2}$Mn$_{0.3}$O$_2$/Li half cells with the cathode prepared at 1100 °C is mainly attributed to the excessive size of the sample particles. An increase in the sintering temperature within a certain range is beneficial to the development of crystals and the electrochemical performance of the material. However, an excessively high sintering temperature may cause problems, such as material compaction and excessively coarse crystal grains, which adversely affect the electrochemical properties of the materials.

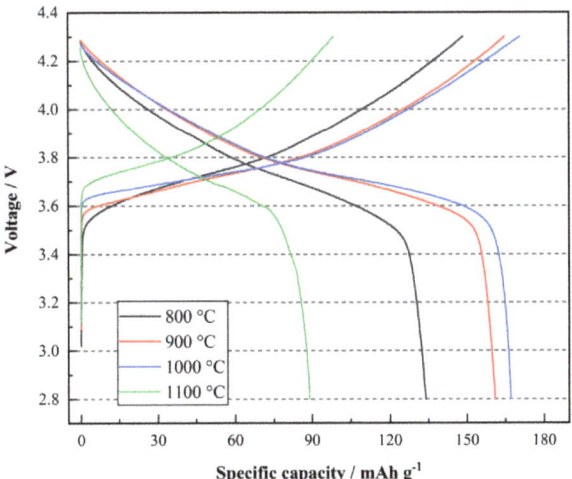

Figure 5. Initial charge/discharge curves of the half cells with LiNi$_{0.5}$Co$_{0.2}$Mn$_{0.3}$O$_2$ synthesized at 800 °C, 900 °C, 1000 °C and 1100 °C.

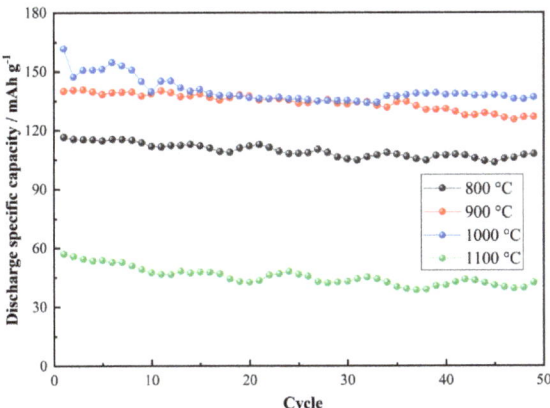

Figure 6. Cycling property of the half cells with LiNi$_{0.5}$Co$_{0.2}$Mn$_{0.3}$O$_2$ synthesized at 800 °C, 900 °C, 1000 °C and 1100 °C.

The high-rate charge/discharge capacity is one of the most significant indicators of battery quality. Increasing the current density from 0.5 to 1 C, 1.5 C and 2 C, the rate performance of LiNi$_{0.5}$Co$_{0.2}$Mn$_{0.3}$O$_2$/Li half cells with cathode materials synthesized at different temperatures was examined in a voltage range of 2.8~4.3 V at room temperature (Figure 7). At the rate of 2 C, the discharge capacities of the four samples with synthesized temperatures of 800 °C, 900 °C, 1000 °C and 1100 °C are 107.1 mAh g^{-1}, 138.9 mAh g^{-1}, 126.2 mAh g^{-1} and 20.6 mAh g^{-1}, respectively. Compared with the condition at the rate of 0.5 C, capacity retention rates reach 86.16%, 93.79%, 88.50% and 47.69%. Upon recovering the rate of 0.5 C, the specific capacity is restored to 124 mAh g^{-1}, 149.8 mAh g^{-1}, 140.1 mAh g^{-1} and 35 mAh g^{-1}. The samples synthesized at 900 °C and 1000 °C exhibit a better rate performance and a stronger structural stability under high-rate conditions. Owing to the oversized crystal grains, lithium ions have to transport via a long pathway in solid LiNi$_{0.5}$Co$_{0.2}$Mn$_{0.3}$O$_2$ prepared at 1100 °C. It is widely believed that a longer pathway for lithium ions in the charge/discharge processes means a worse high-rate property of lithium-ion batteries. On the other hand, the as-prepared sample at 800 °C contains more impurities than that at 900 °C and 1000 °C, so it also delivers a poor high-rate performance. Regardless of the grain sizes or lattice structure, the samples synthesized at 900 °C and 1000 °C all exhibit good properties to support an excellent high-rate capacity.

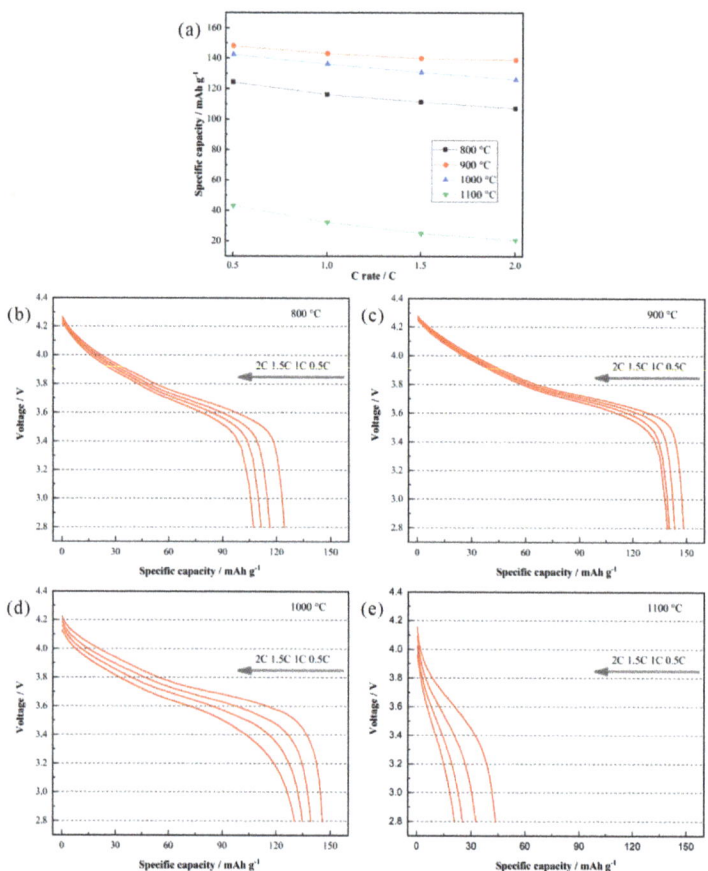

Figure 7. (**a**) Rate performances of the half cells with LiNi$_{0.5}$Co$_{0.2}$Mn$_{0.3}$O$_2$ synthesized at different temperatures and the corresponding discharge curves at the rates of (**b**) 0.5 C, (**c**) 1 C, (**d**) 1.5 C, and (**e**) 2 C.

4. Conclusions

In this paper, using four carbonates as raw materials, the layered LiNi$_{0.5}$Co$_{0.2}$Mn$_{0.3}$O$_2$ cathode material with high specific capacity was successfully synthesized with the solid-phase synthesis method. LiNi$_{0.5}$Co$_{0.2}$Mn$_{0.3}$O$_2$ synthesized at 1000 °C provides an optimal cycling performance due to its high crystallinity and uniform particle size of about 2 μm. The initial discharge specific capacity and coulomb efficiency reached 167.2 mAh g^{-1} and 97.89%, respectively. On the other hand, the high-rate charge/discharge performance of LiNi$_{0.5}$Co$_{0.2}$Mn$_{0.3}$O$_2$ synthesized at 900 °C was particularly noteworthy, with its capacity still maintained at 93.78% at the rate of 2 C (at 0.5 C ≈ 148.1 mAh g^{-1}). Restoring the current density to 0.5 C rate, its specific capacity still recovered to 149.8 mAh g^{-1}. These results indicate that the LiNi$_{0.5}$Co$_{0.2}$Mn$_{0.3}$O$_2$ cathode material for lithium-ion batteries prepared using carbonate raw materials and the solid-phase synthesis method has broad prospects.

This synthesis method not only obtain $LiNi_{0.5}Co_{0.2}Mn_{0.3}O_2$ with excellent electrochemical properties, but also greatly avoid high production costs and environmental pollution. Owing to its simplicity and efficiency, this synthetic process is well suited for large-scale commercial production.

Author Contributions: Formal analysis, X.L.; Investigation, X.L., B.S., J.Z.; Supervision, W.X.; Visualization, X.L.; Writing—original draft, X.L.; Writing—review & editing, X.L., B.S., W.X. and J.Z. All authors have read and agreed to the published version of the manuscript.

Funding: This research was funded by National Key Research and Development Plan (2016YFE0111500).

Institutional Review Board Statement: Not applicable.

Informed Consent Statement: Not applicable.

Data Availability Statement: Not applicable.

Acknowledgments: The authors acknowledge that this work was supported by the National Key Research and Development Plan (2016YFE0111500).

Conflicts of Interest: The authors declare no conflict of interest.

References

1. Chen, Y.; Kang, Y.; Zhao, Y.; Wang, L.; Liu, J.; Li, Y.; Liang, Z.; He, X.; Li, X.; Tavajohi, N.; et al. A review of lithium-ion battery safety concerns: The issues, strategies, and testing standards. *J. Energy Chem.* **2021**, *59*, 83–99. [CrossRef]
2. Huang, B.; Pan, Z.; Su, X.; An, L. Recycling of lithium-ion batteries: Recent advances and perspectives. *J. Power Sources* **2018**, *399*, 274–286. [CrossRef]
3. Choi, S.; Wang, G. Advanced lithium-Ion batteries for practical applications: Technology, development, and future perspectives. *Adv. Mater. Technol.* **2018**, *3*, 1700376. [CrossRef]
4. Dang, R.; Qu, Y.; Ma, Z.; Yu, L.; Duan, L.; Lü, W. The effect of elemental doping on nickel-rich NCM cathode materials of lithium ion batteries. *J. Phys. Chem. C* **2021**, *126*, 151–159. [CrossRef]
5. Chen, C.; Tao, T.; Qi, W.; Zeng, H.; Wu, Y.; Liang, B.; Yao, Y.; Lu, S.; Chen, Y. High-performance lithium ion batteries using SiO_2-coated $LiNi_{0.5}Co_{0.2}Mn_{0.3}O_2$ microspheres as cathodes. *J. Alloys Compd.* **2017**, *709*, 708–716. [CrossRef]
6. Zhao, R.; Miao, J.; Lan, W.; Wu, Z.; Hung, I.M.; Lv, D.; Zeng, R.; Shi, G.; Chen, H. Synthesis of layered materials by ultrasonic/microwave-assisted coprecipitation method: A case study of $LiNi_{0.5}Co_{0.2}Mn_{0.3}O_2$. *Sustain. Mater. Technol.* **2018**, *18*, e00083. [CrossRef]
7. Jung, S.K.; Gwon, H.; Hong, J.; Park, K.Y.; Seo, D.H.; Kim, H.; Hyun, J.; Yang, W.; Kang, K. Understanding the degradation mechanisms of $LiNi_{0.5}Co_{0.2}Mn_{0.3}O_2$ cathode material in lithium ion batteries. *Adv. Energy Mater.* **2014**, *4*, 1300787. [CrossRef]
8. Du, K.; Hua, C.; Tan, C.; Peng, Z.; Cao, Y.; Hu, G. A high-powered concentration-gradient $Li(Ni_{0.85}Co_{0.12}Mn_{0.03})O_2$ cathode material for lithium ion batteries. *J. Power Sources* **2014**, *263*, 203–208. [CrossRef]
9. Zhang, S.; Qiu, X.; He, Z.; Weng, D.; Zhu, W. Nanoparticled $Li(Ni_{1/3}Co_{1/3}Mn_{1/3})O_2$ as cathode material for high-rate lithium-ion batteries. *J. Power Sources* **2006**, *153*, 350–353. [CrossRef]
10. Hu, S.K.; Cheng, G.H.; Cheng, M.Y.; Hwang, B.J.; Santhanam, R. Cycle life improvement of ZrO_2-coated spherical $LiNi_{1/3}Co_{1/3}Mn_{1/3}O_2$ cathode material for lithium ion batteries. *J. Power Sources* **2009**, *188*, 564–569. [CrossRef]
11. Kim, H.S.; Kim, Y.; Kim, S.I.; Martin, S.W. Enhanced electrochemical properties of $LiNi_{(1/3)}Co_{(1/3)}Mn_{(1/3)}O_2$ cathode material by coating with $LiAlO_2$ nanoparticles. *J. Power Sources* **2006**, *161*, 623–627. [CrossRef]
12. Yao, C.; Mo, Y.; Jia, X.; Chen, X.; Xia, J.; Chen, Y. $LiMnPO_4$ surface coating on $LiNi_{0.5}Co_{0.2}Mn_{0.3}O_2$ by a simple sol-gel method and improving electrochemical properties. *Solid State Ion.* **2018**, *317*, 156–163. [CrossRef]
13. Shi, Y.; Zhang, M.; Fang, C.; Meng, Y.S. Urea-based hydrothermal synthesis of $LiNi_{0.5}Co_{0.2}Mn_{0.3}O_2$ cathode material for Li-ion battery. *J. Power Sources* **2018**, *394*, 114–121. [CrossRef]
14. Liu, J.; Qiu, W.; Yu, L.; Zhao, H.; Li, T. Synthesis and electrochemical characterization of layered $Li(Ni_{1/3}Co_{1/3}Mn_{1/3})O_2$ cathode materials by low-temperature solid-state reaction. *J. Alloys Compd.* **2008**, *449*, 326–330. [CrossRef]
15. Tan, L.; Liu, H. High rate charge–discharge properties of $LiNi_{1/3}Co_{1/3}Mn_{1/3}O_2$ synthesized via a low temperature solid-state method. *Solid State Ion.* **2010**, *181*, 1530–1533. [CrossRef]
16. Noh, H.J.; Youn, S.; Yoon, C.S.; Sun, Y.K. Comparison of the structural and electrochemical properties of layered $Li[Ni_xCo_yMn_z]O_2$ (x = 1/3, 0.5, 0.6, 0.7, 0.8 and 0.85) cathode material for lithium-ion batteries. *J. Power Sources* **2013**, *233*, 121–130. [CrossRef]
17. Li, L.; Xia, L.; Yang, H.; Zhan, X.; Chen, J.; Chen, Z.; Duan, J. Solid-state synthesis of lanthanum-based oxides Co-coated $LiNi_{0.5}Co_{0.2}Mn_{0.3}O_2$ for advanced lithium ion batteries. *J. Alloys Compd.* **2020**, *832*, 154959. [CrossRef]
18. He, Y.S.; Ma, Z.F.; Liao, X.Z. Synthesis and characterization of submicron-sized by a simple self-propagating solid-state metathesis method. *J. Power Sources* **2007**, *163*, 1053–1058. [CrossRef]

19. Javanmardi, M.; Emadi, R.; Ashrafi, H. Synthesis of nickel aluminate nanoceramic compound from aluminum and nickel carbonate by mechanical alloying with subsequent annealing. *Trans. Nonferrous Met. Soc. China* **2016**, *26*, 2910–2915. [CrossRef]
20. Kim, J.W.; Lee, H.G. Thermal and carbothermic decomposition of Na_2CO_3 and Li_2CO_3. *Metall. Mater. Trans. B* **2001**, *32*, 17–24. [CrossRef]
21. Ouyang, Z.; Wen, P.; Chen, Y.; Ye, L. Study on thermodynamic equilibrium and character inheritance of cobalt carbonate decomposition. *Vacuum* **2020**, *179*, 109559. [CrossRef]
22. Zaki, M.I.; Nohman, A.K.H.; Kappenstein, C.; Wahdan, T.M. Temperature-programmed characterization studies of thermochemical events occurring in the course of decomposition of Mn oxysalts. *J. Mater. Chem.* **1995**, *5*, 1081–1088. [CrossRef]
23. Wang, L.; Hu, Y.H. Surface modification of $LiNi_{0.5}Co_{0.2}Mn_{0.3}O_2$ cathode materials with Li_2O-B_2O_3-LiBr for lithium-ion batteries. *Int. J. Energy Res.* **2019**, *43*, 4644–4651. [CrossRef]
24. Chen, Y.; Li, Y.; Li, W.; Cao, G.; Tang, S.; Su, Q.; Deng, S.; Guo, J. High-voltage electrochemical performance of $LiNi_{0.5}Co_{0.2}Mn_{0.3}O_2$ cathode material via the synergetic modification of the Zr/Ti elements. *Electrochim. Acta* **2018**, *281*, 48–59. [CrossRef]
25. Hao, J.; Yu, Z.; Liu, H.; Song, W.; Liu, J.; Kong, L.; Li, C. Enhancing electrochemical performances of $LiNi_{0.5}Co_{0.2}Mn_{0.3}O_2$ cathode materials derived from NiF_2 artificial interface at elevated voltage. *J. Alloys Compd.* **2019**, *806*, 814–822. [CrossRef]
26. Zhang, Y.; Wang, Z.; Zhong, Y.; Wu, H.; Li, S.; Cheng, Q.; Guo, P. Coating for improving electrochemical performance of NCM523 cathode for lithium-ion batteries. *Ionics* **2020**, *27*, 13–20. [CrossRef]
27. Jia, X.; Yan, M.; Zhou, Z.; Chen, X.; Yao, C.; Li, D.; Chen, D.; Chen, Y. Nd-doped $LiNi_{0.5}Co_{0.2}Mn_{0.3}O_2$ as a cathode material for better rate capability in high voltage cycling of Li-ion batteries. *Electrochim. Acta* **2017**, *254*, 50–58. [CrossRef]
28. Breuer, O.; Chakraborty, A.; Liu, J.; Kravchuk, T.; Burstein, L.; Grinblat, J.; Kauffman, Y.; Gladkih, A.; Nayak, P.; Tsubery, M.; et al. Understanding the role of minor molybdenum doping in $LiNi_{0.5}Co_{0.2}Mn_{0.3}O_2$ electrodes: From structural and surface analyses and theoretical modeling to practical electrochemical cells. *ACS Appl. Mater. Interfaces* **2018**, *10*, 29608–29621. [CrossRef]
29. Wang, Z.; Sun, Y.; Chen, L.; Huang, X. Electrochemical characterization of positive electrode material $LiNi_{[1/3]}Co_{[1/3]}Mn_{[1/3]}O_{[2]}$ and compatibility with electrolyte for lithium-ion batteries. *J. Electrochem. Soc.* **2004**, *151*, A914. [CrossRef]
30. Trevisanello, E.; Ruess, R.; Conforto, G.; Richter, F.H.; Janek, J. Polycrystalline and single crystalline NCM cathode materials—Quantifying particle cracking, active surface area, and lithium diffusion. *Adv. Energy Mater.* **2021**, *11*, 2003400. [CrossRef]
31. He, L.P.; Li, K.; Zhang, Y.; Liu, J. Substantial doping engineering in layered $LiNi_{0.5+x}Co_{0.2-x}Mn_{0.3}O_2$ materials for lithium-ion batteries. *J. Electrochem. Soc.* **2021**, *168*, 060534. [CrossRef]

Article

Boosting Lithium Storage of a Metal-Organic Framework via Zinc Doping

Wenshan Gou [1], Zhao Xu [1], Xueyu Lin [2,*], Yifei Sun [1], Xuguang Han [1], Mengmeng Liu [1] and Yan Zhang [1,*]

1. Institute of Advanced Cross-Field Science, College of Life Sciences, Qingdao University, Qingdao 200671, China; 2019025181@qdu.edu.cn (W.G.); 2019025203@qdu.edu.cn (Z.X.); 2020025520@qdu.edu.cn (Y.S.); 2020025501@qdu.edu.cn (X.H.); 2020025500@qdu.edu.cn (M.L.)
2. Beijing National Laboratory for Molecular Sciences and State Key Laboratory of Rare Earth Materials Chemistry and Applications, College of Chemistry and Molecular Engineering, Peking University, Beijing 100871, China
* Correspondence: 1801110301@pku.edu.cn (X.L.); yzhang_iacs@qdu.edu.cn (Y.Z.)

Abstract: Lithium-ion batteries (LIBs) as a predominant power source are widely used in large-scale energy storage fields. For the next-generation energy storage LIBs, it is primary to seek the high capacity and long lifespan electrode materials. Nickel and purified terephthalic acid-based MOF (Ni-PTA) with a series amounts of zinc dopant (0, 20, 50%) are successfully synthesized in this work and evaluated as anode materials for lithium-ion batteries. Among them, the 20% atom fraction Zn-doped Ni-PTA ($Zn_{0.2}$-Ni-PTA) exhibits a high specific capacity of 921.4 mA h g^{-1} and 739.6 mA h g^{-1} at different current densities of 100 and 500 mA g^{-1} after 100 cycles. The optimized electrochemical performance of $Zn_{0.2}$-Ni-PTA can be attributed to its low charge transfer resistance and high lithium-ion diffusion rate resulting from expanded interplanar spacing after moderate Zn doping. Moreover, a full cell is fabricated based on the $LiFePO_4$ cathode and as-prepared MOF. The $Zn_{0.2}$-Ni-PTA shows a reversible specific capacity of 97.9 mA h g^{-1} with 86.1% capacity retention (0.5 C) after 100 cycles, demonstrating the superior electrochemical performance of $Zn_{0.2}$-Ni-PTA anode as a promising candidate for practical lithium-ion batteries.

Keywords: metal-organic frameworks; zinc-ions doped; energy storage and conversion; lithium-ion batteries

Citation: Gou, W.; Xu, Z.; Lin, X.; Sun, Y.; Han, X.; Liu, M.; Zhang, Y. Boosting Lithium Storage of a Metal-Organic Framework via Zinc Doping. *Materials* 2022, 15, 4186. https://doi.org/10.3390/ma15124186

Academic Editor: Alessandro Dell'Era

Received: 2 May 2022
Accepted: 9 June 2022
Published: 13 June 2022

Publisher's Note: MDPI stays neutral with regard to jurisdictional claims in published maps and institutional affiliations.

Copyright: © 2022 by the authors. Licensee MDPI, Basel, Switzerland. This article is an open access article distributed under the terms and conditions of the Creative Commons Attribution (CC BY) license (https://creativecommons.org/licenses/by/4.0/).

1. Introduction

Lithium-ion batteries (LIBs), due to their high energy density, low cost and long lifespan, have been regarded as critical energy storage devices for electric vehicles, large-scale electricity storage, etc. [1]. The scientific concern with lithium-ion batteries is developing cathodes, anodes, and electrolytes [2,3]. The performance of anode material has become a restriction for high-energy LIBs. As we know, an ideal electrode material should possess both high lithium-ions storage capacity and stable electrochemical performance. The commercial anode material graphite, which can be easily produced, is limited by an insufficient capacity of 372 mA h g^{-1} [4]. Other types of anode materials such as alloy [5,6] and conversion reaction-based transition metal oxides [7,8] possess ultrahigh specific capacity. Nevertheless, the dramatic volume expansion during the charge/discharge process and poor cycle performance restrict the broad use of those materials. Therefore, it is significant to explore novel anode materials with excellent performances for the further development of LIBs.

Metal-organic frameworks (MOFs), as a class of porous materials combining metal ions or clusters with organic linkers through coordination bonds, with a huge variety of structures, large surface areas and adjustable porosity, are widely used in many fields such as gas storage [9], chemical sensors [10], catalysis [11], and drug delivery [12]. Over the past several years, different MOFs have been applied in the secondary battery field, especially in LIBs [13,14]. Generally, the use of MOFs material in the LIBs field can be classified into two aspects, using MOFs as templates to produce homogeneous metal oxide materials and carbon materials for LIBs [14–16], and using MOFs as electrode materials directly [17–19]. Chen's group first explored MOF-177 with different morphologies as an anode material for LIBs even though the electrochemical performance of MOF-177 was not promising [18]. After that, more and more MOFs have been investigated as electrode materials for LIBs. For instance, Wang et al. synthesized a Co-based coordination polymer nanowire with a specific capacity of 1132 mA h g^{-1} at a current density of 100 mA g^{-1} [19]. However, most of the works mentioned above focus on introducing novel MOF electrodes for LIBs. Research about the modification of MOFs to improve their inherent property (such as poor electronic conductivity) and electrochemical performances are still scarce [20]. As we all know, element doping is a regular modification to enhance the electronic conductivity and improve the electrochemical performances of electrode materials. Previous similar works could be found in LiFePO$_4$ [21,22] and several metal–oxides systems [23]. In this respect, the modification of MOFs by doping should be an effective and practical way to enhance their electrochemical performance. Purified terephthalic acid (PTA) is an optimized ligand used in the synthesis of MOFs due to its easy availability from cheap poly-ethylene terephthalate (PET) plastic products. The purified terephthalic acid-based MOF (Ni-PTA) is a prospective anode material for LIBs due to its unique layered structure. However, the performances of Ni-PTA need to be further elevated by optimizing its fine structure (such as interlayer spacing) to realize the high diffusion of Li$^+$ [16].

Compared to Ni^{2+} (0.065 nm), Zn^{2+} owns a larger radius (0.074 nm), and the doping of Zn^{2+} can expand the interplanar distances of Ni-PTA, which facilitates the diffusion of Li$^+$. Herein, we successfully synthesized a series of nickel and purified terephthalic acid-based MOF (Ni-PTA) with different amounts of zinc doping (the x% atom fraction Zn-doped Ni-PTA was denoted as Zn$_x$-Ni-PTA) through a simple solvothermal method and explored them as anode materials for LIBs. The result shows that Zn$_{0.2}$-Ni-PTA achieves more excellent cycle stability and higher specific capacity than the Ni-PTA without Zn-doped and the 50% atom fraction Zn-doped (Zn$_{0.5}$-Ni-PTA) one. The further kinetics information reveals that the Zn$_{0.2}$-Ni-PTA demonstrates a faster Li$^+$ diffusion rate and lower charge transfer resistance, which also explains the brilliant electrochemical performance of Zn$_{0.2}$-Ni-PTA. In addition, we fabricated a full cell based on a LiFePO$_4$ cathode to test the electrochemical performance of Zn$_{0.2}$-Ni-PTA as anode.

2. Materials and Methods

2.1. Synthesis of Zn$_x$-Ni-PTA

The Zn$_x$-Ni-PTA was prepared by a simple one-pot solvothermal route. Briefly, 0.166 g of purified terephthalic acid (PTA, Aladdin, 99%), 1.5/1.2/0.75 mmol of Ni(NO$_3$)$_2$·6H$_2$O (Aladdin, analytically pure) and different amounts of Zn(NO$_3$)$_2$·6H$_2$O (Aladdin, analytically pure) (0, 0.3, 0.75 mmol) were, respectively, dissolved in a mixed solvent consisting of 15 mL absolute ethanol (Macklin, analytically pure) and 15 mL deionized H$_2$O, which was then stirred for 30 min. Then, the mixture was transformed into a 40 mL Teflon-lined stainless steel autoclave and reacted at 180 °C for 24 h. After the autoclave cooled down to room temperature, the green precipitates were washed by N,N-dimethylformamide (DMF, Aladdin, analytically pure) and absolute ethanol several times. Ultimately, this product was dried at 60 °C in air for 24 h. The 0%, 20% and 50% atom fraction Zn-doped Ni-PTA denoted Zn$_0$-Ni-PTA, Zn$_{0.2}$-Ni-PTA, and Zn$_{0.5}$-Ni-PTA, respectively.

2.2. Materials Characterization

The powder X-ray diffraction (XRD) patterns were recorded by a Rigaku II X-ray diffraction spectrometer (Japan Science Co., Tokyo, Japan) using Cu-Kα radiation. Fourier-transform infrared (FTIR) transmission spectra were performed by FTIR-65 IR spectrophotometer (Tianjin Port East Technology Co., Tianjin, China). Thermogravimetric–Differential Thermal Analysis (TG-DTA) was performed by a Labsys Evo thermogravimetric differential thermal analyzer from the Setaram Instrumentation (France) with a rate of 10 K min^{-1} under ambient condition. The JSM-JSM7500 instrument (Japan Electronics Co., Tokyo, Japan) obtained scanning electron microscopy (SEM) images. X-ray photoelectron spectroscopy (XPS) was taken on the PHI5000VersaProbe instrument (Shanghai Yuzhong Industrial Co., Shanghai, China). Inductively Coupled Plasma–Atomic Emission Spectrometry (ICP-AES) record was obtained by ICP-9000(N+M) atomic emission spectroscopy (Thermo Jarrel-Ash Co., Boston, MA, USA).

2.3. Electrochemical Measurements

To carry out the electrochemical measurements, the active material, Super-P carbon black and polyvinyl difluoride (PVDF, Macklin, M_w = 1,000,000) were blended in a weight ratio of 6:3:1 with several drops of N-methyl pyrrolidone (NMP) added and stirred until the mixture became homogeneous. Afterward, the mixed slurry was pasted onto a copper foil current collector with a diameter of 10 mm and dried in vacuum at 80 °C for 12 h. The average loading of active materials was about 1.2 mg cm^{-1}. The CR 2016-type cells were assembled in an Ar-filled glove box (water and oxygen concentration less than 0.1 ppm), using lithium foil as the counter electrode and Celgard 2300 polypropylene as separators (diameter—16 mm). The electrolyte was 1 M LiPF$_6$ dissolved in a solvent mixed with EC-DMC-EMC (1:1:1 vol %). The galvanostatic charge–discharge (GCD) tests were performed by a LAND CT2001 battery test system (Wuhan Kingnuo Electronic Co., Wuhan, China) at room temperature (25 °C). Cycle voltammetry (CV) and electrochemical impedance spectroscopy (EIS) were taken on a CHI-660B electrochemical station (Shanghai Chenhua Instrument Co., Shanghai, China) at a full charge state of batteries after the 20th cycle. The current density of test is 0.1 A g^{-1}. The CV tests were carried out with a scan rate of 0.1 mV s^{-1}, while the EIS data were recorded in the frequency range of 0.01–100 kHz. The full cell was similarly assembled as above-mentioned: 20 at % Zn-doped Ni-PTA as anode, and the cathode was obtained by homogeneously mixing the LiFePO$_4$ (Canrd, D-1, 80 wt %), PVDF (10 wt %) and NMP (10 wt %). The loading density was about 0.4 mg cm^{-2}, and Al foil was used as the current collector. The specific capacity was calculated based on the cathode material.

3. Results and Discussion

3.1. Characterization of Zn_x-Ni-PTA

To investigate the crystalline phase of as-prepared products, XRD measurement was performed. Figure 1a shows the XRD patterns of Zn_x-Ni-PTA with different amounts of zinc doping and the standard card of $Ni_3(OH)_2(C_8H_4O_4)_2 \cdot (H_2O)_4 \cdot 2H_2O$ (CCDC 638866) belonging to the space group of P-1(2)-triclinic. Although all the XRD patterns of synthesized MOFs shared similarities, with the increasing Zn^{2+} amount, the peaks of (010) and (020) tend to become invisible. This phenomenon illustrates that the crystal structure order along the *b*-axis decreased. In addition to the absence of some reflections, with the amount of zinc dopant increasing, the peaks of (100) and (200) shifted to a lower angle, which can be attributed to the partial replacement of doped larger Zn^{2+} (0.074 nm) to Ni^{2+} (0.065 nm) in MOFs, expanding the interplanar distances along the *a*-axis [24,25]. The SEM images of Zn_x-Ni-PTA are presented in Figure S1. The Zn_0-Ni-PTA showed layered micro-sheet like morphology features, and the flower-like layered structure was observed after Zn^{2+} doping. TGA curves (Figure S2) indicates that the initial weight loss is due to the loss of solvated water molecules and the weight loss in the range of 350–420 °C is attributed to the thermal decomposition of the Zn_x-Ni-PTA.

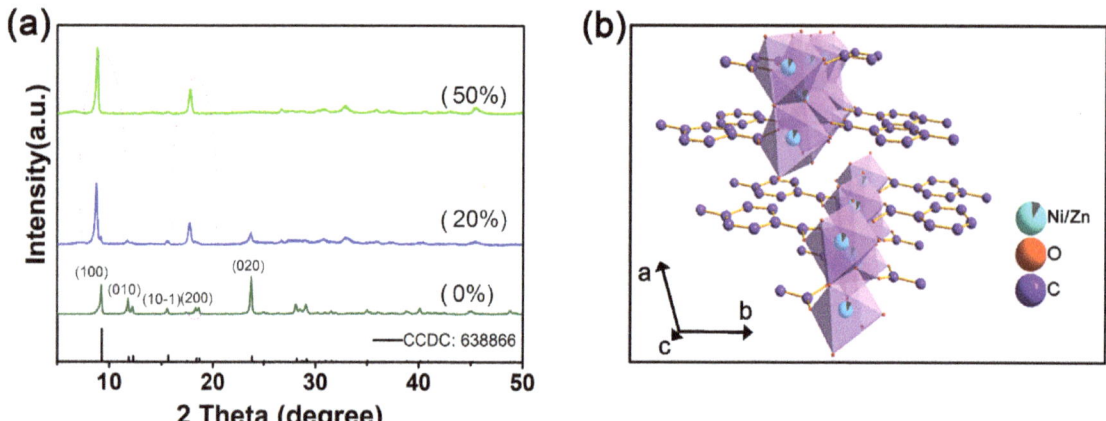

Figure 1. (a) Powder XRD pattern of the different atom fraction Zn_x-Ni-PTA samples. (b) The structure mode of Zn_x-Ni-PTA.

In addition, the similar FTIR peaks also proved that these synthesized MOFs have similar layered topology crystal structures, as shown in Figure 2a. According to the patterns, bands at 1501 cm^{-1} reveal the para-aromatic C-H stretching mode. The bending stretching vibration of -OH is observed at 3613 cm^{-1}. The bands at 3070 cm^{-1}, 3344 cm^{-1} and 3431 cm^{-1} are assigned to stretching vibrations of H_2O in MOFs. In addition, the asymmetric and symmetric vibration of COO^- are located at 1581 cm^{-1} and 1400 cm^{-1}, respectively. The more Zn content, the larger of COO^- groups separation that demonstrated the impact of doped zinc ions on the structure of MOF. The absorption peaks at 522 cm^{-1} in $Zn_{0,0.2,0.5}$-Ni-PTA are associated to Ni-O, and peaks at 437 cm^{-1} in $Zn_{0.2,0.5}$-Ni-PTA are associated to Zn-O vibration bonds, respectively [26,27]. This also confirms the successful doping of Zn^{2+} to the Ni-PTA. The content of dopant Zn^{2+} in Zn_x-Ni-PTA samples (0%, 22.06% and 50.83%) was detected by induced coupled plasma atomic emission spectroscopy (ICP-AES), respectively (Table S1) and matched well with the theoretical value. The above results indicated that different amounts of Zn^{2+} successfully replaced the Ni^{2+} in Ni-MOFs.

For further information about elemental composition and valance state, an XPS test was performed. As observed from the XPS spectra of Zn 2p (Figure 2b), no characteristic peaks were detected in the Zn_0-Ni-PTA sample, and two prominent doublet peaks corresponding to Zn $2p_{1/2}$ and Zn $2p_{3/2}$ can be observed in $Zn_{0.2}$-Ni-PTA and $Zn_{0.5}$-Ni-PTA, indicating the successful doping of Zn in these two samples [28]. In the Ni 2p spectra (Figure 2c) of Zn_x-Ni-PTA, all samples share similar peak types attributed to Ni^{2+}, but the binding energy shows the obvious difference before and after doping. The binding energy of Ni $2p_{3/2}$ in pristine Ni-PTA (Zn_0-Ni-PTA) is 855.8 eV; by contrast, the value shifts to 858.2 and 857.9 eV in $Zn_{0.2}$-Ni-PTA and $Zn_{0.5}$-Ni-PTA. Such difference perhaps originates from that the doped Zn^{2+} with different electronegativity influences the electronic interactions of Ni^{2+} [29,30]. Based on the discussion above, it can be inferred that Zn^{2+} has strong bands with organic ligands in MOFs. All of the evidence discussed above proved the successful doping of zinc. The surface area (BET) date of different Zn_x-Ni-MOF are display in Table S2.

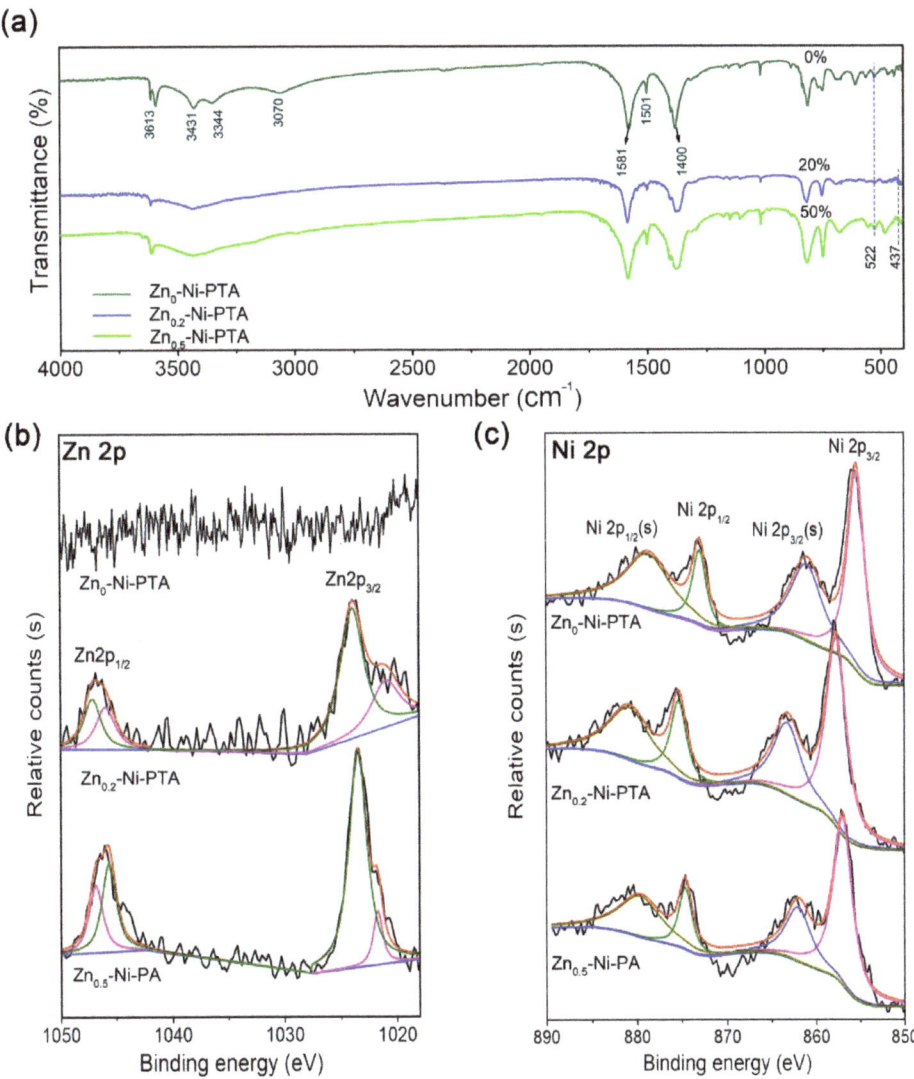

Figure 2. (a) FTIR pattern of the different atom fraction Zn_x-Ni-PTA samples. (b) Zn 2p and (c) Ni 2p spectra of Zn_x-Ni-PTA.

3.2. Half-Cell Test

To investigate the Li-ion storage capacity of $Zn_{0.2}$-Ni-PTA, electrochemical tests were performed and are shown in Figure 3. The CV curves, as shown in Figure 3a, exhibit a weak reduction peak at about 1.30 V and a sharp reduction peak at 0.75 V followed by a broad oxidation peak around 1.25 V during the initial anodic scan, which can be ascribed to the lithiation/de-lithiation of carboxylate groups and aromatic rings and the formation of a solid-electrolyte interphase (SEI) layer [31,32]. Compared with the first cycle, the intensity of the redox peaks in the second and third cycles are growing weaker, which is related to structural or morphological evolution [33]. The sharp redox peaks shift toward higher/lower potential (≈0.75/1.40 V), which corresponds to the Li-ions insertion/extraction into/from the carboxylate groups and the benzene rings of Zn_x-Ni-PTA [34]. Notably, Zn_0-Ni-PTA, $Zn_{0.2}$-Ni-PTA, and $Zn_{0.5}$-Ni-PTA showed similar CV curves

(Figure S3 left). This phenomenon implies that zinc dopant did not change the topology of pristine MOF; instead, it only just boosted lithium storage performance. The charge–discharge curves of the $Zn_{0.2}$-Ni-PTA are shown in Figure 3b. Two discharge voltage plateaus at about 1.30 V and 0.75 V could be observed in the first discharge curve. Then, following the charging process, two charge ranges are demonstrated, which agree well with the CV curves. Combining with the subsequent cycles, we can assume that the reversible discharge plateau of $Zn_{0.2}$-Ni-PTA at 1.0–0.5 V contributes the most capacity.

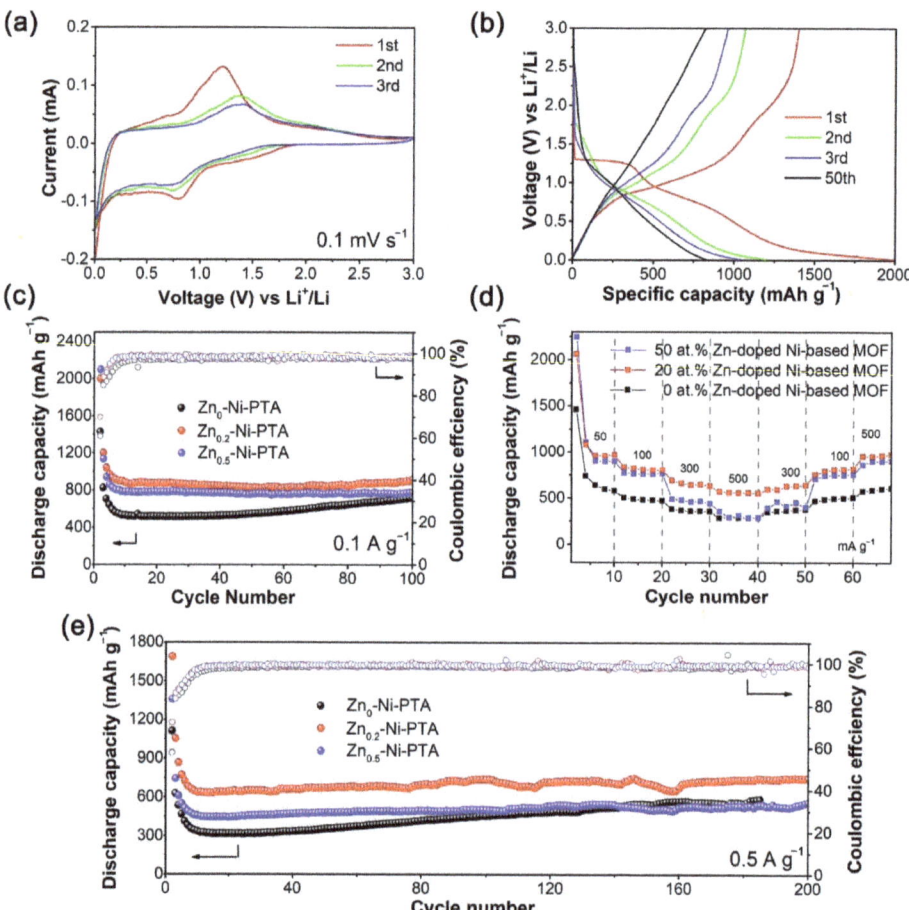

Figure 3. Electrochemical performance of Zn_x-Ni-PTA material: (**a**) CV curves and (**b**) voltage profiles of 20% atom fraction Zn-doped Ni-PTA. (**c**,**e**) Cycle and (**d**) rate performance of different MOFs.

The cycle performance of different amounts of Zn-doped Ni-PTA at the current density 0.1 A g^{-1} and 0.5 A g^{-1} is displayed in Figure 3c,e. The 20% atom fraction Zn-doped Ni-PTA delivers the initial discharge capacity of 2000 mA h g^{-1} at 0.1 A g^{-1} and 1690 mA h g^{-1} at 0.5 A g^{-1}. After 100/200 cycles, the discharge capacity retains up to 921 and 747 mA h g^{-1} at 0.1 A g^{-1}/0.5 A g^{-1}, respectively, with a Coulombic efficiency of nearly 100%. We are noticing that the capacity of three samples mildly increases after about the 10th cycle, which might be due to the electrochemical activation process related to the repeated insertion/extraction of Li-ions in MOFs. In simple terms, during charging and discharging, the inserted Li$^+$ expanded the interlayer spacing of samples and exposed a larger number of active sites, which could be beneficial for gradually increasing capacity. In addition, it is clear that the cycle performance of $Zn_{0.2}$-Ni-PTA is superior to the other two MOFs. In addition, the better rate performance of $Zn_{0.2}$-Ni-PTA is exhibited in Figure 3d. In summary, the reversible capacity at current density values of 50, 100, 300, and 500 mA g^{-1} is 958, 806, 646, and 556 mA h g^{-1}, respectively. Moreover, when the current density decreases to 50 mA g^{-1}, the discharge capacity remains stable.

Based on the above electrochemical tests, combined with the cycle performance information of different atom fractions of (0, 10, 20, 30, 40, 50%) Zn-doped Ni-PTA (Figure S5), we can prove that $Zn_{0.2}$-Ni-PTA displays the best electrochemical performance among different atom fractions of zinc doped Ni-PTA samples. For comparison, their electrochemical performances are presented along with the other MOFs in Table S3. In addition, the XRD patterns and SEM images of $Zn_{0.2}$-Ni-PTA electrodes before and after the cycle are displayed in Figure S4. The SEM images show that $Zn_{0.2}$-Ni-PTA maintains its pristine microplate-like characteristics after cycles, indicating its good morphological stability. Interestingly, an amorphization process after cycles is determined by XRD. This phenomenon perhaps originates from the electrochemical powderization effect due to the repeat insertion/extraction of Li^+ in its crystal [35]. Such an amorphization process was also observed in other lithium storage cases of MOFs materials [33,35]. Moreover, the decrease in redox peaks in the first few cycles observed from CV curves may be also related to this amorphization process.

3.3. Electrochemical Analysis

To better understand why $Zn_{0.2}$-Ni-PTA possesses brilliant electrochemical performance, the effect of doped Zn^{2+} on the migration of Li^+ was also investigated. It is well known that the high migration rate of lithium-ions can improve the electrochemical performance of the battery [36–38]. Therefore, electrochemical impedance spectroscopy (EIS) experiments are conducted to explore the kinetics. From the Nyquist diagrams, as shown in Figure 4a, it is shown that $Zn_{0.2}$-Ni-PTA possesses a lower electrochemical impedance than Zn_0-Ni-PTA. Furthermore, the chemical diffusion coefficient of Li^+ (D_{Li^+}) was calculated by Equation (1). R, T, A, F, n, and C are the ideal gas constant, the thermal–dynamic temperature, the surface area of electrodes, the Faraday's constant, the number of electrons per molecule during oxidation, and the Li^+ concentration in the cathode, respectively. The σ_W is the Warburg coefficient, which has a linear relation with Z' (Equation (2)):

$$D_{Li^+} = \left(\frac{2RT}{\sqrt{2}n^2F^2\sigma_W AC}\right)^2 = \frac{2R^2T^2}{n^4F^4\sigma_W^2 A^2C^2} \quad (1)$$

$$Z' = R_s + R_{ct} + \sigma_W\omega^{-1/2} \quad (2)$$

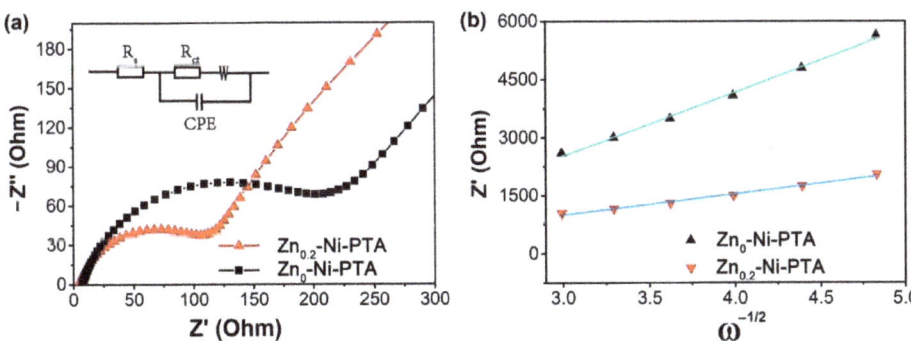

Figure 4. (**a**) Nyquist plot of $Zn_{0.2}$-Ni-PTA and Zn_0-Ni-PTA. (**b**) Z' vs. $\omega^{-1/2}$ plots in low frequency region.

In Equation (2), R_s and R_{ct} represent the resistance of solution and the charge transfer, respectively [25]. Figure 4b illustrates the linear relationship plot of $\omega^{-1/2}$ (reciprocal square root of angular frequency) vs. Z' (actual impedance) in the low frequency of Zn_x-Ni-PTA. The slope of this linear plot is equal to the value of σ_W. As shown clearly in the plot, by contrast with Zn_0-Ni-PTA, $Zn_{0.2}$-Ni-PTA possess a smaller slope, which demonstrates a higher Li$^+$ diffusion coefficient and better electrochemical performance. Compared to Ni^{2+} (0.065 nm), Zn^{2+} has a larger radius (0.074 nm), and the doped Zn^{2+} expands the interplanar distances of $Zn_{0.2}$-Ni-PTA, which facilitates the diffusion of Li$^+$. Thus, $Zn_{0.2}$-Ni-PTA displays a higher Li$^+$ diffusion coefficient and improved lithium storage performances [39].

3.4. Full-Cell Test

To further investigate the application prospects of the sample, we constructed the full cell using Zn_x-Ni-PTA as anode coupled with LiFePO$_4$ as cathode. The electrochemical test is performed at a voltage range of 0.5-4.0 V. Figure 5a shows the charge and discharge curves of the $Zn_{0.2}$-Ni-PTA/LiFePO$_4$ full cell. It yields a discharge capacity of 113.7 mA h g^{-1} with an average operation voltage of around 2.6 V at 0.5 C. At 0.1 C, 0.5 C, 1 C, 2 C, and 5 C, the reversible capacity remains 124.5, 105.9, 82.2, 62.4, and 50.3 mA h g^{-1}, respectively (Figure 5b). After 200 cycles, the capacity still retains 94.7 mA h g^{-1} with a high capacity retention of 83.24%, which contrasted sharply with the terrible cycle performance of Ni-PTA without Zn doped (Figure 5c). It should be mentioned that the mass loading of LiFePO$_4$ was relatively lower in this work compared to that in commercial LIBs. For most lab-level research, the mass loading of active materials is also lower [40]. In consideration of practical application, we will try modifying the loading of LFP in our later research to achieve a higher energy density.

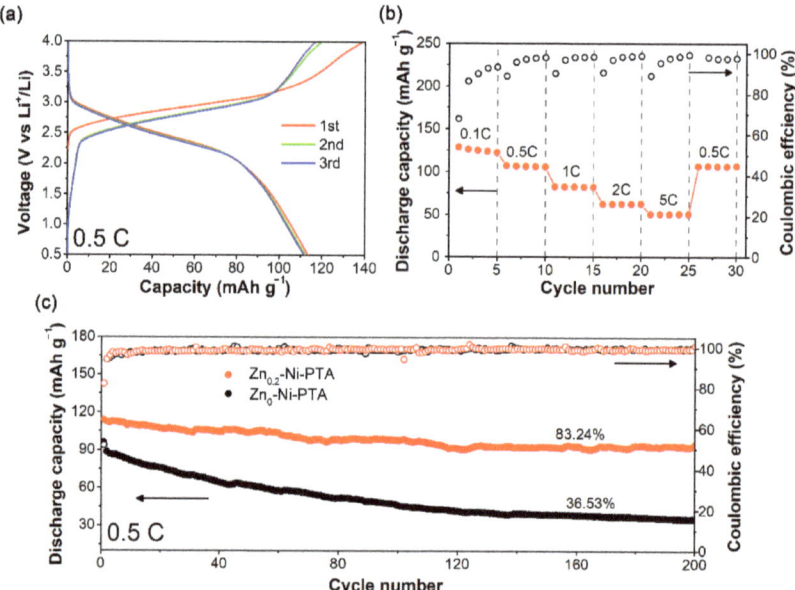

Figure 5. (**a**) Charge–discharge curves, (**b**) rate performance and (**c**) cycle performance of $Zn_{0.2}$-Ni-PTA/LiFePO$_4$ full cell.

The above electrochemical result of both half and full cells based on $Zn_{0.2}$-Ni-PTA with high reversible capacity, safe operation potential, excellent rate and cycle stability is superior to the previous report of synthesized Ni-MOF [16]. This could be attributed to the following aspects: First, the zinc with a large radius can hinder the destruction of the electrode material and enlarge the interplanar distances [21,24,31,32]. Meanwhile, the EIS results showed that $Zn_{0.2}$-Ni-PTA delivered less impedance and charge-transfer resistance, bringing a faster Li-ion transmission rate. In summary, $Zn_{0.2}$-Ni-PTA is a valuable anode material for LIBs.

4. Conclusions

In this work, we successfully synthesized different amounts of Zn doped Ni-PTA and used it as an electrode material for LIBs. In contrast with Ni-PTA, the visible enhancement of lithium storage ability (initial discharge capacity is 2000 mA h g^{-1} at the current density of 100 mA g^{-1}) and cycle stability ($\approx$85% reversible discharge capacity retained after 100 cycles) are observed in 20% Zn-doped Ni-PTA. The following characterizations indicated that the better performance of samples is because of the higher structural stability and lower impedance and charge-transfer resistance after zinc ions are doped. The work brings new perspectives to the modification of present MOFs electrode materials for LIBs. It may lead us to think about the potential use of $Zn_{0.2}$-Ni-PTA as anode materials for lithium-ion batteries in the future.

Supplementary Materials: The following supporting information can be downloaded at: https://www.mdpi.com/article/10.3390/ma15124186/s1. Figure S1: SEM of the samples, Figure S2: TG-DTA of the (a) 0% Zn-doped Ni-MOF, (b) 20% Zn-doped Ni-MOF, (c) 50% Zn-doped Ni-MOF, Figure S3: CV (left) and charge/discharge curves (right) of the samples, Figure S4: (a,b,c,d) SEM micrograph of electrode, (e) XRD patterns of the electrodes, Figure S5: The cycle performance of different amounts of Zn-doped Ni-PTA, Table S1: Induced coupled plasma atomic emission spectroscopy (ICP-AES) results of the Zn-doped Ni-MOF, Table S2: The surface area (BET) of the Zn_x-Ni-MOF, Table S3: Summary of metal-organic frameworks reported as anode materials in rechargeable lithium-ions batteries. References [41–51] are cited in the supplementary materials.

Author Contributions: Conceptualization, W.G. and Y.Z.; methodology, W.G.; software, W.G. and Z.X.; validation, W.G., Z.X. and Y.S.; formal analysis, W.G. and X.H.; investigation, Z.X. and M.L.; resources, X.L.; data curation, Y.S. and X.H.; writing—original draft preparation, W.G.; writing—review and editing, W.G. and Y.Z.; visualization, X.L. and M.L.; supervision, X.L. and Y.Z.; project administration, Y.Z.; funding acquisition, Y.Z. All authors have read and agreed to the published version of the manuscript.

Funding: This work was supported by Research Start-up Funds of Young Talents from Qingdao University (DC2000003363).

Institutional Review Board Statement: Not applicable.

Informed Consent Statement: Not applicable.

Data Availability Statement: Not applicable.

Conflicts of Interest: The authors declare no conflict of interest.

References

1. Zhu, P.C.; Gastol, D.; Marshall, J.; Sommerville, R.; Goodship, V.; Kendrick, E. A Review of Current Collectors for lithium-ion Batteries. *J. Power Sources* **2021**, *485*, 229321229321. [CrossRef]
2. Chen, Y.Q.; Kang, Y.Q.; Zhao, Y.; Wang, L.; Liu, J.L.; Li, Y.X.; Liang, Z.; He, X.M.; Li, X.; Tavajohi, N.; et al. A Review of Lithium-ion Battery Safety Concerns: The Issues, Strategies, and Testing Standards. *J. Energy Chem.* **2021**, *59*, 83–99. [CrossRef]
3. Shen, L.; Wu, H.B.; Liu, F.; Brosmer, J.L.; Shen, G.; Wang, X.; Zink, J.I.; Xiao, Q.; Cai, M.; Wang, G.; et al. Creating Lithium-Ion Electrolytes with Biomimetic Ionic Channels in Metal-Organic Frameworks. *Adv. Mater.* **2018**, *30*, 17074761707476. [CrossRef] [PubMed]
4. Chen, J.J.; Mao, Z.Y.; Zhang, L.X.; Tang, Y.H.; Wang, D.J.; Bie, L.J.; Fahlman, B.D. Direct Production of Nitrogen-doped Porous Carbon from Urea via Magnesiothermic Reduction. *Carbon* **2018**, *130*, 41–47. [CrossRef]
5. Corsi, J.S.; Welborn, S.S.; Stach, E.A.; Detsi, E. Insights into the Degradation Mechanism of Nanoporous Alloy-Type Li-Ion Battery Anodes. *ACS Energy Lett.* **2021**, *6*, 1749–1756. [CrossRef]
6. He, J.; Wei, Y.Q.; Hu, L.T.; Li, H.Q.; Zhai, T.Y. Aqueous Binder Enhanced High-Performance GeP5 Anode for Lithium-Ion Batteries. *Front. Chem.* **2018**, *6*, 2121. [CrossRef]

7. Jiang, T.; Ma, S.Y.; Deng, J.B.; Yuan, T.; Lin, C.F.; Liu, M.L. Partially Reduced Titanium Niobium Oxide: A High-Performance Lithium-Storage Material in a Broad Temperature Range. *Adv. Sci.* **2021**, *9*, 2105119. [CrossRef]
8. Li, Q.; Li, H.S.; Xia, Q.T.; Hu, Z.Q.; Zhu, Y.; Yan, S.S.; Ge, C.; Zhang, Q.H.; Wang, X.X.; Shang, X.T.; et al. Extra Storage Capacity in Transition Metal Oxide Lithium-ion Batteries Revealed by in Situ Magnetometry. *Nat. Mater.* **2021**, *20*, 76–83. [CrossRef]
9. Chakraborty, G.; Park, I.H.; Medishetty, R.; Vittal, J.J. Two-Dimensional Metal-Organic Framework Materials: Synthesis, Structures, Properties and Applications. *Chem. Rev.* **2021**, *121*, 3751–3891. [CrossRef]
10. Yao, M.S.; Li, W.H.; Xu, G. Metal-organic Frameworks and Their Derivatives for Electrically-transduced Gas sensors. *Coord. Chem. Rev.* **2021**, *426*, 213479213479. [CrossRef]
11. Dybtsev, D.N.; Bryliakov, K.P. Asymmetric Catalysis Using Metal-organic Frameworks. *Coord. Chem. Rev.* **2021**, *437*, 213845213845. [CrossRef]
12. Cai, W.; Wang, J.Q.; Chu, C.C.; Chen, W.; Wu, C.S.; Liu, G. Metal Organic Framework-Based Stimuli-Responsive Systems for Drug Delivery. *Adv. Sci.* **2019**, *6*, 18015261801526. [CrossRef] [PubMed]
13. Wang, D.; Zhou, W.W.; Zhang, R.; Huang, X.X.; Zeng, J.J.; Mao, Y.F.; Ding, C.Y.; Zhang, J.; Liu, J.P.; Wen, G.W. MOF-derived Zn-Mn Mixed oxides@carbon Hollow Disks with Robust Hierarchical Structure for High-performance lithium-ion Batteries. *J. Mater. Chem. A* **2018**, *6*, 2974–2983. [CrossRef]
14. Zheng, M.B.; Tang, H.; Li, L.L.; Hu, Q.; Zhang, L.; Xue, H.G.; Pang, H. Hierarchically Nanostructured Transition Metal Oxides for Lithium-Ion Batteries. *Adv. Sci.* **2018**, *5*, 17005921700592. [CrossRef] [PubMed]
15. Wang, Y.Y.; Zhang, M.; Li, S.L.; Zhang, S.R.; Xie, W.; Qin, J.S.; Su, Z.M.; Lan, Y.Q. Diamondoid-structured Polymolybdate-based Metal-organic Frameworks as High-capacity Anodes for lithium-ion Batteries. *Chem. Commun.* **2017**, *53*, 5204–5207. [CrossRef]
16. Zhang, Y.; Niu, Y.B.; Liu, T.; Li, Y.T.; Wang, M.Q.; Hou, J.K.; Xu, M.W. A nickel-based Metal-organic Framework: A novel optimized Anode Material for Li-ion Batteries. *Mater. Lett.* **2015**, *161*, 712–715. [CrossRef]
17. Gong, T.; Lou, X.B.; Gao, E.Q.; Hu, B.W. Pillared-Layer Metal Organic Frameworks for Improved Lithium-Ion Storage Performance. *ACS Appl. Mater. Interfaces* **2017**, *9*, 21839–21847. [CrossRef]
18. Li, X.X.; Cheng, F.Y.; Zhang, S.N.; Chen, J. Shape-controlled Synthesis and lithium-storage Study of Metal-organic Frameworks $Zn_4O(1,3,5\text{-benzenetribenzoate})_2$. *J. Power Sources* **2006**, *160*, 542–547. [CrossRef]
19. Wang, P.; Lou, X.B.; Li, C.; Hu, X.S.; Yang, Q.; Hu, B.W. One-Pot Synthesis of Co-Based Coordination Polymer Nanowire for Li-Ion Batteries with Great Capacity and Stable Cycling Stability. *Nano-Micro Lett.* **2018**, *10*, 1919. [CrossRef]
20. Yang, J.; Zheng, C.; Xiong, P.X.; Li, Y.F.; Wei, M.D. Zn-doped Ni-MOF Material with a High Supercapacitive Performance. *J. Mater. Chem. A* **2014**, *2*, 19005–19010. [CrossRef]
21. Liu, H.; Cao, Q.; Fu, L.J.; Li, C.; Wu, Y.P.; Wu, H.Q. Doping effects of zinc on LiFePO4 Cathode Material for lithium ion Batteries. *Electrochem. Commun.* **2006**, *8*, 1553–1557. [CrossRef]
22. Raju, K.; Venkataiah, G.; Yoon, D.H. Effect of Zn Substitution on the Structural and Magnetic Properties of Ni-Co ferrites. *Ceram. Int.* **2014**, *40*, 9337–9344. [CrossRef]
23. Han, X.Y.; Qing, G.Y.; Sun, J.T.; Sun, T.L. How Many Lithium Ions Can Be Inserted onto Fused C_6 Aromatic Ring Systems. *Angew. Chem.* **2011**, *124*, 5237–5241. [CrossRef]
24. Lee, G.; Varanasi, C.V.; Liu, J. Effects of Morphology and Chemical Doping on Electrochemical Properties of Metal Hydroxides in Pseudocapacitors. *Nanoscale* **2015**, *7*, 3181–3188. [CrossRef] [PubMed]
25. Niu, Y.B.; Xu, M.W.; Cheng, C.J.; Bao, S.J.; Hou, J.K.; Liu, S.G.; Yi, F.L.; He, H.; Li, C.M. $Na_{3.12}Fe_{2.44}(P_2O_7)_2$/multi-walled Carbon Nanotube Composite as a Cathode Material for Sodium-ion Batteries. *J. Mater. Chem. A* **2015**, *3*, 17224–17229. [CrossRef]
26. Rahdar, A.; Aliahmad, M.; Azizi, Y. NiO Nanoparticles: Synthesis and Characterization. *J. Nanostruct.* **2015**, *5*, 145–151. Available online: https://www.sid.ir/en/journal/ViewPaper.aspxid=490392 (accessed on 1 June 2015).
27. Xiong, G.; Pal, U.; Serrano, J.G.; Ucer, K.B.; Williams, R.T. Photoluminesence and FTIR study of ZnO Nanoparticles: The Impurity and Defect Perspective. *Phys. Status Solidi C* **2006**, *3*, 3577–3581. [CrossRef]
28. Chen, J.; Liu, R.; Gao, H.; Chen, L.; Ye, D. Amine-functionalized Metal-organic Frameworks for the Transesterification of Triglycerides. *J. Mater. Chem. A* **2014**, *2*, 7205–7213. [CrossRef]
29. Fomekong, R.L.; Tsobnang, P.K.; Magnin, D.; Hermans, S.; Delcorte, A.; Ngolui, J.L. Coprecipitation of Nickel zinc malonate: A facile and reproducible synthesis route for $Ni_{1-x}Zn_xO$ nanoparticles and $Ni_{1-x}Zn_xO/ZnO$ nanocomposites via pyrolysis. *J. Solid State Chem.* **2015**, *230*, 381–389. [CrossRef]
30. Yan, Y.; Lin, J.; Cao, J.; Guo, S.; Zheng, X.; Feng, J.; Qi, J. Activating and Optimizing the Activity of NiCoP Nanosheets for Electrocatalytic Alkaline Water Splitting Through the V doping Effect Enhanced by P vacancies. *J. Mater. Chem. A* **2019**, *7*, 24486–24492. [CrossRef]
31. Zhang, J.H.; Cai, G.F.; Zhou, D.; Tang, H.; Wang, X.L.; Gu, C.D.; Tu, J.P. Co-doped NiO nanoflake Array Films with Enhanced Electrochromic Properties. *J. Mater. Chem. C* **2014**, *2*, 7013–7021. [CrossRef]
32. Zhao, L.L.; Su, G.; Liu, W.; Cao, L.X.; Wang, J.; Dong, Z.; Song, M.Q. Optical and Electrochemical Properties of Cu-doped NiO Films Prepared by Electrochemical Deposition. *Appl. Surf. Sci.* **2011**, *257*, 3974–3979. [CrossRef]
33. Armand, M.; Grugeon, S.; Vezin, H.; Laruelle, S.; Ribiere, P.; Poizot, P.; Tarascon, J.M. Conjugated Dicarboxylate anodes for Li-ion Batteries. *Nat. Mater.* **2009**, *8*, 120–125. [CrossRef] [PubMed]

34. Li, C.; Hu, X.; Lou, X.; Zhang, L.; Wang, Y.; Amoureux, J.-P.; Shen, M.; Chen, Q.; Hu, B. The Organic-moiety-dominated Li$^+$ InterCalation/deintercalation Mechanism of a Cobalt-based Metal-organic Framework. *J. Mater. Chem. A* **2016**, *4*, 16245–16251. [CrossRef]
35. Wang, L.; Yu, Y.; Chen, P.C.; Zhang, D.W.; Chen, C.H. Electrospinning Synthesis of C/Fe$_3$O$_4$ Composite Nanofibers and their Application for High Performance lithium-ion batteries. *J. Power Sources* **2008**, *183*, 717–723. [CrossRef]
36. Chen, P.; Wu, Z.; Guo, T.; Zhou, Y.; Liu, M.; Xia, X.; Sun, J.; Lu, L.; Ouyang, X.; Wang, X.; et al. Strong Chemical Interaction between Lithium Polysulfides and Flame-Retardant Polyphosphazene for Lithium-Sulfur Batteries with Enhanced Safety and Electrochemical Performance. *Adv. Mater.* **2021**, *33*, e2007549. [CrossRef] [PubMed]
37. Qin, B.; Cai, Y.; Si, X.; Li, C.; Cao, J.; Fei, W.; Qi, J. Ultra-lightweight ion-sieving Membranes for High-rate lithium Sulfur Batteries. *Chem. Eng. J.* **2022**, *430*, 132698. [CrossRef]
38. Zhou, G.; Tian, H.; Jin, Y.; Tao, X.; Liu, B.; Zhang, R.; Seh, Z.W.; Zhuo, D.; Liu, Y.; Sun, J.; et al. Catalytic Oxidation of Li2S on the Surface of Metal Sulfides for Li-S Batteries. *Proc. Natl. Acad. Sci. USA* **2017**, *114*, 840–845. [CrossRef]
39. Qi, J.; Yan, Y.; Cai, Y.; Cao, J.; Feng, J. Nanoarchitectured Design of Vertical-Standing Arrays for Supercapacitors: Progress, Challenges, and Perspectives. *Adv. Funct. Mater.* **2020**, *31*, 2006030. [CrossRef]
40. Rui, Z.; Jie, L.; Gu, J. The Effects of Electrode Thickness on the Electrochemical and Thermal Characteristics of Lithium-ion Battery. *Appl. Energy* **2015**, *139*, 220–229. [CrossRef]
41. Senthil Kumar, R.; Nithya, C.; Gopukumar, S.; Anbu Kulandainathan, M. Diamondoid-Structured Cu-Dicarboxylate-based Metal-Organic Frameworks as High-Capacity Anodes for Lithium-Ion Storage. *Energy Technol.* **2014**, *2*, 921–927. [CrossRef]
42. Tang, B.; Huang, S.; Fang, Y.; Hu, J.; Malonzo, C.; Truhlar, D.G.; Stein, A. Mechanism of electrochemical lithiation of a metal-organic framework without redox-active nodes. *J. Chem. Phys.* **2016**, *144*, 194702. [CrossRef]
43. Han, X.; Yi, F.; Sun, T.; Sun, J. Synthesis and electrochemical performance of Li and Ni 1,4,5,8-naphthalenetetracarboxylates as anodes for Li-ion batteries. *Electrochem. Commun.* **2012**, *25*, 136–139. [CrossRef]
44. Lin, Y.; Zhang, Q.; Zhao, C.; Li, H.; Kong, C.; Shen, C.; Chen, L. An exceptionally stable functionalized metal-organic framework for lithium storage. *Chem. Commun.* **2015**, *51*, 697–699. [CrossRef] [PubMed]
45. Li, G.; Li, F.; Yang, H.; Cheng, F.; Xu, N.; Shi, W.; Cheng, P. Graphene oxides doped MIL-101(Cr) as anode materials for enhanced electrochemistry performance of lithium ion battery. *Inorg. Chem. Commun.* **2016**, *64*, 63–66. [CrossRef]
46. Hu, L.; Lin, X.M.; Mo, J.T.; Lin, J.; Gan, H.L.; Yang, X.L.; Cai, Y.P. Lead-Based Metal-Organic Framework with Stable Lithium Anodic Performance. *Inorg. Chem.* **2017**, *56*, 4289–4295. [CrossRef]
47. Saravanan, K.; Nagarathinam, M.; Balaya, P.; Vittal, J.J. Lithium storage in a metal organic framework with diamondoid topology—A case study on metal formates. *J. Mater. Chem.* **2010**, *20*, 8329–8335. [CrossRef]
48. Liu, Q.; Yu, L.; Wang, Y.; Ji, Y.; Horvat, J.; Cheng, M.L.; Jia, X.; Wang, G. Manganese-based layered coordination polymer: Synthesis, structural characterization, magnetic property, and electrochemical performance in lithium-ion batteries. *Inorg. Chem.* **2013**, *52*, 2817–2822. [CrossRef]
49. Li, H.; Su, Y.; Sun, W.; Wang, Y. Carbon Nanotubes Rooted in Porous Ternary Metal Sulfide@N/S-Doped Carbon Dodecahedron: Bimetal-Organic-Frameworks Derivation and Electrochemical Application for High-Capacity and Long-Life Lithium-Ion Batteries. *Adv. Funct. Mater.* **2016**, *26*, 8345–8353. [CrossRef]
50. Zhou, D.; Ni, J.; Li, L. Self-supported multicomponent CPO-27 MOF nanoarrays as high-performance anode for lithium storage. *Nano Energy* **2019**, *57*, 711–717. [CrossRef]
51. Xia, S.-B.; Yu, S.-W.; Yao, L.-F.; Li, F.-S.; Li, X.; Cheng, F.-X.; Shen, X.; Sun, C.-K.; Guo, H.; Liu, J.-J. Robust hexagonal nut-shaped titanium(IV) MOF with porous structure for ultra-high performance lithium storage. *Electrochim. Acta* **2019**, *296*, 746–754. [CrossRef]

Article

The Effects of Ru^{4+} Doping on LiNi$_{0.5}$Mn$_{1.5}$O$_4$ with Two Crystal Structures

Xinli Li [1], Ben Su [1], Wendong Xue [1,*] and Junnan Zhang [2]

[1] School of Materials Science and Engineering, University of Science and Technology Beijing, Beijing 100083, China; xinbattery@163.com (X.L.); suben111@163.com (B.S.)
[2] Shandong Wina Green Power Technology Co., Ltd., Weifang 261000, China; jnzhang@winabattery.com
* Correspondence: xuewendong@ustb.edu.cn; Tel.: +86-185-0134-1077

Abstract: Doping of Ru has been used to enhance the performance of LiNi$_{0.5}$Mn$_{1.5}$O$_4$ cathode materials. However, the effects of Ru doping on the two types of LiNi$_{0.5}$Mn$_{1.5}$O$_4$ are rarely studied. In this study, Ru^{4+} with a stoichiometric ratio of 0.05 is introduced into LiNi$_{0.5}$Mn$_{1.5}$O$_4$ with different space groups (Fd$\bar{3}$m, P4$_3$32). The influence of Ru doping on the properties of LiNi$_{0.5}$Mn$_{1.5}$O$_4$ (Fd$\bar{3}$m, P4$_3$32) is comprehensively studied using multiple techniques such as XRD, Raman, and SEM methods. Electrochemical tests show that Ru^{4+}-doped LiNi$_{0.5}$Mn$_{1.5}$O$_4$ (P4$_3$32) delivers the optimal electrochemical performance. Its initial specific capacity reaches 132.8 mAh g^{-1}, and 97.7% of this is retained after 300 cycles at a 1 C rate at room temperature. Even at a rate of 10 C, the capacity of Ru^{4+}-LiNi$_{0.5}$Mn$_{1.5}$O$_4$ (P4$_3$32) is still 100.7 mAh g^{-1}. Raman spectroscopy shows that the Ni/Mn arrangement of Ru^{4+}-LiNi$_{0.5}$Mn$_{1.5}$O$_4$ (Fd$\bar{3}$m) is not significantly affected by Ru^{4+} doping. However, LiNi$_{0.5}$Mn$_{1.5}$O$_4$ (P4$_3$32) is transformed to semi-ordered LiNi$_{0.5}$Mn$_{1.5}$O$_4$ after the incorporation of Ru^{4+}. Ru^{4+} doping hinders the ordering process of Ni/Mn during the heat treatment process, to an extent.

Keywords: solid-state reactions; LiNi$_{0.5}$Mn$_{1.5}$O$_4$; Ru^{4+}; space group; comparative study

1. Introduction

The continuous development of lithium-ion batteries has driven the progress of the electric vehicle industry. However, the cruising mileage of electric vehicles still has a great deal of room for improvement. Energy density has always been one of the key factors limiting the range. From the formula $W = Q \bullet U$, we know that increasing the specific capacity and discharge voltage of the cathode material is the main way to increase the power density of lithium-ion batteries (LIBs). Spinel LiNi$_{0.5}$Mn$_{1.5}$O$_4$ (LNMO) is considered to be one of the most promising candidates for high-power lithium-ion battery systems, and this is attributed to its ultra-fast 3D Li$^+$ diffusion speed, good high-rate capability, excellent cyclic stability, low cost, and environmental friendliness. LiNi$_{0.5}$Mn$_{1.5}$O$_4$ has a high discharge platform (vs. Li/Li$^+$ ≈ 4.7 V) [1–3] and a theoretical discharge capacity of 146.8 mAh g^{-1}; therefore, it has a fairly high theoretical energy density [4].

Built on the arrangement of Mn^{4+} and Ni^{2+} in the crystal lattice, LiNi$_{0.5}$Mn$_{1.5}$O$_4$ can have two crystal structures: a face-centered cubic structure (Fd$\bar{3}$m) or a simple cubic structure (P4$_3$32). The former is a disordered structure in which Mn^{4+} and Ni^{2+} ions are randomly distributed on the octahedral 16d sites. The latter has an ordered structure in which Mn^{4+} ions occupy the 12d positions and Ni^{2+} ions occupy the 4a sites [5,6]. In addition, almost no Mn^{3+} ions are generated in the crystal structure. It is generally believed that disordered LiNi$_{0.5}$Mn$_{1.5}$O$_4$ containing Mn^{3+} has higher electronic conductivity and lithium-ion conductivity, and thus has better high-rate performance [7–10]. However, the presence of too many Mn^{3+} ions in LiNi$_{0.5}$Mn$_{1.5}$O$_4$ will damage the cyclic stability of the cathode material, because Mn^{3+} is prone to the disproportionation reaction

$2Mn^{3+} = Mn^{2+} + Mn^{4+}$. Generated Mn^{2+} ions dissolve in the electrolyte, migrate to the anode electrode, and further deposit on the surface of the anode electrode. The continuous disproportionation–dissolution–migration–deposition behavior causes irreversible loss of battery capacity [11–13]. However, the ordered $LiNi_{0.5}Mn_{1.5}O_4$ also has the problem of poor cycling stability. Owing to the two-step phase transition occurring during the charge/discharge process, the structural stability of $LiNi_{0.5}Mn_{1.5}O_4$ (P4$_3$32) is highly susceptible and prone to irreversible structural damage. Thus, $LiNi_{0.5}Mn_{1.5}O_4$ with a small amount Mn^{3+} and a structure that is not completely ordered may have optimal properties in actual usage. When the calcination temperature is increased above 700 °C, a disordered phase is formed accompanied by the loss of oxygen and the appearance of Mn^{3+} inside the crystal lattice. The arrangement state of the transition metal ions can be controlled by an annealing process [14]. The oxygen defects and Mn^{3+} can be eliminated by annealing at 700 °C, and a disordered state can be transformed into an ordered state [15–18].

Due to poor uniformity of the raw material mixture, $LiNi_{0.5}Mn_{1.5}O_4$ synthesized by the conventional solid-state method may contain rock-salt impurities and deliver a low specific capacity. To enhance the electrochemical performance of $LiNi_{0.5}Mn_{1.5}O_4$, one efficient strategy is to dope with transition metal ions such as Na [19], Mg [20], Er [21], Fe [22], Ga [23], Ru [24], Cr [25], and Al [26] in the framework, to improve conductivity. Jianbing G. et al. studied the effects of Ru doping and the doping amount on the properties of $LiNi_{0.5}Mn_{1.5}O_4$ cathode material [27,28]. The electrochemical performance of $LiNi_{0.5}Mn_{1.5}O_4$ improved greatly after Ru doping; the initial discharge specific capacity increased from 103 mAh g^{-1} to 125.3 mAh g^{-1} and the 5 C-rate capacity increased by about 30 mAh g^{-1}. However, previous studies have usually been carried out on $LiNi_{0.5}Mn_{1.5}O_4$ (Fd$\bar{3}$m) [24,29], and few studies have compared the effects of transition metal doping on the two types of cathode materials. In this study, $LiNi_{0.5}Mn_{1.5}O_4$ cathode materials with Fd$\bar{3}$m and P4$_3$32 space groups were synthesized through a combination of typical solid-state reactions and heat treatment processes. Multiple techniques such as XRD, Raman and SEM methods were utilized to comprehensively study the influence of Ru^{4+} doping modification on the microstructure, micromorphology, and electrochemical performance of $LiNi_{0.5}Mn_{1.5}O_4$ with the two space groups.

2. Materials and Methods

2.1. Material Synthesis

$LiNi_{0.5}Mn_{1.5}O_4$ with different structures was synthesized via traditional solid-state reactions. A typical route was as follows: (1) Li_2CO_3, $NiCO_3$, and $MnCO_3$ in a stoichiometric ratio of 0.525:0.5:1.5 were mixed in alcohol; (2) a process of ball milling was performed at a speed of 400 r/min for 4 h, and the obtained mixture was completely dried at 60 °C and subsequently pulverized; (3) the powder was heated to 900 °C in a tube furnace at a rate of 5 °C/min and held for 12 h, followed by cooling naturally to room temperature to obtain the $LiNi_{0.5}Mn_{1.5}O_4$ with space group of Fd$\bar{3}$m, denoted LNMO (Fd$\bar{3}$m). Alternatively, the heated material was insulated at 700 °C for 48 h and cooled naturally to room temperature to give the P4$_3$32-structured $LiNi_{0.5}Mn_{1.5}O_4$, denoted LNMO (P4$_3$32). Li_2CO_3, $NiCO_3$, $MnCO_3$, and RuO_2 were mixed in a stoichiometric ratio of 0.525:0.45:1.5:0.05, and the above steps were repeated to obtain Ru-doped $LiNi_{0.5}Mn_{1.5}O_4$ with different structures, denoted Ru^{4+}-LNMO (Fd$\bar{3}$m) and Ru^{4+}-LNMO (P4$_3$32), respectively.

2.2. Characterization

To investigate the influence of the Ru^{4+} doping on the crystal structure of $LiNi_{0.5}Mn_{1.5}O_4$ with different space groups, X-ray diffraction (XRD, Ultima IV, Tokyo, Japan) was carried out using Cu Kα radiation in the range $10° \leq 2\theta \leq 70°$. The morphologies of Ru-doped $LiNi_{0.5}Mn_{1.5}O_4$ with different structures were recorded using scanning electron microscopy (SEM, FESEM Quanta TEG 450, Hillsboro, OR, USA). The phase structures of $LiNi_{0.5}Mn_{1.5}O_4$ with and without doping for the different structures were investigated using a Renishaw inVia plus-type micro-Raman spectrometer.

2.3. Preparation of Electrodes and Construction of Cells

LiNi$_{0.5}$Mn$_{1.5}$O$_4$ (80 wt%) was mixed with 10 wt% acetylene black and 10 wt% polyvinyli-dene fluoride (PVDF) in the appropriate amount of N-Methyl pyrrolidone (NMP). The obtained slurry was coated onto aluminum foil, then dried at 80 °C for 2 h in air and at 120 °C for 8 h in a vacuum oven. Finally, a Celgard 2400 argon-filled glove box was used as the separator to assemble coin-type LiNi$_{0.5}$Mn$_{1.5}$O$_4$/Li cells.

2.4. Electrochemical Measurements

The electrochemical properties of LiNi$_{0.5}$Mn$_{1.5}$O$_4$ before and after Ru^{4+} doping were measured with CR2025-type coin cells (HF-Kejing, Hefei, China). All the charge-discharge behaviors were evaluated at a rate of 1 C at room temperature, utilizing a LAND battery testing system. The rate of 1 C was set at 147 mA g^{-1}, and the current densities for testing were determined on the basis of the weight of cathode material. In the evaluation of cycle performance, the cells were charged and discharged in the voltage range of 3.5 V to 5.0 V for 300 cycles. The rate tests were conducted at rates of 0.2 C, 0.5 C, 1 C, 2 C, 5 C, and 10 C and then reversed successively back to 0.2 C.

3. Results and Discussion

The X-ray diffraction patterns of the LiNi$_{0.5}$Mn$_{1.5}$O$_4$ (Fd$\bar{3}$m, P4$_3$32), with and without Ru^{4+} doping, are compared in Figure 1. Clearly, all peaks of the four as-prepared samples are in agreement with the XRD patterns of typical spinel LiNi$_{0.5}$Mn$_{1.5}$O$_4$. Ru^{4+} doping does not change the primary lattice framework of the LiNi$_{0.5}$Mn$_{1.5}$O$_4$. Since the radius of Ru^{4+} is comparable to that of Ni^{2+} [30], a slight distortion of the lattice will be caused by the introduction of Ru^{4+}. This is also reflected in the diffraction patterns, with the (111) peaks of Ru^{4+}-LiNi$_{0.5}$Mn$_{1.5}$O$_4$ (Fd$\bar{3}$m, P4$_3$32) all shifting slightly towards the low-angle region. Furthermore, the larger lattice parameters a and c facilitate the diffusion of Li$^+$ and subsequently enhance the high-rate capability of the battery. The weak peaks (2θ at ≈37.5°, 43.6°, 47.5°, and 63.5°) correspond to the rock-salt impurity phase Li$_x$Ni$_{1-x}$O, which is mainly caused by oxygen loss [31,32]. In terms of traditional solid-state methods, insufficient mixing and high sintering temperatures inevitably contribute to the volatilization of Ni/Li and the formation of Li$_x$Ni$_{1-x}$O components. Amplifying a partial area (2θ = 40°~50°), it can be found that the weak peaks of the impurity basically disappear in the spectrum of Ru^{4+}-LiNi$_{0.5}$Mn$_{1.5}$O$_4$ (P4$_3$32), but tiny impurity peaks still emerge in the spectrum of Ru^{4+}-LiNi$_{0.5}$Mn$_{1.5}$O$_4$ (Fd$\bar{3}$m). This demonstrates that Ru^{4+} doping has a noticeable effect in eliminating Li$_x$Ni$_{1-x}$O-like impurity phases for LiNi$_{0.5}$Mn$_{1.5}$O$_4$ (both Fd$\bar{3}$m and P4$_3$32) and stabilizing the spinel crystal structure.

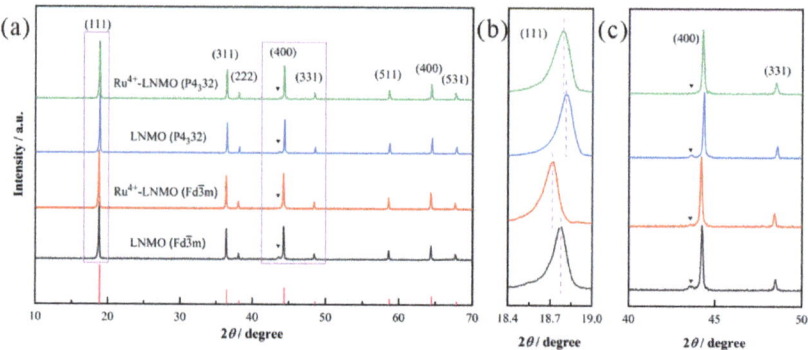

Figure 1. (a) XRD patterns of LiNi$_{0.5}$Mn$_{1.5}$O$_4$ (Fd$\bar{3}$m, P4$_3$32) before and after Ru^{4+} doping; (b) the enlarged (111) diffraction peak in the XRD patterns; (c) the enlarged impurities diffraction peak in the XRD patterns.

The Raman spectra of LiNi$_{0.5}$Mn$_{1.5}$O$_4$ with different structures before and after doping are shown in Figure 2. The peaks at around 630 cm^{-1} are assigned to the symmetric Mn-O stretching vibration, and the peaks at around 482 cm^{-1} correspond to the Ni-O stretching mode. The splitting of the F$_{2g}^{(1)}$ vibration mode near 580–600 cm^{-1} is clear evidence for the ordered structure, while a lack of F$_{2g}^{(1)}$ splitting corresponds to the disordered state. It is generally accepted that a higher degree of disorder indicates a higher Mn^{3+} content. It can be observed that Ru^{4+}-LiNi$_{0.5}$Mn$_{1.5}$O$_4$ (Fd$\bar{3}$m) and LiNi$_{0.5}$Mn$_{1.5}$O$_4$ (Fd$\bar{3}$m) have highly similar F$_{2g}^{(1)}$ vibration modes, in which no obvious splitting peaks appear. Ru^{4+} doping does not affect the arrangement of Ni/Mn in the lattice. Ru^{4+}-LiNi$_{0.5}$Mn$_{1.5}$O$_4$ (Fd$\bar{3}$m) still maintains a highly Ni/Mn disordered state. Nevertheless, the splitting degree of Ru^{4+}-LiNi$_{0.5}$Mn$_{1.5}$O$_4$ (P4$_3$32) is much less than the obvious splitting in the LiNi$_{0.5}$Mn$_{1.5}$O$_4$ (P4$_3$32) sample, indicating that Ru doping enhances the degree of disorder of LiNi$_{0.5}$Mn$_{1.5}$O$_4$ (P4$_3$32) and enhances the content of Mn^{3+} ions to some extent. During the process of heat treatment, the introduction of Ru^{4+} inhibits the Ni/Mn ordering process so that ordered LiNi$_{0.5}$Mn$_{1.5}$O$_4$ is transformed to semi-ordered LiNi$_{0.5}$Mn$_{1.5}$O$_4$. Previous studies have demonstrated that the Ni/Mn arrangement is mainly related to the radius and valence of nickel/manganese ions. LiNi$_{0.5}$Mn$_{1.5}$O$_4$ may have disordered (space group: Fd$\bar{3}$m) or ordered (space group: P4$_3$32) spinel structures. In the ordered structure, Ni occupies the 4b octahedral sites and Mn occupies the 12d octahedral sites. Ni and Mn are randomly distributed among the 16d octahedral sites in the disordered LiNi$_{0.5}$Mn$_{1.5}$O$_4$ structure. In this study, Ru^{4+} was utilized to replace the corresponding content of Ni^{2+} in the lattice structures. However, Ru^{4+} and Mn^{4+} have same valence state and similar ionic radii [30], and this can easily cause cationic mixing of Ru/Mn. Ru^{4+} can tend to occupy the 12c sites of Mn^{4+}, causing the substituted Mn^{4+} to compete with Ni^{2+} for the 4b sites. By substituting Ru^{4+} with a high valence state for Ni^{2+} with a low valence state, some of the Mn^{4+} ions are reduced to Mn^{3+} for the sake of charge compensation, thereby increasing the cationic mixing degree.

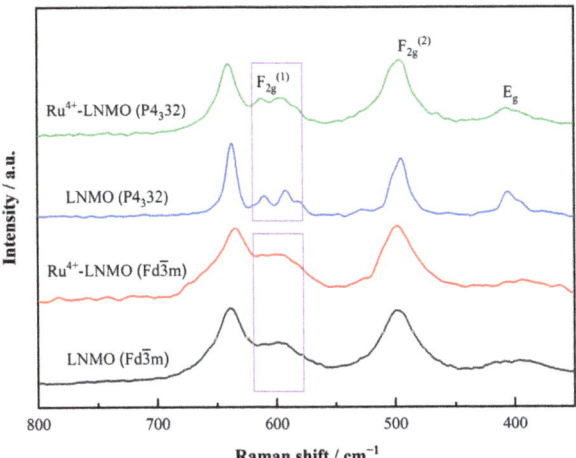

Figure 2. Raman spectra of the LiNi$_{0.5}$Mn$_{1.5}$O$_4$ (Fd$\bar{3}$m, P4$_3$32) before and after Ru^{4+} doping.

It is generally believed that the crystal plane in contact with the electrolyte and the particle size both have an important influence on the electrochemical performance of LiNi$_{0.5}$Mn$_{1.5}$O$_4$ cathode material [33–36]. The dissolution of the transition metal is closely linked to the stability of the interface. For example, Mn^{2+} derived from the disproportionation reaction is most easily dissolved from the {110} crystal plane into the electrolyte. Therefore, inhibiting the growth of the {110} crystal plane can effectively reduce the dissolution of the transition metal and thus improve the cyclic stability [37]. Compared with the {110} planes, the {100} crystal planes have a positive effect on the electrochemical performance [38,39]. Particle size is another important factor affecting stability. Nanoscale $LiNi_{0.5}Mn_{1.5}O_4$ has shorter Li^+ diffusion paths, but at the same time, the larger specific surface area also leads to more serious side reactions at the interface and instability of the battery system [40,41]. Large (micron level) $LiNi_{0.5}Mn_{1.5}O_4$ particles have a smaller specific surface area, which effectively reduces the degree of side reactions to enhance cycling performance.

The micro-morphology of the $LiNi_{0.5}Mn_{1.5}O_4$ ($Fd\overline{3}m$, $P4_332$) before and after Ru^{4+} doping is shown in Figure 3. Ru^{4+}-$LiNi_{0.5}Mn_{1.5}O_4$ ($Fd\overline{3}m$) particles have a truncated octahedral morphology, and Ru^{4+}-$LiNi_{0.5}Mn_{1.5}O_4$ ($P4_332$) particles have a spherical truncated polyhedron morphology. Additionally, both of them have a certain particle size distribution. Some particle diameters are about 2 μm, and a large number of particle diameters are about 1 μm. After doping, grains of $LiNi_{0.5}Mn_{1.5}O_4$ with different structures have no obvious distortion or morphological changes. However, strictly speaking, the particle growth of the active material seems to be repressed by Ru^{4+} cooperation during the calcination process, which reduces the final particle size to some extent.

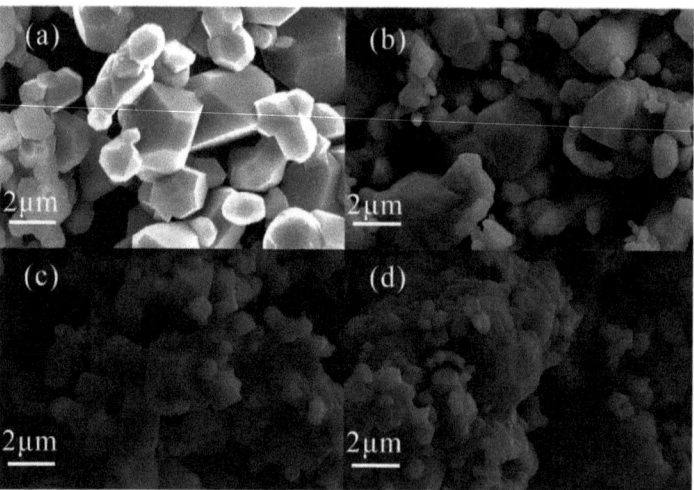

Figure 3. SEM images of the $LiNi_{0.5}Mn_{1.5}O_4$ ($Fd\overline{3}m$, $P4_332$) before and after Ru^{4+} doping. (**a**) $LiNi_{0.5}Mn_{1.5}O_4$ ($Fd\overline{3}m$); (**b**) $LiNi_{0.5}Mn_{1.5}O_4$ ($P4_332$); (**c**) Ru^{4+}-$LiNi_{0.5}Mn_{1.5}O_4$ ($Fd\overline{3}m$); (**d**) Ru^{4+}-$LiNi_{0.5}Mn_{1.5}O_4$ ($P4_332$).

The initial charge–discharge curves of the $LiNi_{0.5}Mn_{1.5}O_4$ ($Fd\overline{3}m$, $P4_332$) before and after doping in the voltage range of 3.5~5.0 V at a rate of 0.2 C are shown in Figure 4. Compared with the discharge capacity (123.0 mAh g^{-1}) of $LiNi_{0.5}Mn_{1.5}O_4$ ($Fd\overline{3}m$), the capacity of Ru^{4+}-$LiNi_{0.5}Mn_{1.5}O_4$ ($Fd\overline{3}m$) is increased to 127.7 mAh g^{-1}. Meanwhile, the 4.1 V platform capacity and the capacity ratio are reduced from 18.9 mAh g^{-1}/15.4% to 14.1 mAh g^{-1}/11.0%, respectively. Compared to $LiNi_{0.5}Mn_{1.5}O_4$ ($P4_332$), the 4.1V platform capacity (9.3 mAh g^{-1}, 7.07%) of Ru^{4+}-$LiNi_{0.5}Mn_{1.5}O_4$ ($P4_332$) has more than doubled, and the discharge capacity has reached the maximum value. In addition, the average valence of the transition metal nickel ions may decrease with the introduction of Ru^{4+}, resulting in an increase in the 4.7 V platform discharge capacity. On the other hand, Ru^{4+} doping can also prevent the sample from reacting with oxygen during the heat treatment process, which is manifested by the presence of a certain amount of Mn^{3+}. The presence of Mn^{3+} can further increase the discharge capacity of the material. Therefore, the discharge capacity of Ru^{4+}-$LiNi_{0.5}Mn_{1.5}O_4$ ($P4_332$) reaches the maximum value among the four samples.

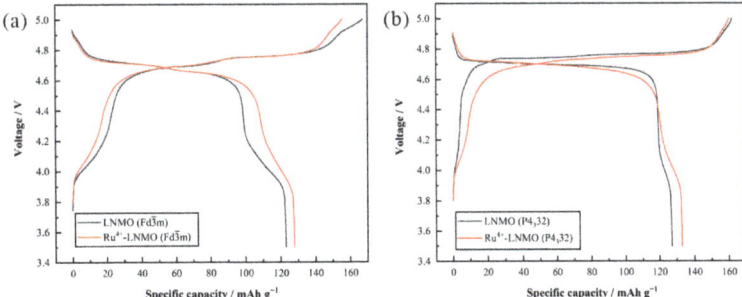

Figure 4. Initial charge–discharge curves of (**a**) LiNi$_{0.5}$Mn$_{1.5}$O$_4$ (Fd$\bar{3}$m) and (**b**) LiNi$_{0.5}$Mn$_{1.5}$O$_4$ (P4$_3$32) before and after Ru^{4+} doping in the voltage range of 3.5 V~5.0 V at a rate of 0.2 C.

The LiNi$_{0.5}$Mn$_{1.5}$O$_4$/Li half cells were all pre-cycled at 0.2 C for three cycles. Figure 5 shows the cycling performance of cells at the 1 C rate (147.0 mA g^{-1}) in the voltage range of 3.5 V to 5.0 V (vs. Li/Li$^+$) at room temperature. It can be observed that the Ru^{4+}-LiNi$_{0.5}$Mn$_{1.5}$O$_4$ (Fd$\bar{3}$m) delivers a higher discharge capacity than LiNi$_{0.5}$Mn$_{1.5}$O$_4$ (Fd$\bar{3}$m), with an initial specific capacity increase from 121.7 mAh g^{-1} to 124.8 mAh g^{-1}. In addition, the capacity retention after 300 cycles is also increased from 97.3% to 98.5%. On the other hand, the LiNi$_{0.5}$Mn$_{1.5}$O$_4$ (P4$_3$32) sample shows an initial discharge capacity of 125.4 mAh g^{-1} but delivers a lower retention of 92.0%. With Ru^{4+} doping, the initial capacity of Ru^{4+}-LiNi$_{0.5}$Mn$_{1.5}$O$_4$ (P4$_3$32) reaches the highest value of 132.8 mAh g^{-1}, and the capacity retention after 300 cycles also recovers to 97.8%.

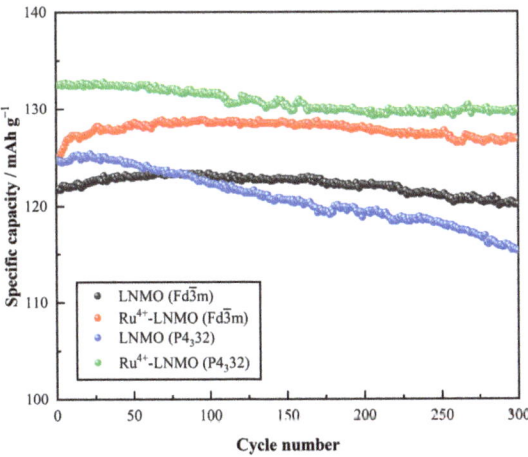

Figure 5. Cycling performance of the LiNi$_{0.5}$Mn$_{1.5}$O$_4$ (Fd$\bar{3}$m, P4$_3$32) before and after Ru^{4+} doping at a 1 C rate in the voltage range of 3.5 V~5.0 V at 25 °C.

The discharge voltage curves in the 1st and 300th cycles of the four samples at the 1C rate (147.0 mAh g^{-1}) at room temperature are displayed in Figure 6. Before and after 300 cycles, the curves for LiNi$_{0.5}$Mn$_{1.5}$O$_4$ (Fd$\bar{3}$m) have a high degree of coincidence, but careful observation shows that both the 4.7 V and 4.1 V platforms are shortened slightly. Ni^{2+} and Mn^{3+} dissolve into the electrolyte, causing a loss of battery capacity. The introduction of Ru^{4+} inhibits the dissolution of Ni^{2+} and Mn^{3+} during the charge–discharge process by stabilizing the crystal structure. Thus, the plateau-shortening degree of Ru^{4+}-LiNi$_{0.5}$Mn$_{1.5}$O$_4$ (Fd$\bar{3}$m) is slightly less than that of LiNi$_{0.5}$Mn$_{1.5}$O$_4$ (Fd$\bar{3}$m). In addition, LiNi$_{0.5}$Mn$_{1.5}$O$_4$ (P4$_3$32) as a cathode material suffers severe capacity attenuation. The discharge voltage curves in the 1st and 300th cycles are significantly dissimilar, reflecting the observable shortening of the 4.7 V platform. Compared to $LiNi_{0.5}Mn_{1.5}O_4$ (P4$_3$32), Ru^{4+}-$LiNi_{0.5}Mn_{1.5}O_4$ (P4$_3$32) delivers excellent cyclic stability after 300 cycles. Ru^{4+} plays a major role in improving the cyclic stability. The Ru-O bond energy is higher than those of the Ni-O bond and the Mn-O bond, which can stabilize the crystal structure to reduce the lattice damage. Based on the analysis of the Raman spectroscopy results, Ru^{4+}-$LiNi_{0.5}Mn_{1.5}O_4$ (P4$_3$32) transforms to semi-ordered $LiNi_{0.5}Mn_{1.5}O_4$ and contains a certain amount of Mn^{3+}. The octahedral distortion caused by Mn^{3+} promotes the generation of more Li^+ diffusion channels in active materials, which is beneficial to the cycling stability of the electrode. Therefore, Ru^{4+}-$LiNi_{0.5}Mn_{1.5}O_4$ (P4$_3$32) not only has superior capacity but also has considerable cyclic stability.

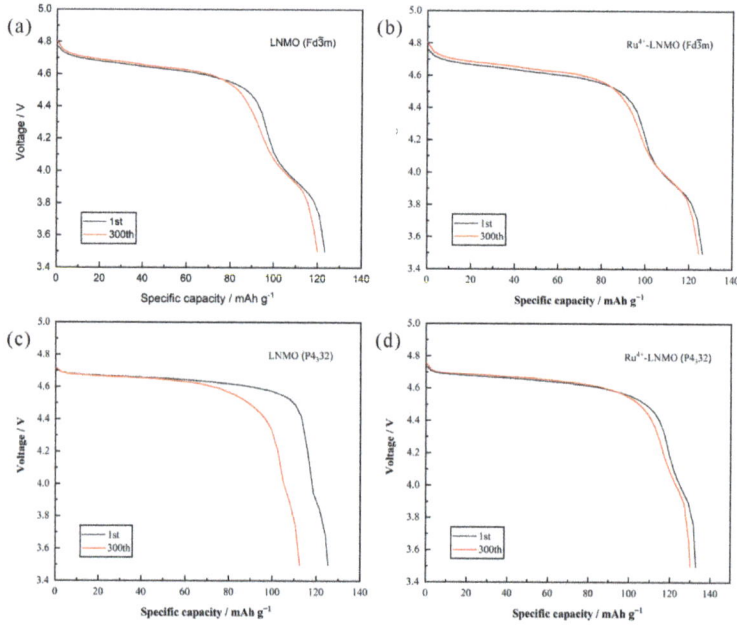

Figure 6. Discharge profiles of the $LiNi_{0.5}Mn_{1.5}O_4$ (Fd$\bar{3}$m, P4$_3$32) before and after Ru^{4+} doping from 3.5 V to 5.0 V at a 1 C rate in the 1st and 300th cycle. (**a**) $LiNi_{0.5}Mn_{1.5}O_4$ (Fd$\bar{3}$m); (**b**) $LiNi_{0.5}Mn_{1.5}O_4$ (P4$_3$32); (**c**) Ru^{4+}-$LiNi_{0.5}Mn_{1.5}O_4$ (Fd$\bar{3}$m); (**d**) Ru^{4+}-$LiNi_{0.5}Mn_{1.5}O_4$ (P4$_3$32).

The rate tests for $LiNi_{0.5}Mn_{1.5}O_4$ (Fd$\bar{3}$m, P4$_3$32) before and after Ru^{4+} doping at different current densities were carried out in the voltage range of 3.5 V to 5.0 V, as shown in Figure 7. At rates of 0.2 C, 0.5 C, 1 C, 2 C, 5 C, and 10 C, Ru^{4+}-$LiNi_{0.5}Mn_{1.5}O_4$ (Fd$\bar{3}$m) and Ru^{4+}-$LiNi_{0.5}Mn_{1.5}O_4$ (P4$_3$32) both demonstrate higher discharge capacities than $LiNi_{0.5}Mn_{1.5}O_4$ (Fd$\bar{3}$m, P4$_3$32), and Ru^{4+}-$LiNi_{0.5}Mn_{1.5}O_4$ (P4$_3$32) delivers the optimum rate capacity. The effect of Ru^{4+} doping on the high-rate performance for ordered $LiNi_{0.5}Mn_{1.5}O_4$ is more noticeable at high rates. Upon increasing the current intensity, the superiority becomes particularly evident. All samples deliver a decrease in specific discharge capacity when the imposed current density increases from a rate of 0.2 C (29.4 mA g^{-1}) to a rate of 5 C (735 mA g^{-1}). Ru^{4+}-$LiNi_{0.5}Mn_{1.5}O_4$ (P4$_3$32) shows a discharge capacity retention of ≈89.4% (≈118 mAh g^{-1} at 5 C; ≈132 mAh g^{-1} at 0.2 C), compared with ≈85.3% retention (≈108.3 mAh g^{-1} at 5 C; ≈126.2 mAh g^{-1} at 0.2 C) for Ru^{4+}-$LiNi_{0.5}Mn_{1.5}O_4$ (Fd$\bar{3}$m). Upon increasing the current intensity to a rate of 10 C (1470 mA g^{-1}), the capacity retention of Ru^{4+}-$LiNi_{0.5}Mn_{1.5}O_4$ (P4$_3$32) still remains at ≈75.8%, compared with ≈67.2% for Ru^{4+}-$LiNi_{0.5}Mn_{1.5}O_4$ (Fd$\bar{3}$m). Meanwhile, the capacities decay to only ≈63.4% and 52.8% of the initial capacities for $LiNi_{0.5}Mn_{1.5}O_4$ (Fd$\bar{3}$m) and $LiNi_{0.5}Mn_{1.5}O_4$ (P4$_3$32), respectively.

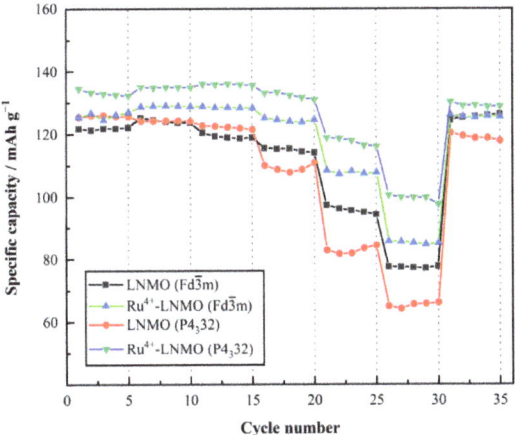

Figure 7. Rate performance of LiNi$_{0.5}$Mn$_{1.5}$O$_4$ (Fd$\bar{3}$m, P4$_3$32) before and after Ru^{4+} doping.

The superior rate performance of Ru^{4+}-LiNi$_{0.5}$Mn$_{1.5}$O$_4$ (P4$_3$32) is mainly related to rapid migration rate of Li$^+$ during the charge/discharge process. Firstly, owing to the incorporation of Ru^{4+}, the lattice parameters of LiNi$_{0.5}$Mn$_{1.5}$O$_4$ increase, which may widen the Li$^+$ movement path and reduce the activation energy required for the migration of Li$^+$. Secondly, the electron popping path changes from an O-Ni-O-Ni route to an O-Ru/Ni-O-Ru/Ni route, which facilitates the transfer of electrons. As a result, Ru^{4+}-LiNi$_{0.5}$Mn$_{1.5}$O$_4$ (Fd$\bar{3}$m, P4$_3$32) all have a better electronic conductivity than LiNi$_{0.5}$Mn$_{1.5}$O$_4$ (Fd$\bar{3}$m, P4$_3$32). Thirdly, compared with Ni (3d^8, 2 vacancies), Ru (4d^4, 6 vacancies) has more outer vacancies and has a wider conduction band overlapping with the O 2p orbitals, which both contribute to enhancing the movement of electrons and lithium ions. It is worth noting that the LiNi$_{0.5}$Mn$_{1.5}$O$_4$ cathode material has a tendency to be disordered as a result of Ru^{4+} doping. Disordered LiNi$_{0.5}$Mn$_{1.5}$O$_4$ remains in a disordered state, and ordered LiNi$_{0.5}$Mn$_{1.5}$O$_4$ transforms to semi-ordered LiNi$_{0.5}$Mn$_{1.5}$O$_4$. Therefore, Ru^{4+}-LiNi$_{0.5}$Mn$_{1.5}$O$_4$ (P4$_3$32) without an impurity phase has better high-rate properties than Ru^{4+}-LiNi$_{0.5}$Mn$_{1.5}$O$_4$ (Fd$\bar{3}$m).

4. Conclusions

With the introduction of Ru^{4+}, the samples all deliver a higher discharge capacity, greater cycling stability, and better rate performance (especially at high charge–discharge current). Compared with LiNi$_{0.5}$Mn$_{1.5}$O$_4$ (Fd$\bar{3}$m), the P4$_3$32-structured cathode material shows the most obvious improvement in electrochemical performance after doping. After 300 cycles at a 1 C rate, the capacity retention of Ru^{4+}-LiNi$_{0.5}$Mn$_{1.5}$O$_4$ (P4$_3$32) is still at 97.7% (at 1st cycle ≈132.8 mAh g^{-1}; at 300th cycle ≈129.8 mAh g^{-1}), which is larger than for undoped samples. Intriguingly, the specific capacity of Ru^{4+}-LiNi$_{0.5}$Mn$_{1.5}$O$_4$ (P4$_3$32) remains at 100 mAh g^{-1} even at the extreme charge–discharge rate of 10 C (1470 mAh g^{-1}). The introduction of Ru^{4+} hinders the ordering process of nickel/manganese ions during annealing treatment to suppress the lattice damage in the charge–discharge process. The stronger Ru-O bonding in the Ru^{4+}-doped samples stabilizes the cathode structure and further improves the cycling stability. The greater number of O-Ru/Ni-O-Ru/Ni electron movement paths in the Ru^{4+}-doped LiNi$_{0.5}$Mn$_{1.5}$O$_4$ contribute to increasing the electron movement and enhancing the high-rate performance. After doping with Ru, disordered LiNi$_{0.5}$Mn$_{1.5}$O$_4$ remains in a disordered state, and ordered LiNi$_{0.5}$Mn$_{1.5}$O$_4$ is transformed into semi-ordered LiNi$_{0.5}$Mn$_{1.5}$O$_4$ with no impurities, which explains why the electrochemical performance of Ru^{4+}-LiNi$_{0.5}$Mn$_{1.5}$O$_4$ (P4$_3$32) is better than that of Ru^{4+}-LiNi$_{0.5}$Mn$_{1.5}$O$_4$ (Fd$\bar{3}$m). In terms of the low-cost solid-state synthesis method, Ru^{4+} doping of ordered LiNi$_{0.5}$Mn$_{1.5}$O$_4$ provides a potential method of improving the electrochemical characteristics of high-voltage cathode materials for lithium-ion batteries.

Author Contributions: Data curation, X.L.; Formal analysis, X.L.; Investigation, X.L.; Software, X.L.; Supervision, W.X.; Writing—original draft, X.L., B.S. and J.Z.; Writing—review & editing, B.S., W.X. and J.Z. All authors have read and agreed to the published version of the manuscript.

Funding: The authors acknowledge that this work was supported by the National Key Research and Development Plan (2016YFE0111500).

Institutional Review Board Statement: Not applicable.

Informed Consent Statement: Not applicable.

Data Availability Statement: Not applicable.

Conflicts of Interest: The authors declare no conflict of interest.

References

1. Risthaus, T.; Wang, J.; Friesen, A.; Wilken, A.; Berghus, D.; Winter, M.; Li, J. Synthesis of spinel LiNi$_{0.5}$Mn$_{1.5}$O$_4$ with secondary plate morphology as cathode material for lithium ion batteries. *J. Power Sources* **2015**, *293*, 137–142. [CrossRef]
2. Liu, H.; Kloepsch, R.; Wang, J.; Winter, M.; Li, J. Truncated octahedral LiNi$_{0.5}$Mn$_{1.5}$O$_4$ cathode material for ultralong-life lithium-ion battery: Positive (100) surfaces in high-voltage spinel system. *J. Power Sources* **2015**, *300*, 430–437. [CrossRef]
3. Zhong, Q.; Bonakdarpour, A.; Zhang, M.; Gao, Y.; Dahn, J.R. ChemInform Abstract: Synthesis and Electrochemistry of LiNi$_x$Mn$_{20-x}$O$_4$. *ChemInform* **1997**, *144*, 205–213. [CrossRef]
4. Liu, D.; Zhu, W.; Trottier, J.; Gagnon, C.; Barray, F.; Guerfi, A.; Mauger, A.; Groult, H.; Julien, C.M.; Goodenough, J.B.; et al. Spinel materials for high-voltage cathodes in Li-ion batteries. *RSC Adv.* **2014**, *4*, 154–167. [CrossRef]
5. Song, J.; Shin, D.W.; Lu, Y.; Amos, C.D.; Manthiram, A.; Goodenough, J.B. Role of oxygen vacancies on the performance of Li[Ni$_{0.5-x}$Mn$_{1.5+x}$]O$_4$ (x = 0, 0.05, and 0.08) spinel cathodes for lithium-ion batteries. *Chem. Mater.* **2012**, *24*, 3101–3109. [CrossRef]
6. Deng, Y.-F.; Zhao, S.-X.; Xu, Y.-H.; Nan, C.-W. Effect of temperature of Li$_2$O–Al$_2$O$_3$–TiO$_2$–P$_2$O$_5$ solid-state electrolyte coating process on the performance of LiNi$_{0.5}$Mn$_{1.5}$O$_4$ cathode materials. *J. Power Sources* **2015**, *296*, 261–267. [CrossRef]
7. Xiao, J.; Chen, X.; Sushko, P.V.; Sushko, M.L.; Kovarik, L.; Feng, J.; Deng, Z.; Zheng, J.; Graff, G.L.; Nie, Z.; et al. High-performance LiNi$_{0.5}$Mn$_{1.5}$O$_4$ spinel controlled by Mn^{3+} concentration and site disorder. *Adv. Mater.* **2012**, *24*, 2109–2116. [CrossRef]
8. Patoux, S.; Daniel, L.; Bourbon, C.; Lignier, H.; Pagano, C.; LE Cras, F.; Jouanneau, S.; Martinet, S. High voltage spinel oxides for Li-ion batteries: From the material research to the application. *J. Power Sources* **2009**, *189*, 344–352. [CrossRef]
9. Santhanam, R.; Rambabu, B. Research progress in high voltage spinel LiNi$_{0.5}$Mn$_{1.5}$O$_4$ material. *J. Power Sources* **2010**, *195*, 5442–5451. [CrossRef]
10. Kunduraci, M.; Al-Sharab, J.F.; Amatucci, G.G. High-power nanostructured LiMn$_{2-x}$Ni$_x$O$_4$ high-voltage lithium-ion battery electrode materials: Electrochemical impact of electronic conductivity and morphology. *Chem. Mater.* **2006**, *18*, 3585–3592. [CrossRef]
11. Li, B.; Xing, L.; Xu, M.; Lin, H.; Li, W. New solution to instability of spinel LiNi$_{0.5}$Mn$_{1.5}$O$_4$ as cathode for lithium ion battery at elevated temperature. *Electrochem. Commun.* **2013**, *34*, 48–51. [CrossRef]
12. Park, O.K.; Cho, Y.; Lee, S.; Yoo, H.-C.; Song, H.-K.; Cho, J. Who will drive electric vehicles, olivine or spinel? *Energy Environ. Sci.* **2011**, *4*, 1621–1633. [CrossRef]
13. Jarry, A.; Gottis, S.; Yu, Y.S.; Roque-Rosell, J.; Kim, C.; Cabana, J.; Kerr, J.; Kostecki, R. The formation mechanism of fluorescent metal complexes at the Li$_x$Ni$_{0.5}$Mn$_{1.5}$O$_4$-delta/carbonate ester electrolyte interface. *J. Am. Chem. Soc.* **2015**, *137*, 3533–3539. [CrossRef] [PubMed]
14. Liu, G.Q.; Wen, L.; Liu, Y.M. Spinel LiNi$_{0.5}$Mn$_{1.5}$O$_4$ and its derivatives as cathodes for high-voltage Li-ion batteries. *J. Solid State Electrochem.* **2010**, *14*, 2191–2202. [CrossRef]
15. Zhu, X.; Li, X.; Zhu, Y.; Jin, S.; Wang, Y.; Qian, Y. LiNi$_{0.5}$Mn$_{1.5}$O$_4$ nanostructures with two-phase intergrowth as enhanced cathodes for lithium-ion batteries. *Electrochim. Acta* **2014**, *121*, 253–257. [CrossRef]
16. Kim, J.-H.; Huq, A.; Chi, M.; Pieczonka, N.P.W.; Lee, E.; Bridges, C.A.; Tessema, M.M.; Manthiram, A.; Persson, K.A.; Powell, B.R. Integrated nano-domains of disordered and ordered spinel phases in LiNi$_{0.5}$Mn$_{1.5}$O$_4$ for Li-ion batteries. *Chem. Mater.* **2014**, *26*, 4377–4386. [CrossRef]
17. Liu, G.; Park, K.-S.; Song, J.; Goodenough, J.B. Influence of thermal history on the electrochemical properties of Li[Ni$_{0.5}$Mn$_{1.5}$]O$_4$. *J. Power Sources* **2013**, *243*, 260–266. [CrossRef]
18. Kim, J.H.; Myung, S.T.; Yoon, C.S.; Kang, S.G.; Sun, Y.K. Comparative study of LiNi$_{0.5}$Mn$_{1.5}$O$_{4-\delta}$ and LiNi$_{0.5}$Mn$_{1.5}$O$_4$ cathodes having two crystallographic structures: Fd$\bar{3}$m and P4$_3$32. *Chem. Mater.* **2004**, *16*, 906–914. [CrossRef]
19. Wang, J.F.; Chen, D.; Wu, W.; Wang, L.; Liang, G.C. Effects of Na$^+$ doping on crystalline structure and electrochemical performances of LiNi$_{0.5}$Mn$_{1.5}$O$_4$ cathode material. *Trans. Nonferr. Metal. Soc.* **2017**, *27*, 2239–2248. [CrossRef]
20. Liu, G.; Zhang, L.; Lu, S.; Lun, W. A new strategy to diminish the 4 V voltage plateau of LiNi$_{0.5}$Mn$_{1.5}$O$_4$. *Mater. Res. Bull.* **2013**, *48*, 4960–4962. [CrossRef]
21. Liu, S.; Zhao, H.; Tan, M.; Hu, Y.; Shu, X.; Zhang, M.; Chen, B.; Liu, X. Er-Doped LiNi$_{0.5}$Mn$_{1.5}$O$_4$ Cathode Material with Enhanced Cycling Stability for Lithium-Ion Batteries. *Materials* **2017**, *10*, 859. [CrossRef] [PubMed]

22. Liu, G.; Zhang, J.; Zhang, X.; Du, Y.; Zhang, K. Study on oxygen deficiency in spinel $LiNi_{0.5}Mn_{1.5}O_4$ and its Fe and Cr-doped compounds. *J. Alloys Compd.* **2017**, *725*, 580–586. [CrossRef]
23. Lan, L.; Li, S.; Li, J.; Lu, L.; Lu, Y.; Huang, S.; Xu, S.; Pan, C.; Zhao, F. Enhancement of the electrochemical performance of the spinel structure $LiNi_{0.5-x}Ga_xMn_{1.5}O_4$ cathode material by Ga doping. *Nanoscale Res. Lett.* **2018**, *13*, 251. [CrossRef] [PubMed]
24. Wang, H.; Tan, T.A.; Yang, P.; Lai, M.O.; Lu, L. High-rate performances of the Ru-doped spinel $LiNi_{0.5}Mn_{1.5}O_4$: Effects of doping and particle size. *J. Phys. Chem. C* **2011**, *115*, 6102–6110. [CrossRef]
25. Wang, S.; Li, P.; Shao, L.; Wu, K.; Lin, X.; Shui, M.; Long, N.; Wang, D.; Shu, J. Preparation of spinel $LiNi_{0.5}Mn_{1.5}O_4$ and Cr-doped $LiNi_{0.5}Mn_{1.5}O_4$ cathode materials by tartaric acid assisted sol-gel method. *Ceram. Int.* **2015**, *41*, 1347–1353. [CrossRef]
26. Sun, J.; Li, P.; Wang, K.; Tan, Y.; Xue, B.; Niu, J. Effect of Al^{3+} and Al_2O_3 co-modification on electrochemical characteristics of the 5-V cathode material $LiNi_{0.5}Mn_{1.5}O_4$. *Ionics* **2020**, *26*, 3725–3736. [CrossRef]
27. Zhou, D.; Li, J.; Chen, C.; Lin, F.; Wu, H.; Guo, J. A hydrothermal synthesis of Ru-doped $LiMn_{1.5}Ni_{0.5}O_4$ cathode materials for enhanced electrochemical performance. *RSC Adv.* **2021**, *11*, 12549–12558. [CrossRef]
28. Zhou, D.; Li, J.; Chen, C.; Chen, C.; Wu, H.; Lin, F.; Guo, J. Ruthenium doped $LiMn_{1.5}Ni_{0.5}O_4$ microspheres with enhanced electrochemical performance as lithium-ion battery cathode. *J. Mater. Sci. Mater. Electron.* **2021**, *32*, 23786–23797. [CrossRef]
29. Chae, J.S.; Jo, M.R.; Kim, Y.-I.; Han, D.-W.; Park, S.-M.; Kang, Y.-M.; Roh, K.C. Kinetic favorability of Ru-doped $LiNi_{0.5}Mn_{1.5}O_4$ for high-power lithium-ion batteries. *J. Ind. Eng. Chem.* **2015**, *21*, 731–735. [CrossRef]
30. Oh, S.H.; Chung, K.Y.; Jeon, S.H.; Kim, C.S.; Cho, W.I.; Cho, B.W. Structural and electrochemical investigations on the $LiNi_{0.5-x}Mn_{1.5-y}M_{x+y}O_4$ (M = Cr, Al, Zr) compound for 5 V cathode material. *J. Alloys Compd.* **2009**, *469*, 244–250. [CrossRef]
31. Manthiram, A.; Chemelewski, K.; Lee, E.-S. A perspective on the high-voltage $LiMn_{1.5}Ni_{0.5}O_4$ spinel cathode for lithium-ion batteries. *Energy Environ. Sci.* **2014**, *7*, 1339–1350. [CrossRef]
32. Yi, T.-F.; Mei, J.; Zhu, Y.-R. Key strategies for enhancing the cycling stability and rate capacity of $LiNi_{0.5}Mn_{1.5}O_4$ as high-voltage cathode materials for high power lithium-ion batteries. *J. Power Sources* **2016**, *316*, 85–105. [CrossRef]
33. Cabana, J.; Zheng, H.; Shukla, A.K.; Kim, C.; Battaglia, V.S.; Kunduraci, M. Comparison of the performance of $LiNi_{1/2}Mn_{3/2}O_4$ with different microstructures. *J. Electrochem. Soc.* **2011**, *158*, A997–A1004. [CrossRef]
34. Chen, Z.; Qiu, S.; Cao, Y.; Ai, X.; Xie, K.; Hong, X.; Yang, H. Surface-oriented and nanoflake-stacked $LiNi_{0.5}Mn_{1.5}O_4$ spinel for high-rate and long-cycle-life lithium ion batteries. *J. Mater. Chem.* **2012**, *22*, 17768–17772. [CrossRef]
35. Kim, J.-S.; Kim, K.; Cho, W.; Shin, W.H.; Kanno, R.; Choi, J.W. A Truncated Manganese Spinel Cathode for Excellent Power and Lifetime in Lithium-Ion Batteries. *Nano Lett.* **2012**, *12*, 6358–6365. [CrossRef]
36. Zhang, X.; Cheng, F.; Yang, J.; Chen, J. $LiNi_{0.5}Mn_{1.5}O_4$ Porous Nanorods as high-rate and long-life cathodes for Li-ion batteries. *Nano. Lett.* **2013**, *13*, 2822–2825. [CrossRef]
37. Hirayama, M.; Sonoyama, N.; Ito, M.; Minoura, M.; Mori, D.; Yamada, A.; Tamura, K.; Mizuki, J.I.; Kanno, R. Characterization of electrode/electrolyte interface with X-Ray reflectometry and epitaxial-film $LiMn_2O_4$ electrode. *J. Electrochem. Soc.* **2007**, *154*, A1065–A1072. [CrossRef]
38. Benedek, R.; Thackeray, M.M. Simulation of the surface structure of lithium manganese oxide spinel. *Phys. Rev. B* **2011**, *83*, 195439. [CrossRef]
39. Fang, C.C.M.; Parker, S.; De With, G. Atomistic Simulation of the Surface Energy of Spinel $MgAl_2O_4$. *J. Am. Ceram. Soc.* **2000**, *83*, 2082–2084. [CrossRef]
40. Huang, F.; Gilbert, B.; Zhang, H.; Banfield, J. Reversible, Surface-Controlled Structure Transformation in Nanoparticles Induced by an Aggregation State. *Phys. Rev. Lett.* **2004**, *92*, 155501. [CrossRef]
41. Lafont, U.; Locati, C.; Kelder, E. Nanopowders of spinel-type electrode materials for Li-ion batteries. *Solid State Ionics* **2006**, *177*, 3023–3029. [CrossRef]

Article

Hierarchical and Heterogeneous Porosity Construction and Nitrogen Doping Enabling Flexible Carbon Nanofiber Anodes with High Performance for Lithium-Ion Batteries

Jun Liu [1], Yuan Liu [1], Jiaqi Wang [1], Xiaohu Wang [1,2], Xuelei Li [1,3,*], Jingshun Liu [1,3], Ding Nan [4,5] and Junhui Dong [1]

1. Inner Mongolia Key Laboratory of Graphite and Graphene for Energy Storage and Coating, School of Materials Science and Engineering, Inner Mongolia University of Technology, Hohhot 010051, China; clxylj@163.com (J.L.); ly2547068790@163.com (Y.L.); wjq024466341811@163.com (J.W.); wxh20220208@163.com (X.W.); jingshun_liu@163.com (J.L.); jhdong@imut.edu.cn (J.D.)
2. Rising Graphite Applied Technology Research Institute, Chinese Graphite Industrial Park—Xinghe, Ulanqab 013650, China
3. Collaborative Innovation Center of Non-ferrous Metal Materials and Processing Technology Co-Constructed by the Province and Ministry, Inner Mongolia Autonomous Region, Inner Mongolia University of Technology, Hohhot 010051, China
4. College of Chemistry and Chemical Engineering, Inner Mongolia University, West University, Street 235, Hohhot 010021, China; nd@imu.edu.cn
5. Inner Mongolia Enterprise Key Laboratory of High Voltage and Insulation Technology, Inner Mongolia Power Research Institute Branch, Inner Mongolia Power (Group) Co., Ltd., Hohhot 010020, China
* Correspondence: lglixuelei@163.com

Citation: Liu, J.; Liu, Y.; Wang, J.; Wang, X.; Li, X.; Liu, J.; Nan, D.; Dong, J. Hierarchical and Heterogeneous Porosity Construction and Nitrogen Doping Enabling Flexible Carbon Nanofiber Anodes with High Performance for Lithium-Ion Batteries. *Materials* 2022, *15*, 4387. https://doi.org/10.3390/ma15134387

Academic Editor: Satyam Panchal

Received: 3 May 2022
Accepted: 17 June 2022
Published: 21 June 2022

Publisher's Note: MDPI stays neutral with regard to jurisdictional claims in published maps and institutional affiliations.

Copyright: © 2022 by the authors. Licensee MDPI, Basel, Switzerland. This article is an open access article distributed under the terms and conditions of the Creative Commons Attribution (CC BY) license (https://creativecommons.org/licenses/by/4.0/).

Abstract: With the rapid development of flexible electronic devices, flexible lithium-ion batteries are widely considered due to their potential for high energy density and long life. Anode materials, as one of the key materials of lithium-ion batteries, need to have good flexibility, an excellent specific discharge capacity, and fast charge–discharge characteristics. Carbon fibers are feasible as candidate flexible anode materials. However, their low specific discharge capacity restricts their further application. Based on this, N-doped carbon nanofiber anodes with microporous, mesoporous, and macroporous structures are prepared in this paper. The hierarchical and heterogeneous porosity structure can increase the active sites of the anode material and facilitate the transport of ions, and N-doping can improve the conductivity. Moreover, the N-doped flexible carbon nanofiber with a porous structure can be directly used as the anode for lithium-ion batteries without adding an adhesive. It has a high first reversible capacity of 1108.9 mAh g^{-1}, a stable cycle ability (954.3 mAh g^{-1} after 100 cycles), and excellent rate performance. This work provides a new strategy for the development of flexible anodes with high performance.

Keywords: lithium-ion battery; flexible anode; hierarchical and heterogeneous porosity structure; N-doping; high performances

1. Introduction

Lithium-ion batteries have become the most widely used energy storage devices due to their long service life, high safety, low self-discharge, rapid charge and discharge, high power/energy densities, and environmental friendliness [1,2]. Nowadays, the application of lithium-ion batteries can be seen everywhere, such as in the aerospace industry, satellite communication, mobile phones, watches, and other portable electronic devices. In recent years, flexible and portable electronic devices have gradually entered the market, including flexible display screens, wearable electronic devices, flexible electronic medical devices, etc. The flexible lithium-ion battery is one of the important parts of flexible electronic devices, which has the advantages of a light weight, small volume, large energy storage capacity, long service life, wide temperature adaptation range, and high specific energy compared to other flexible energy devices [3]. Flexible electrode materials are the main factor determining the electrochemical performance of flexible lithium-ion batteries [4,5]. Therefore, they have become a research hotspot for scholars to develop flexible electrode materials with excellent electrochemical properties.

Carbon fibers are used extensively as flexible lithium-ion anode materials due to their good conductivity and mechanical strength [6]. However, carbon fiber anodes face the problems of poor flexibility, low reversible capacity, and insufficient rate performance when used directly. To improve their electrochemical performance, an effective measure is to fabricate holes of different sizes in the carbon fibers or dope the carbon fibers with some active substances. Zhang et al. reported that micropores could provide a larger surface area, which is beneficial to the distribution of reaction sites [7]. Mesopores were reported to greatly enhance the rate capability and facilitate the transport of ions [8]. Additionally, Zhu et al. reported that macropores could accommodate great volume variation during cycling [9]. At present, N-doping is also widely considered due to the advantages of low cost, stable combination of carbon and nitrogen, and improved performance [10,11]. However, it is difficult for a single mending strategy to fully satisfy the requirements of high electrochemical performances. Therefore, it may be an effective strategy to coordinate various modification methods for improving the performance of carbon nanofiber anodes.

Based on the above analysis, we produced a flexible N-doped carbon nanofiber anode with microporous, mesoporous, and macroporous structures using gas–electric co-spinning technology in this work. Compared to commonly used electrospinning, this technology can not only greatly improve the production efficiency but can also reduce the power consumption caused by a high-voltage power supply. In addition, this technology has a series of advantages such as device simplicity, easy operation, and low cost. The hierarchical and heterogeneous porosity structure can increase the number of reactive active sites of lithium ions and facilitate the transport of ions to effectively improve the reversible capacity of the anode, and N-doping can improve the conductivity. The N-doped carbon nanofiber anode with hierarchical and heterogeneous porosity structure can be directly used as the anode of lithium-ion batteries without adding adhesives and any substrate. Moreover, it has a high first reversible capacity of 1108.9 mAh g^{-1}, capacity retention of 86.1% after 100 cycles, and excellent rate performance, which is much higher than the results reported for other carbon fiber anodes for lithium-ion batteries. This work provides a new way to manufacture flexible carbon nanofiber anodes with high performance for lithium-ion batteries.

2. Experimental

2.1. Materials

The reagents used in this paper included polyacrylonitrile (PAN) (solid, molecular weight 150,000; Macklin), graphene (solid, 5 μm; Jiangnan Graphene Research Institute, Changzhou, China), N, N-dimethylformamide (DMF) (liquid, analytical purity; Macklin), and melamine ($C_3H_6N_6$) (solid, analytically pure; Macklin), which were of analytical purity to be used directly.

2.2. Preparation of NH_3-Activated Carbon Nanofibers

Gas–electric co-spinning technology combines electrostatic spinning technology and gas spinning technology. First, 2 g polyacrylonitrile (PAN) were added to N, N-dimethylformamide (DMF) and stirred for 10 h at 70 °C in a water bath to form a 10% PAN/DMF solution. The prepared precursor solution was put into a 50-milliliter syringe which had a coaxial stainless steel needle with an inner diameter of 1.11 mm and an outer diameter of 1.49 mm. An airflow nozzle was connected to the needle, and then the syringe was placed on the propeller. The collecting plate was 15 × 20 cm wire mesh. The distance between the needle and the collecting plate was controlled at 10~15 cm, and a high-voltage generator was used to connect the syringe needle and the steel wire mesh. Then, the prepared precursor solution was continuously spun on the collecting plate for 1 h to obtain a fiber cloth. In the process of gas–electric co-spinning, the feed rate was 4 mL h^{-1}, the voltage was 5 kV, and the airflow was 10 psi. For pre-oxidation, the obtained fiber cloth was pre-oxidized in a blast drying oven at 280 °C at a heating rate of 2 °C min^{-1} for 6 h. The purpose of pre-oxidation was to make PAN undergo three chemical reactions, namely cyclization, dehydrogenation, and oxidation, so as to make the carbon nanofiber cloth more stable before carbonization. For carbonization, the pre-oxidized fiber cloth was placed into a high-temperature tubular furnace, and the temperature was raised from room temperature to 950 °C at a rate of 3 °C min^{-1} under the protection of a high-purity nitrogen atmosphere for 1 h. Then, after carbonization, the high-purity nitrogen was replaced by ammonia and kept for 0–40 min for activation. Finally, the ammonia was changed into high-purity nitrogen gas and cooled to room temperature to obtain NH_3-activated carbon nanofibers. In the process of preparing the above materials, the NH_3-activated carbon nanofibers were named CNFs-0NH_3, CNFs-10NH_3, CNFs-20NH_3, CNFs-30NH_3, and CNFs-40NH_3 according to the activation times of 0, 10, 20, 30, and 40 min, respectively.

2.3. Preparation of N-Doped Carbon Nanofibers

The preparation process of N-doped carbon nanofibers with the hierarchical and heterogeneous porosity structure is shown in Figure 1. First, 0.1 g graphene was dissolved in 20 g DMF and sonicated for 20 min. Then, 2 g PAN and 1 g $C_3H_6N_6$ were added to the above solution and stirred for 10 h at 70 °C in a water bath to form a uniform spinning solution. Then, the prepared precursor solution was spun continuously for 1 h to obtain a nanofiber cloth by the gas–electric co-spinning device. The conditions for the gas–electric co-spinning and subsequent pre-oxidation were the same as above. The activation time of ammonia was 30 min. In the process of preparing the above materials, according to the mass ratios of PAN to $C_3H_6N_6$ of 2:0, 2:1, 2:2, and 2:3, the final products were named CNFs-0N, CNFs-1N, CNFs-2N, and CNFs-3N, respectively.

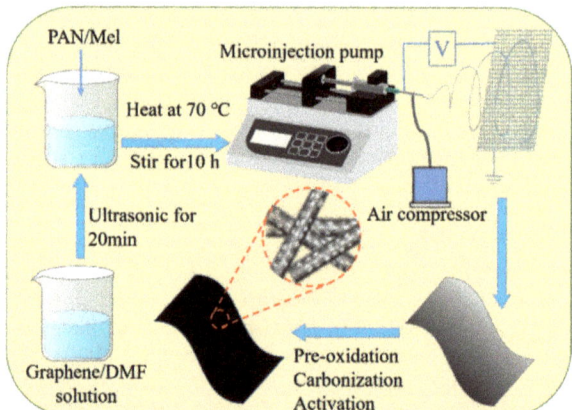

Figure 1. Flow chart of the preparation of N-doped carbon nanofiber anodes with the hierarchical and heterogeneous porosity structure.

2.4. Materials Characterization

The sample morphology was characterized using scanning electron microscopy (SEM, HITACHI-SU8220, HITACHI, Tokyo, Japan), and the corresponding element mapping on the surface of the materials was analyzed using an energy-dispersive spectrometer (EDS). The composition, content, and chemical valence of surface elements were studied by X-ray photoelectron spectroscopy (XPS, ESCALAB 250Xi, Thermo Fisher, Waltham, MA, USA). The pore size distribution of samples was measured at 77.2 K by the Brunauer–Emmett–Teller method (BET, BELSORP-mini II, BEL Japan Inc., Osaka, Japan). Before the pore size distribution test, the samples needed to be degassed for 8 h at 200 °C under an N_2 atmosphere to remove the adsorbed water molecules and low volatile content in the material.

2.5. Electrochemical Measurements

The prepared anode sheet of lithium-ion cells was a flexible carbon nanofiber which did not need the addition of binders or to be coated on copper foil as with traditional electrodes. The prepared N-doped carbon nanofiber anodes with the hierarchical and heterogeneous porosity structure after being sliced can be used directly as an anode. The electrode loading for the cells was ~1.2 mg cm^{-2}, and the diameter of this electrode was ~0.9 cm. A metal lithium sheet was used as the counter electrode, the separator was porous polypropylene, and the electrolyte was an EC:DEC:EMC = 1:1:1 (v/v) solvent with added 20% fluoroethylene carbonate (FEC) and lithium hexafluorophosphate (LiPF$_6$). The volume of the electrolyte used in each of the cells was about 0.4 mL. The electrochemical experiments were carried out using CR2032 cells at room temperature. The blue electric system (Land CT2001A, Blue Power Company) was set at constant current charge–discharge between the voltage range of 0.01 and 3 V. The rate performance of cells was tested at different current densities (50, 100, 200, 500, and 1000 mA g^{-1}). A Princeton (PMC1000A) electrochemical workstation was used to test the cyclic voltammetry (CV) in the voltage range of 0.01~3 V with a scanning rate of 0.01 mV s^{-1}. Additionally, electrochemical impedance spectroscopy (EIS) tests were carried out between 0.1 Hz and 100 kHz with an amplitude of 5 mV.

3. Results and Discussion

Figure 2 show the microscopic morphology of carbon nanofibers before activation and after activation for 30 min by ammonia. It can be seen from the images that the carbon nanofiber has a good fiber shape, and its diameter is about 300 nm. The SEM images in Figure 2a show that some particulate structures emerge on the surface of the carbon nanofibers before activation, which is caused by pre-oxidation and carbonization. In Figure 2b, the surface of the carbon nanofibers after activation for 30 min by ammonia not only has no small agglomerations but also shows the appearance of abundant pores, including the microporous, mesoporous, and macroporous structures produced by ammonia activation. The hierarchical and heterogeneous porosity structure is instrumental in increasing the active sites and facilitates the transport of ions of flexible carbon anode materials [7–9].

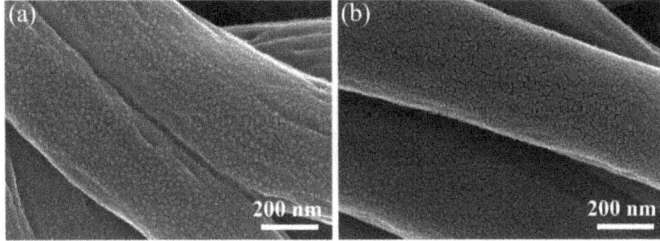

Figure 2. SEM images of carbon nanofibers (**a**) before activation and (**b**) after activation for 30 min by ammonia.

To analyze the types and states of surface elements of the carbon nanofibers after different ammonia activation times, all NH$_3$-activated carbon nanofibers were tested by XPS. The full spectrum in Figure 3a shows that these carbon nanofibers after ammonia activation contained three elements: C, N, and O; the corresponding three orbits C_{1s}, N_{1s}, and O_{1s}, were located at 285, 400, and 533 eV, respectively. The C/N atomic ratios of CNFs-10NH$_3$, CNFs-20NH$_3$, CNFs-30NH$_3$, and CNFs-40NH$_3$ were 4.57%, 5.64%, 5.76%, and 6.76%, respectively. The results show that ammonia activation can not only make pores but can also dope N, and with the increase in ammonia activation time, the N content in the carbon nanofibers also increased. Furthermore, Figure 3b–e show the N_{1s} spectra of the carbon nanofibers after different ammonia activation times. It can clearly be seen that all samples have two peaks of pyridinic nitrogen (N-6) and pyrrolic/pyridone nitrogen (N-5), which are located near 398.5 and 400.1 eV, respectively. It is reported that the existence of pyridine nitrogen and pyrrole nitrogen in carbon nanofibers can increase the conductivity [12,13]. Simultaneously, pyridine nitrogen and pyrrole nitrogen have a strong ability to adsorb lithium, which can improve the lithium storage performance of carbon nanofibers.

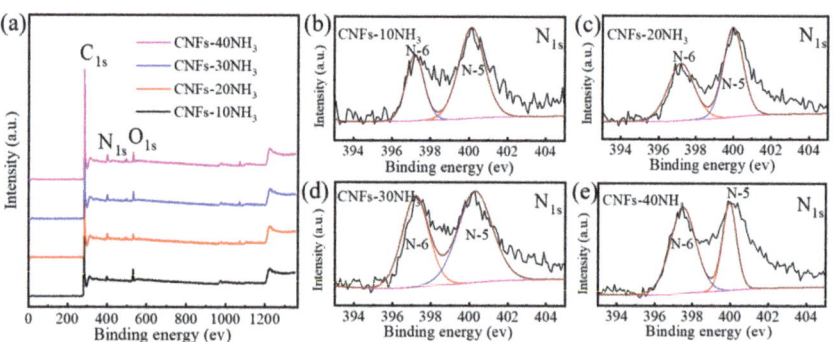

Figure 3. XPS analysis of carbon nanofibers with different ammonia activation times: (**a**) Full spectra of CNFs-10NH$_3$, CNFs-20NH$_3$, CNFs-30NH$_3$, and CNFs-40NH$_3$. N_{1s} spectra of (**b**) CNFs-10NH$_3$, (**c**) CNFs-20NH$_3$, (**d**) CNFs-30NH$_3$, and (**e**) CNFs-40NH$_3$.

The effect of the different ammonia activation times on the electrochemical properties of the carbon nanofiber anode was tested in lithium-ion batteries. Figure 4a show the cycle performance curves of the CNFs-0NH$_3$, CNFs-10NH$_3$, CNFs-20NH$_3$, CNFs-30NH$_3$, and CNFs-40NH$_3$ anodes at the current density of 50 mA g^{-1} from 0.01 to 3 V. The digital picture in Figure 4a shows the anode sheet activated by ammonia, which is flexible and foldable and can be used directly as a flexible lithium-ion anode without the addition of binders. It should be noted that the capacity vs. cycle number curves (and the CE curves as well) exhibit a lot of fluctuations (or wavy nature), especially at the slow rate of 50 mA g^{-1}. This is mainly because the temperature of the laboratory was affected by the ambient temperature, which led to the temperature fluctuation of the test battery near the room temperature. However, in the whole test process, this performance change trend was basically correct, and the data can explain the performance of the materials. With the increase in the ammonia activation time, the initial capacity of the carbon nanofiber anode gradually increased. After 80 cycles, the reversible capacity of the CNFs-0NH$_3$, CNFs-10NH$_3$, CNFs-20NH$_3$, CNFs-30NH$_3$, and CNFs-40NH$_3$ anodes was 406.3, 424.4, 440.2, 728.1, and 689.2 mAh g^{-1}, respectively, indicating that the CNFs-30NH$_3$ anode had the best cycle stability. Figure 4b show the rate performance curves of the CNFs-0NH$_3$, CNFs-10NH$_3$, CNFs-20NH$_3$, CNFs-30NH$_3$, and CNFs-40NH$_3$ anodes. Similarly, with the increase in the ammonia activation time from the CNFs-0NH$_3$, CNFs-10NH$_3$, and CNFs-20NH$_3$ anodes to the CNFs-30NH$_3$ anode, the rate performance of the carbon nanofiber anodes also gradually increased. The CNFs-30NH$_3$ anode showed the strongest rate performance, with an average specific capacity of 818.4, 666.5, 623.8, 508.9, and 452.4 mAh g^{-1} at the current density of 50, 100, 200, 500, and 1000 mA g^{-1}, respectively. When the current density returned to 50 mA g^{-1}, the specific capacity of the CNFs-30NH$_3$ anode could still reach 736.7 mAh g^{-1}. However, with the further increase in the ammonia activation time, the rate performance of the CNFs-40NH$_3$ anode was attenuated. These results illustrate that a moderate ammonia activation time is beneficial to improving the electrochemical properties of carbon nanofibers, as excessive activation may make the anode sheet brittle due to the presence of too many holes, resulting in poor rate performance.

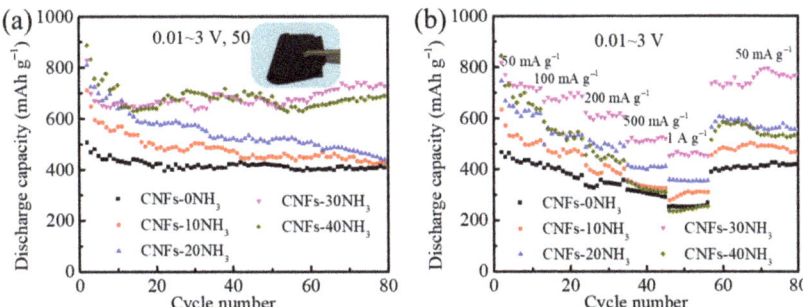

Figure 4. (a) Cycle performance curves and (b) rate performance curves of CNFs-0NH$_3$, CNFs-10NH$_3$, CNFs-20NH$_3$, CNFs-30NH$_3$, and CNFs-40NH$_3$ anodes. The digital picture in (a) is the anode sheet activated by ammonia.

Based on the analysis of the above results, ammonia activation can make carbon nanofibers obtain a hierarchical and heterogeneous porosity structure, which is instrumental in promoting electrochemical properties. In addition, it is worth noting that a small amount of nitrogen was doped in the carbon nanofibers in the process of manufacturing the microporous, mesoporous, and macroporous structures. Therefore, to further explore the effect of N-doping on the electrochemical performance of carbon nanofiber anodes, we chose CNFs-30NH$_3$ with the best performance as the N-doping object to reveal the influence mechanism of nitrogen on carbon nanofibers with the hierarchical and heterogeneous porosity structure and to further improve its electrochemical performance.

Figure 5a–d show the micromorphology of the hierarchical and heterogeneous porosity structure of carbon nanofibers with different N-doping contents. The images show that all N-doped porous carbon nanofibers have a good nanofiber shape, with a diameter of about 300 nm. With the increase in the N-doping content, there was little difference in the micromorphology of the carbon nanofibers. However, there are many cross-linking phenomena in the CNFs-3N anode. This is because when the added content of melamine is 3 g, the spinnability of the nanofiber becomes poor, and the spinning rate decreases due to the high viscosity of the solution, so the CNFs-3N anode will be locally uneven after pre-oxidation, carbonization, and activation. The SEM image of CNFs-2N and the corresponding EDS mapping of elements are shown in Figure 5e. The C and N elements are evenly distributed, indicating that N was effectively doped in the nanofibers. Figure 5f show the nitrogen adsorption–desorption curves. The hysteresis loop of each sample appears at P/P$_0$ > 0.4, belonging to type IV isotherms. This indicates that all samples contain mesopores. Furthermore, the corresponding pore size distribution curves of CNFs-0N, CNFs-1N, CNFs-2N, and CNFs-3N in Figure 5g indicate that all samples have micropores (pore size less than 2 nm), mesopores (pore size between 2 and 50 nm), and macropores (pore size greater than 50 nm), which meets the expectation of preparing N-doped carbon nanofibers with a hierarchical and heterogeneous porosity structure.

To further analyze the types and states of surface elements on the porous carbon nanofibers with different nitrogen contents, CNFs-0N, CNFs-1N, CNFs-2N, and CNFs-3N were tested by XPS. The full spectra in Figure 6a show that the prepared carbon nanofibers with different nitrogen contents contained three elements: C, N, and O; the corresponding three orbits, C$_{1s}$, N$_{1s}$, and O$_{1s}$, were located at 285, 400, and 533 eV, respectively. The C/N atomic ratios of CNFs-0N, CNFs-1N, CNFs-2N, and CNFs-3N were 5.8%, 6.6%, 8.4%, and 9.9%, respectively, obtained by XPS data analysis. The C/N atomic ratios of all samples were higher than those of CNFs-30NH$_3$, indicating that the amount of nitrogen was effectively increased by adding melamine during preparation. Figure 6b–e show the N$_{1s}$ spectra of

CNFs-0N, CNFs-1N, CNFs-2N, and CNFs-3N. All samples have two peaks of pyridinic nitrogen (N-6) at 398.5 eV and pyrrolic/pyridone nitrogen (N-5) at 400.1 eV. These results explain that nitrogen was effectively doped into CNFs-30NH$_3$, and the chemical bond formed by doping melamine is consistent with that formed by ammonia activation.

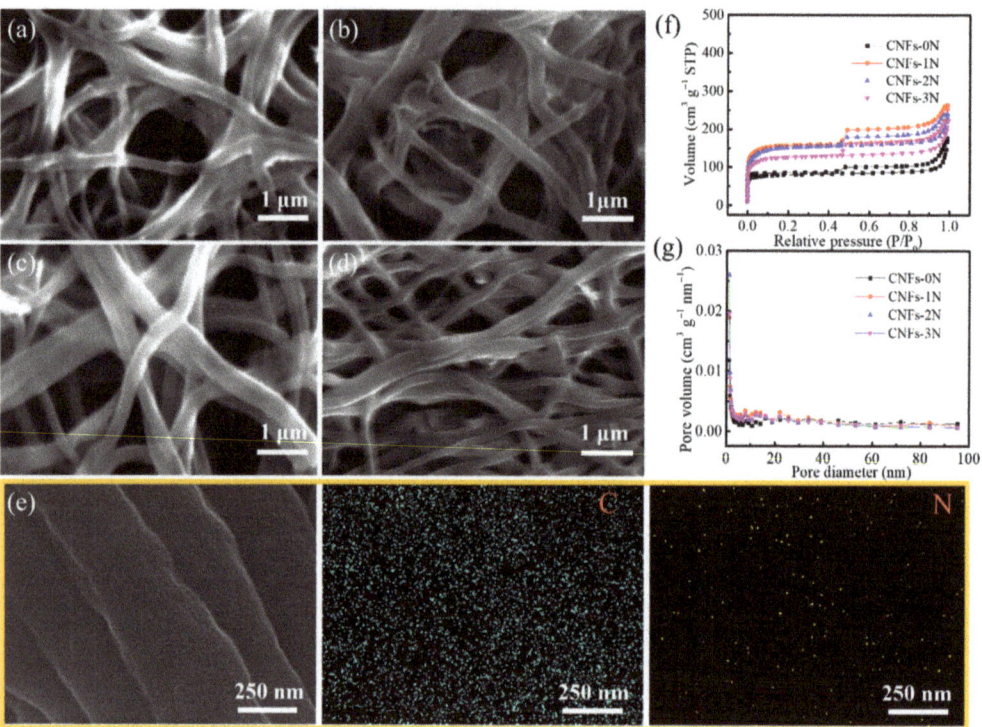

Figure 5. SEM images of porous carbon nanofibers with different nitrogen doping contents: (**a**) CNFs-0N, (**b**) CNFs-1N, (**c**) CNFs-2N, and (**d**) CNFs-3N. (**e**) SEM image and the corresponding EDS mapping of C and N elements in partial CNFs-2N. (**f**) Nitrogen adsorption–desorption curves and (**g**) pore size distribution curves of CNFs-0N, CNFs-1N, CNFs-2N, and CNFs-3N.

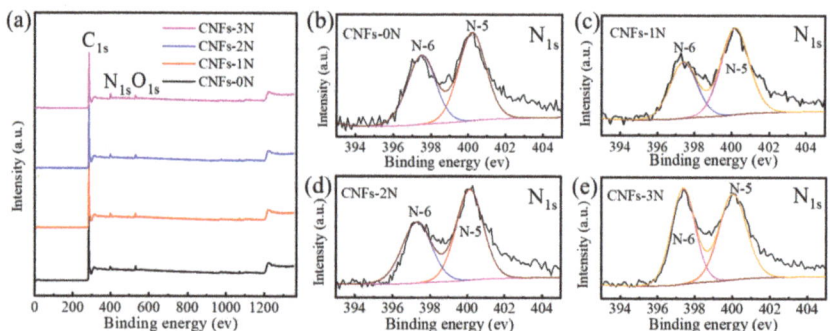

Figure 6. XPS spectra of porous carbon nanofibers with different nitrogen contents: (**a**) Full spectra of CNFs-0N, CNFs-1N, CNFs-2N, and CNFs-3N. Spectra of N$_{1s}$: (**b**) CNFs-0N, (**c**) CNFs-1N, (**d**) CNFs-2N, and (**e**) CNFs-3N.

Figure 7a show the cycle performance curves of the CNFs-0N, CNFs-1N, CNFs-2N, and CNFs-3N anodes at 50 mA g^{-1} from 0.01 to 3 V in lithium-ion batteries. After 100 cycles, the reversible capacity of each anode changed little. In particular, the CNFs-2N anode had a high initial capacity of 1108.9 mAh g^{-1} and still maintained a high reversible capacity of 954.3 mAh g^{-1} after 100 cycles, which was the highest capacity among the four anodes. However, the initial Coulombic efficiency of the CNFs-0N, CNFs-1N, CNFs-2N, and CNFs-3N anodes was only 49.72%, 50.24%, 46.29%, and 45.71%, respectively. The low initial Coulombic efficiency is due to the fact that after ammonia activation, the materials developed a number of micropores, mesopores, and macroporous structures, providing many reactive sites and forming a high irreversible capacity with the formation of a large area of SEI film during the first charge and discharge process. After the first charge and discharge, the Coulomb efficiency of each anode was stable at basically more than 90%, which is attributed to the formation of stable SEI film.

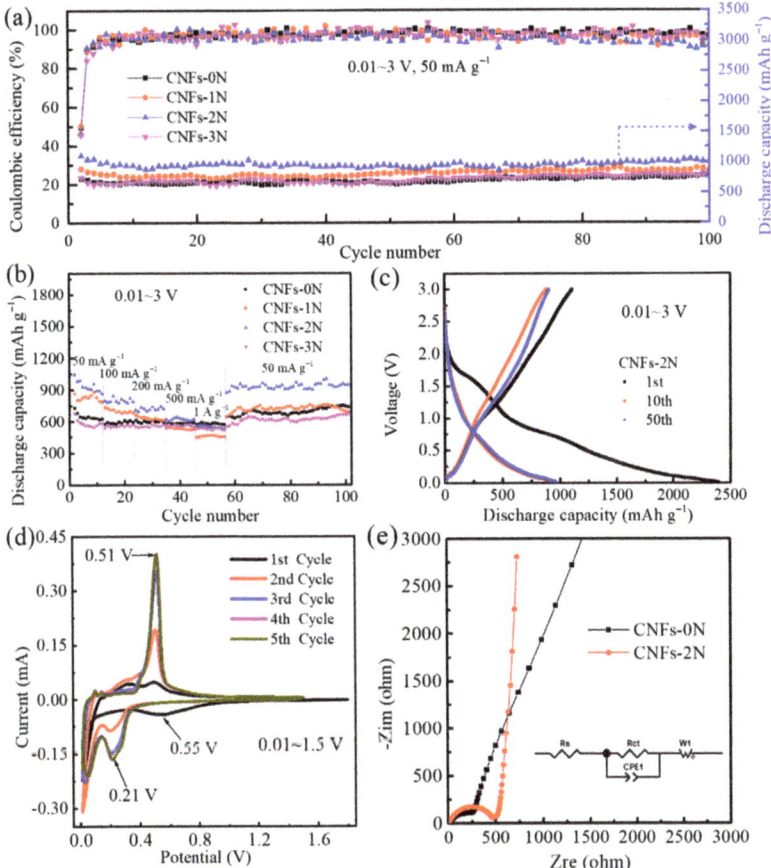

Figure 7. (**a**) Cycle performance curves and (**b**) rate performance curves of CNFs-0N, CNFs-1N, CNFs-2N, and CNFs-3N anodes. (**c**) The charge–discharge curves of the CNFs-2N anode, (**d**) CV curves of the CNFs-2N anode, and (**e**) Nyquist plots of CNFs-0N and CNFs-2N anodes in lithium-ion batteries. The insertion diagram in (**e**) is the equivalent circuit.

Figure 7b show the rate performance curves of the CNFs-0N, CNFs-1N, CNFs-2N, and CNFs-3N anodes in lithium-ion batteries at different current densities. The CNFs-2N anode still exhibited excellent specific capacities of 1047.2, 796.9, 709.6, 619.9, and 558.3 mAh g^{-1} at 50, 100, 200, 500, and 1000 mA g^{-1}, respectively. When the current density changed to

50 mA g^{-1} again, the specific capacity of the CNFs-2N anode was higher than 921.2 mAh g^{-1}, indicating a good rate performance. This is because N-doping and the hierarchical and heterogeneous porosity structure can improve the conductivity, increase the reactive sites, and improve the Li-ion transport rate. To the best of our knowledge, the outstanding electrochemical performance of the CNFs-2N anode is much better than that of other reported carbon fiber anodes for lithium-ion batteries (Table 1).

Table 1. Electrochemical performances of flexible carbon fiber anodes for lithium-ion batteries reported elsewhere and in this work.

Anodes	Mass Loading (mg cm^{-2})	First Reversible Capacity (mAh g^{-1})	Cycle Performance (mAh g^{-1})	Rate Performance (mAh g^{-1})	Reference
V$_2$O$_3$/MCCNFs	1.5~2.5	790.6 (0.1 A g^{-1})	487.7 (5 A g^{-1}, 5000 cycles)	456.8 (5 A g^{-1})	[7]
Sn@C@CNF	2.0	891.2 (0.1 A g^{-1})	610.8 (0.2 A g^{-1}, 180 cycles)	305.1 (2 A g^{-1})	[9]
SnS/CNFs	/	898 (0.05 A g^{-1})	548 (0.5 A g^{-1}, 500 cycles)	206 (4 A g^{-1})	[14]
γ-Fe$_2$O$_3$/C films	1.0	923.97 (0.2 A g^{-1})	1088 (0.2 A g^{-1}, 300 cycles)	380 (5 A g^{-1})	[15]
am-Fe$_2$O$_3$/rGO/CNFs	1.5~2.0	825 (0.1 A g^{-1})	739 (1 A g^{-1}, 400 cycles)	570 (2 A g^{-1})	[16]
In$_2$O$_3$@CF	1.4	510 (0.1 A g^{-1})	435 (0.1 A g^{-1}, 500 cycles)	190 (1.5 A g^{-1})	[17]
MnSe@C-700	1.6	614.6 (0.1 A g^{-1})	684 (0.1 A g^{-1}, 100 cycles)	/	[18]
NCNFs	7.64	752.3 (0.05 A g^{-1})	411.9 (0.1 A g^{-1}, 160 cycles)	148.8 (2 A g^{-1})	[19]
CNF@SnO$_2$	1.77~3.54	793 (0.5 A g^{-1})	485 (0.1 A g^{-1}, 850 cycles)	359 (4 A g^{-1})	[20]
G/Si@CFs	0.65~1	1036 (0.1 A g^{-1})	896.8 (0.1 A g^{-1}, 200 cycles)	543 (1 A g^{-1})	[21]
C/CuO/rGO	1.30~1.95	550 (0.1 A g^{-1})	400 (1 A g^{-1}, 600 cycles)	300 (2 A g^{-1})	[22]
FeCo@NCNFs-600	1.77~2.65	736.3 (0.1 A g^{-1})	566.5 (0.1 A g^{-1}, 100 cycles)	130 (2 A g^{-1})	[23]
SnO$_2$/TiO$_2$@CNFs	/	1061.2 (0.1 A g^{-1})	729.6 (0.1 A g^{-1}, 150 cycles)	206.2 (3 A g^{-1})	[24]
MoO$_2$/C	85.7	752.5 (0.2 A g^{-1})	450 (2 A g^{-1}, 500 cycles)	432 (2 A g^{-1})	[25]
FCNF-3/4	1.0	775 (0.2 A g^{-1})	630 (0.2 A g^{-1}, 100 cycles)	250 (5 A g^{-1})	[26]
Fe$_3$O$_4$/NCNFs	1.33	686 (0.1 A g^{-1})	522 (0.1 A g^{-1}, 200 cycles)	407 (5 A g^{-1})	[27]
Fe$_2$O$_3$/SnO$_x$/CNF	/	797 (0.1 A g^{-1})	756 (0.1 A g^{-1}, 55 cycles)	540 (1 A g^{-1})	[28]
V$_2$O$_3$/CNF	/	415.3 (0.2 A g^{-1})	420 (0.2 A g^{-1}, 100 cycles)	80 (10 A g^{-1})	[29]
ZnSe@CNFs-2.5	0.8~1.2	737.5 (0.1 A g^{-1})	426.1 (5 A g^{-1}, 3000 cycles)	547.6 (5 A g^{-1})	[30]
SiOC/C fibers-NH	0.8~1.5	518 (0.1 A g^{-1})	595 (0.2 A g^{-1}, 100 cycles)	195 (4 A g^{-1})	[31]
10-SnO$_2$@CNFs/CNT	1.5~2.5	500.9 (0.1 A g^{-1})	460.3 (0.1 A g^{-1}, 200 cycles)	222.2 (3.2 A g^{-1})	[32]
γ-Fe$_2$O$_3$@CNFs	2.0	1065 (0.5 A g^{-1})	430 (6 A g^{-1}, 1000 cycles)	222 (60 A g^{-1})	[33]
CNFs-2N	1.2	1108.9 (0.05 A g^{-1})	954.3 (0.05 A g^{-1}, 100 cycles)	549.7 (1 A g^{-1})	This work

Figure 7c show the constant-current charge–discharge curves of the CNFs-2N anode during the 1st, 10th, and 50th cycles in the voltage range of 0.01~3 V. In the first cycle, the discharge and charge capacities were observed to be 2395 and 1108.9 mAh g^{-1}, respectively, corresponding to 46.3% of the initial Coulombic efficiency. For the 10th and 50th cycles, the charge capacity was 879 and 906 mAh g^{-1}, respectively, and the discharge capacity was 929 and 963 mAh g^{-1}, respectively. Compared with the first cycle, the decrease in capacity was due to the formation of SEI film and the irreversible lithiation of carbon in the initial discharge process [14]. To further estimate the lithium storage performance of the CNFs-2N anode, CV curves of the first five cycles were obtained from 0.01 to 1.5 V at a scan rate of 0.1 mV s^{-1}, as shown in Figure 7d. During the first lithiation, the generation of the SEI film and the decomposition of the electrolyte led to a broad irreversible peak at about 0.55 V, which disappeared in the following cycles. For the anodic peak at 0.51 V and the reduction peak at 0.21 V, the current intensity gradually increased with the increase in the number of cycles, which corresponds to the activation process of more active materials reacting with lithium ions [34].

To further explore the effect of N-doping on the performance of lithium-ion batteries, the EIS of the CNFs-0N and CNFs-2N anodes was tested. Figure 7e show the Nyquist plots of the CNFs-0N and CNFs-2N anodes in lithium-ion batteries. The curve consists of a semicircle in the high-frequency region and a slash in the low-frequency region. The semicircle in the high-frequency region represents the charge transfer impedance. The oblique line in the low-frequency region is related to lithium-ion diffusion [35]. The semicircle in the high-frequency region of the CNFs-2N anode is larger than that of the CNFs-0N anode—that is, the charge transfer impedance increases after N-doping. The increase in the charge transfer impedance after N-doping may be due to the formation of a larger SEI film after forming pores, which makes CNFs-2N possess a larger specific surface area. However, the ideal pore size distribution can shorten the ion diffusion path and provide more active sites. Moreover, the porous structure can accommodate the volume strain and inhibit the volume expansion in the process of Li$^+$ insertion/de-insertion to achieve long-term cycle stability. Therefore, the CNFs-2N anode has a high specific capacity, an excellent rate performance, and cycle stability. In addition, compared with CNFs-0N, CNFs-2N has a higher slope in the low-frequency region. The larger the slope is, the lower the lithium-ion diffusion resistance is. This indicates that CNFs-2N has high conductivity after N-doping.

4. Conclusions

In this work, flexible N-doped carbon nanofiber anodes with a hierarchical and heterogeneous porosity structure were synthesized, which can be directly used as the anode of lithium-ion batteries without adding adhesives. SEM images and nitrogen adsorption–desorption tests confirmed the existence of microporous, mesoporous, and macroporous structures in the carbon nanofibers. EDS and XPS spectra proved that the nitrogen element was successfully doped in the carbon nanofibers. The hierarchical and heterogeneous porosity structure increased the active sites of the anode materials, improved the ion transport rate, and inhibited the volume expansion in the process of Li$^+$ insertion/de-insertion. The N-doping improved the conductivity of the carbon nanofibers. As expected, the prepared CNFs-2N anode had a high initial reversible specific capacity of 1108.9 mAh g^{-1} and excellent capacity retention and rate performance in lithium-ion batteries. This work provides a new way to develop high-performance flexible anode materials for lithium-ion batteries.

Author Contributions: J.L. (Jun Liu): conceptualization, data interpretation, funding acquisition, project administration; Y.L.: methodology, writing—original draft; J.W.: software, data curation; X.W.: cell fabrication and testing; X.L.: formal analysis, data interpretation, writing—review and editing; J.L. (Jingshun Liu): conceptualization, funding acquisition; D.N.: supervision, funding acquisition; J.D.: supervision. All authors have read and agreed to the published version of the manuscript.

Funding: This work was financially supported by Natural Science Foundation of Inner Mongolia (no. 2019MS05068), Inner Mongolia Major Science and Technology Project (no. 2020ZD0024), Scientific Research Project of Inner Mongolia University of Technology (no. ZZ202106), the Alashan League's Project of Applied Technology Research and Development Fund (no. AMYY2020-01), the research project of Inner Mongolia Electric Power (Group) Co., Ltd. for post-doctoral studies, Program for Innovative Research Team in Universities of Inner Mongolia Autonomous Region (no. NMGIRT2211), Inner Mongolia University of Technology Key Discipline Team Project of Materials Science (no. ZD202012), Inner Mongolia Natural Science Cultivating Fund for Distinguished Young Scholars (no. 2020JQ05), Science and Technology Planning Project of Inner Mongolia Autonomous Region (no. 2020GG0267), and Local Science and Technology Development Project of the Central Government (no. 2021ZY0006).

Institutional Review Board Statement: Not applicable.

Informed Consent Statement: Not applicable.

Data Availability Statement: Not applicable.

Conflicts of Interest: The authors declare no conflict of interest.

References

1. Alain, M.; Christian, M.J.; John, B.G.; Karim, Z. Tribute to michel armand: From rocking chair–Li-ion to solid-state lithium batteries. *J. Electrochem. Soc.* **2020**, *167*, 070507.
2. Alvaro, M.; James, M.; William, A.P. Opportunities and challenges of lithium ion batteries in automotive applications. *ACS Energy Lett.* **2021**, *6*, 621–630.
3. Wang, P.; Hu, M.; Wang, H.; Chen, Z.; Feng, Y.; Wang, J.; Ling, W.; Huang, Y. The evolution of flexible electronics: From nature, beyond nature, and to nature. *Adv. Sci.* **2020**, *7*, 2001116. [CrossRef] [PubMed]

4. Fu, J.; Kang, W.B.; Guo, X.D.; Wen, H.; Zeng, T.B.; Yuan, R.X.; Zhang, C.H. 3D hierarchically porous NiO/graphene hybrid paper anode for long-life and high rate cycling flexible Li-ion batteries. *J. Energy Chem.* **2020**, *47*, 172–179. [CrossRef]
5. Zeng, L.C.; Qiu, L.; Cheng, H.M. Towards the practical use of flexible lithium ion batteries. *Energy Storage Mater.* **2019**, *23*, 434–438. [CrossRef]
6. Joshi, B.; Samuel, E.; Kim, Y.I.; Yarin, A.L.; Swihart, M.T.; Yoon, S.S. Progress and potential of electrospinning-derived substrate-free and binder-free lithium-ion battery electrodes. *Chem. Eng. J.* **2022**, *430*, 132876. [CrossRef]
7. Zhang, T.; Zhang, L.; Zhao, L.N.; Huang, X.X.; Li, W.; Li, T.; Shen, T.; Sun, S.N.; Hou, Y.L. Free-standing, foldable V_2O_3/multichannel carbon nanofibers electrode for flexible Li-Ion batteries with ultralong lifespan. *Small* **2020**, *16*, 2005302. [CrossRef]
8. Li, W.H.; Li, M.S.; Wang, M.; Zeng, L.C.; Yu, Y. Electrospinning with partially carbonization in air: Highly porous carbon nanofibers optimized for high-performance flexible lithium-ion batteries. *Nano Energy* **2015**, *13*, 693–701. [CrossRef]
9. Zhu, S.Q.; Huang, A.M.; Wang, Q.; Xu, Y. MOF-derived porous carbon nanofibers wrapping Sn nanoparticles as flexible anodes for lithium/sodium ion batteries. *Nanotechnology* **2021**, *32*, 165401. [CrossRef]
10. Nan, D.; Huang, Z.H.; Lv, R.T.; Yang, L.; Wang, J.G.; Shen, W.C.; Liu, Y.X.; Yu, X.L.; Ye, L.; Sun, H.Y.; et al. Nitrogen-enriched electrospun porous carbon nanofiber networks as high-performance free-standing electrode materials. *J. Mater. Chem. A* **2014**, *2*, 19678–19684. [CrossRef]
11. Mao, Y.; Duan, H.; Xu, B.; Zhang, L.; Hu, Y.S.; Zhao, C.C.; Wang, Z.X.; Chen, L.Q.; Yang, Y.S. Lithium storage in nitrogen-rich mesoporous carbon materials. *Energy Environ. Sci.* **2012**, *5*, 7950–7955. [CrossRef]
12. Tan, Z.Q.; Ni, K.; Chen, G.X.; Zeng, W.C.; Tao, Z.C.; Ikram, M.; Zhang, Q.B.; Wang, H.J.; Sun, L.T.; Zhu, X.J.; et al. Incorporating pyrrolic and pyridinic nitrogen into a porous carbon made from C_{60} molecules to obtain superior energy storage. *Adv. Mater.* **2017**, *29*, 1603414. [CrossRef]
13. Liu, C.; Xiao, N.; Wang, Y.W.; Zhou, Y.; Wang, G.; Li, H.Q.; Ji, Y.Q.; Qiu, J.S. Electrospun nitrogen-doped carbon nanofibers with tuned microstructure and enhanced lithium storage properties. *Carbon* **2018**, *139*, 716–724. [CrossRef]
14. Xia, J.; Liu, L.; Jamil, S.; Xie, J.J.; Yan, H.X.; Yuan, Y.T.; Zhang, Y.; Nie, S.; Pan, J.; Wang, X.Y.; et al. Free-standing SnS/C nanofiber anodes for ultralong cycle-life lithium-ion batteries and sodium-ion batteries. *Energy Storage Mater.* **2019**, *17*, 1–11. [CrossRef]
15. Chen, Y.J.; Zhao, X.H.; Liu, Y.; Razzaq, A.A.; Haridas, A.K.; Cho, K.K.; Peng, Y.; Deng, Z.; Ahn, J.H. γ-Fe_2O_3 nanoparticles aligned in porous carbon nanofibers towards long life-span lithium ion batteries. *Electrochim. Acta* **2018**, *289*, 264–271. [CrossRef]
16. Zhao, Q.S.; Liu, J.L.; Li, X.X.; Xia, Z.Z.; Zhang, Q.X.; Zhou, M.; Tian, W.; Wang, M.; Hu, H.; Li, Z.T.; et al. Graphene oxide-induced synthesis of button-shaped amorphous Fe_2O_3/rGO/CNFs films as flexible anode for high-performance lithium-ion batteries. *Chem. Eng. J.* **2019**, *369*, 215–222. [CrossRef]
17. Zhao, H.; Yin, H.; Yu, X.X.; Zhang, W.; Li, C.; Zhu, M.Q. In_2O_3 nanoparticles/carbonfiber hybrid mat as free-standing anode for lithium-ion batteries with enhanced electrochemical performance. *J. Alloy. Compd.* **2018**, *735*, 319–326. [CrossRef]
18. Han, Z.S.; Kong, F.J.; Zheng, J.H.; Chen, J.Y.; Tao, S.; Qian, B. MnSe nanoparticles encapsulated into N-doped carbon fibers with a binder-free and free-standing structure for lithium ion batteries. *Ceram. Int.* **2021**, *47*, 1429–1438. [CrossRef]
19. Guo, J.Y.; Liu, J.Q.; Dai, H.H.; Zhou, R.; Wang, T.Y.; Zhang, C.C.; Ding, S.; Wang, H.G. Nitrogen doped carbon nanofiber derived from polypyrrole functionalized polyacrylonitrile for applications in lithium-ion batteries and oxygen reduction reaction. *J. Colloid Interf. Sci.* **2017**, *507*, 154–161. [CrossRef]
20. Abe, J.; Takahashi, K.; Kawase, K.; Kobayashi, Y.; Shiratori, S. Self-standing carbon nanofiber and SnO_2 nanorod composite as a high-capacity and high-rate-capability anode for lithium-ion batteries. *Acs Appl. Nano Mater.* **2018**, *1*, 2982–2989. [CrossRef]
21. Ma, X.X.; Hou, G.M.; Ai, Q.; Zhang, L.; Si, P.C.; Feng, J.K.; Ci, L.J. A heart-coronary arteries structure of carbon nanofibers/graphene/silicon composite anode for high performance lithium ion batteries. *Sci. Rep.* **2017**, *7*, 9642. [CrossRef]
22. Wu, S.H.; Han, Y.D.; Wen, K.C.; Wei, Z.H.; Chen, D.J.; Lv, W.Q.; Lei, T.Y.; Xiong, J.; Gu, M.; He, W.D. Composite nanofibers through in-situ reduction with abundant active sites as flexible and stable anode for lithium ion batteries. *Compos. Part B-Eng.* **2019**, *161*, 369–375. [CrossRef]
23. Li, X.Q.; Xiang, J.; Zhang, X.K.; Li, H.B.; Yang, J.N.; Zhang, Y.M.; Zhang, K.Y.; Chu, Y.Q. Electrospun FeCo nanoparticles encapsulated in N-doped carbon nanofibers as self-supporting flexible anodes for lithium-ion batteries. *J. Alloy. Compd.* **2021**, *873*, 159703. [CrossRef]
24. Mou, H.Y.; Chen, S.X.; Xiao, W.; Miao, C.; Li, R.; Xu, G.L.; Xin, Y.; Nie, S.Q. Encapsulating homogenous ultra-fine SnO_2/TiO_2 particles into carbon nanofibers through electrospinning as high-performance anodes for lithium-ion batteries. *Ceram. Int.* **2021**, *47*, 19945–19954. [CrossRef]
25. Zhang, X.Y.; Gao, M.Z.; Wang, W.; Liu, B.; Li, X.B. Encapsulating MoO_2 nanocrystals into flexible carbon nanofibers via electrospinning for high-performance lithium storage. *Polymers* **2021**, *13*, 22. [CrossRef]
26. Chen, R.Z.; Hu, Y.; Shen, Z.; Pan, P.; He, X.; Wu, K.S.; Zhang, X.W.; Cheng, Z.L. Facile fabrication of foldable electrospun polyacrylonitrile-based carbon nanofibers for flexible lithium-ion batteries. *J. Mater. Chem. A* **2017**, *5*, 12914–12921. [CrossRef]
27. Guo, L.G.; Sun, H.; Qin, C.Q.; Li, W.; Wang, F.; Song, W.L.; Du, J.; Zhong, F.; Ding, Y. Flexible Fe_3O_4 nanoparticles/N-doped carbon nanofibers hybrid film as binder-free anode materials for lithium-ion batteries. *Appl. Surf. Sci.* **2018**, *459*, 263–270. [CrossRef]
28. Joshi, B.N.; An, S.; Yong, Y.I.; Samuel, E.P.; Song, K.Y.; Seong, I.W.; Al-Deyab, S.S.; Swihart, M.T.; Yoon, W.Y.; Yoon, S.S. Flexible freestanding Fe_2O_3-SnO_x-carbon nanofiber composites for Liion battery anodes. *J. Alloy. Compd.* **2017**, *700*, 259–266. [CrossRef]

29. Gao, S.; Zhang, D.; Zhu, K.; Tang, J.A.; Gao, Z.M.; Wei, Y.J.; Chen, G.; Gao, Y. Flexible V_2O_3/carbon nano-felts as free-standing electrode for high performance lithium ion batteries. *J. Alloy. Compd.* **2017**, *702*, 13–19. [CrossRef]
30. Zhang, T.; Qiu, D.P.; Hou, Y.L. Free-standing and consecutive ZnSe@carbon nanofibers architectures as ultra-long lifespan anode for flexible lithium-ion batteries. *Nano Energy* **2022**, *94*, 106909. [CrossRef]
31. Ma, M.B.; Wang, H.J.; Li, X.; Peng, K.; Xiong, L.L.; Du, X.F. Free-standing SiOC/nitrogen-doped carbon fibers with highly capacitive Li storage. *J. Eur. Ceram. Soc.* **2020**, *40*, 5238–5246. [CrossRef]
32. Zhang, S.G.; Yue, L.C.; Wang, M.; Feng, Y.; Li, Z.; Mi, J. SnO_2 nanoparticles confined by N-doped and CNTs-modified carbon fibers as superior anode material for sodium-ion battery. *Solid State Ion.* **2018**, *323*, 105–111. [CrossRef]
33. Su, Y.; Fu, B.; Yuan, G.L.; Ma, M.; Jin, H.Y.; Xie, S.H.; Li, J.Y. Three-dimensional mesoporous γ-Fe_2O_3@carbon nanofiber network as high performance anode material for lithium- and sodium-ion batteries. *Nanotechnology* **2020**, *31*, 155401. [CrossRef] [PubMed]
34. Liu, X.L.; Meng, Y.S.; Li, R.N.; Du, M.Q.; Zhu, F.L.; Zhang, Y. Nitrogen-doped carbon-coated cotton-derived carbon fibers as high-performance anode materials for lithium-ion batteries. *Ionics* **2019**, *25*, 5799–5807. [CrossRef]
35. Chan, C.K.; Peng, H.L.; Liu, G.; McIlwrath, K.; Zhang, X.F.; Huggins, R.A.; Cui, Y. High-performance lithium battery anodes using silicon nanowires. *Nat. Nanotechnol.* **2008**, *3*, 31–35. [CrossRef]

Article

Thermal, Microstructural and Electrochemical Hydriding Performance of a $Mg_{65}Ni_{20}Cu_5Y_{10}$ Metallic Glass Catalyzed by CNT and Processed by High-Pressure Torsion

Ádám Révész [1,*], Marcell Gajdics [2], Miratul Alifah [1], Viktória Kovács Kis [2,3], Erhard Schafler [4], Lajos Károly Varga [5], Stanislava Todorova [6], Tony Spassov [6] and Marcello Baricco [7]

1. Department of Materials Physics, Eötvös University, P.O. Box 32, H-1518 Budapest, Hungary
2. Center of Energy Research, Hungarian Academy of Sciences, H-1121 Budapest, Hungary
3. Department of Mineralogy, Eötvös University, Pázmány Péter Sétány 1/c, H-1119 Budapest, Hungary
4. Physics of Nanostructured Materials, Faculty of Physics, University of Vienna, A-1090 Vienna, Austria
5. Research Institute for Solid State Physics and Optics, Hungarian Academy of Sciences, P.O. Box 49, H-1525 Budapest, Hungary
6. Department of Chemistry, University of Sofia "St. Kl. Ohridski", 1164 Sofia, Bulgaria
7. Dipartimento di Chimica and NIS-INSTM, Università di Torino, Via P. Giuria 7, 10125 Torino, Italy
* Correspondence: revesz.adam@ttk.elte.hu

Abstract: A $Mg_{65}Ni_{20}Cu_5Y_{10}$ metallic glass was produced by melt spinning and was mixed with a 5 wt.% multiwall carbon nanotube additive in a high-energy ball mill. Subsequently, the composite mixture was exposed to high-pressure torsion deformation with different torsion numbers. Complimentary XRD and DSC experiments confirmed the exceptional structural and thermal stability of the amorphous phase against severe plastic deformation. Combined high-resolution transmission electron microscopy observations and fast Fourier transform analysis revealed deformation-induced Mg_2Ni nanocrystals, together with the structural and morphological stability of the nanotubes. The electrochemical hydrogen discharge capacity of the severely deformed pure metallic glass was substantially lower than that of samples with the nanotube additive for several cycles. It was also established that the most deformed sample containing nanotubes exhibited a drastic breakdown in the electrochemical capacity after eight cycles.

Keywords: metallic glass; melt spinning; high-pressure torsion; hydrogen storage

1. Introduction

Alternative energy sources are barely competitive with conventional fossil fuels at the moment; nevertheless, hydrogen as a secondary energy carrier has received rapidly growing attention in the last several decades, mainly due to its very high chemical energy (120–140 MJ/kg) [1]. Hydrogen is clean, renewable and environmentally friendly; however, significant economic and technical challenges should be solved, including efficient production and storage [2]. Recently, a large number of attempts have been made to realize hydrogen storage in the solid state with sufficient storage capacity [3,4].

Among different hydrogen absorbing systems, magnesium-based alloys and compounds have been intensively investigated because of their high H-storage capacities (3700 Wh/L or 2600 Wh/kg), high abundance on Earth, low density, non-toxic nature and low cost, resulting in a potential candidate for future industrial applications [5–9]. Unfortunately, the relatively high hydrogenation enthalpy of Mg and its sluggish sorption kinetics are the main difficulties that still impede the widespread practical utilization of the Mg-H system [5,6]. In order to overcome these limitations of magnesium, different non-equilibrium techniques, such as rapid quenching by melt spinning and copper mold casting and high-energy ball milling (HEBM), have been applied to synthesize Mg-based hydrogen storage materials, including Mg-TM-RE ternary or pseudoternary systems (TM:

transition metal; RE: rare earth element) [10]. The as-quenched alloy can contain fully crystalline phases [11,12], supersaturated solid solution [13], partial amorphous structure [14], or monolithic amorphous glassy phases [15,16].

It was reported that fully amorphous melt-spun $Mg_{60}RE_xNi_{30-x}Cu_{10}$ alloys can absorb 3.0 wt.% H_2 at a temperature as low as 130 °C. The enhanced hydrogenation rate was explained by the formation of hydrogen-induced phase separation [17]. A similar Mg-Ni-Ce system exhibits a hydrogen-induced glass-to-glass transition with a storage capacity of 5 wt.% H_2, which is considerably higher than the value obtained for the crystalline counterpart due to the disordered atomic structure and free volume of the glass [18]. A similar phenomenon was reported for the amorphous $Mg_{87}Ni_{12}Y_1$ alloy, which also exhibits faster H-sorption kinetics than partially or fully crystallized alloys due to the faster diffusion of the H atoms in the amorphous matrix; however, the storage capacity is practically independent of the atomic structure [19]. Excellent hydrogenation/dehydrogenation cycling performance was reported for the rapidly quenched partially crystalline $Mg_{80}Ni_{10}Y_{10}$ alloy [20], while the $Mg_{12}YNi$ solid solution exhibited enhanced H-kinetics due to the catalytic effects of Mg_2Ni and Y [13]. Due to the low mixing enthalpy between Ni and Y ($\Delta H = -25$ kJ/mol), the activation energy of crystallization of melt-spun $Mg_{85}Ni_5Y_{10}$ can exceed 291 kJ/mol, indicating that atomic rearrangement is difficult due to interaction between these two elements [21]. In situ XRD analysis carried out during continuous heating revealed phase separation in the amorphous state prior to the nucleation of Mg_2Ni nanoparticles. The improved H-storage behavior of the $Mg_{86}N_4Y_{10}$ glass is related to the cracking and pulverization of the alloy pieces when a MgH_2 matrix with finely dispersed $MgNiH_4$ and YH_3 particles develops during hydrogen uptake [22]. The hydrogen storage performance of Mg-Ni-Y alloys can be improved by altering the addition of Ni and Y in order to nucleate a ternary eutectic 14H-LPSO phase; however, this long-period stacking ordered lamellar phase containing Ni and Y atoms does not form after dehydrogenation [23]. High-pressure hydrogen sorption experiments revealed that the as-cast fully amorphous $Mg_{54}Cu_{28}Ag_7Y_{11}$ bulk metallic glass exhibits the largest enthalpy of hydrogen desorption compared to its partially and fully crystallized counterparts; therefore, it was assumed that the disordered local atomic structure of the glass is responsible for the hydrogen release [24]. It was found that the hydrogen storage capacity of fully amorphous melt-spun $Mg_{85}Ni_{15-x}M_x$ (M = Y or La) alloys and their partial devitrified state can exceed 5 wt.% at 300 °C [25]. The hydrogen absorption and desorption kinetics of an amorphous Mg-Y-Ni ternary alloy can be improved when a long-period stacking ordered phase nucleates [26].

Hydrogen absorption of a melt-spun Mg-Ni-Mm alloy promotes the nucleation of a metastable cubic Mg_2NiH_4 phase and the vanishing of its typical monolithic counterpart. With increasing spinning velocity, this tendency becomes more pronounced [27]. The precipitation of the $Mg_{12}Mm$ intermetallic phase in the melt-spun Mg-Ni-Mm (Mm = Ce, La) system preferentially occurs at the Mg grain boundaries, which provide pathways for accelerated hydrogen diffusion [28]. The applied quenching rate has a significant influence on the sorption properties of these alloys [28,29]. When La substitutes Ni in the $Mg_{98}Ni_{2-x}Ce_x$ system, the formation of a refined eutectic structure facilitates hydrogenation. High-density LaH_3 nanoparticles nucleated in situ are responsible for the improved desorption processes [30].

An Mg-Mg_2Ni-LaH_x nanocomposite material formed from the hydrogen-induced decomposition of $Mg_{98}Ni_{1.67}La_{0.33}$ exhibits a H-storage capacity as high as 7.2 wt.% H_2. The significantly reduced absorption activation energy is attributed to the LaH_x and Mg_2Ni nanoparticles embedded in the eutectic alloy [31]. In situ X-ray synchrotron radiation of the Mg-Ni-La system during hydrogenation revealed the formation of a previously non-reported $La_2Mg_{17}H_{\sim 1.0}$ solid solution [32]. The electrochemical performance of a $LaMgNi_4$ electrode reveals acceptable cycling stability, decreasing to 47% of its initial capacity after 250 cycles [33]. It was confirmed that considerable H-induced amorphization occurs during the first sorption cycle, while the volume fraction of the amorphous component increases with the cycling number. This phenomenon was ascribed to the hydrogen-induced lattice instability. It was reported very recently that phase boundaries in a ternary Mg-Ni-La eutectic alloy can act as preferential nucleation sites for MgH_2 and apparently promote the hydrogenation process [34]. As an alternative method, DC magnetron sputtering was also applied to produce the fully amorphous $Mg_{85}Ni_{14}Ce_1$ alloy. It was demonstrated that as the thickness of the thin films decreases, the hydrogen sorption kinetics is significantly improved [35]. In addition, these samples possess an excellent cycling ability at 120 °C.

Severe plastic deformation by ball milling of $Mg_{90}Ni_7Ce_3$ not only refines the nanostructured powder but also enhances the creation of an amorphous component, which is beneficial for improving the H-storage kinetics at a temperature as low as 100 °C, at which 3.5 wt.% H_2 can be absorbed within 30 min [36]. A fully crystalline $Mg_{90}Ce_3Ni_4Y_3$ alloy synthesized by induction melting and subsequent HEBM exhibits an altered rate-limiting step of hydrogen desorption with respect to the host $Mg_{90}Ce_{10}$ intermetallic compound; i.e., the surface-controlled mode is transformed into nucleation- and growth-controlled. The addition of Ni and Y induces a large number of interface channels and nucleation sites [37]. Silver addition by HEBM to Mg_2Ni results in the formation of a hyper-eutectic Mg-Ni-Ag alloy with an increased cell parameter with respect to Mg_2Ni, which facilitates the diffusion of hydrogen atoms, reduces the onset of dehydrogenation and decreases its activation energy [38]. Very recently, different nanotube additives added by ball milling were proved to successfully improve the H-storage performance of Mg-based systems. For example, when multiwall carbon nanotubes (CNTs) were milled with a Mg-Ni-La alloy, a significant decrease in the dehydrogenation energy (82 kJ/mol) was observed due to the dispersed distribution of the nanotube sections on the surface of Mg-based particles [39].

The hydrogenation performance of different non-equilibrium alloys can be significantly improved when the material is subjected to a massive severe plastic deformation technique called high-pressure torsion (HPT) [40–46]. Among the different deformation methods, HPT exhibits the largest equivalent strain [47]. A $Mg_{65}Ni_{20}Cu_5Ce_{10}$ metallic glass synthesized by rapid quenching and additional torsional straining by HPT possesses a reduced dehydrogenation temperature and improved H-sorption kinetics that can be attributed to the interfaces between different nanoglass regions developed during the HPT process [48]. HPT on a fully amorphous $Mg_{65}Ni_{20}Cu_5Y_{10}$ glass promotes the nucleation of deformation-induced Mg_2Ni nanocrystals, which lowers the hydrogen absorption temperature and increases the storage capacity [49]. It was observed that these phenomena are more pronounced in the most deformed regions of samples [50].

In the current research, an amorphous $Mg_{65}Ni_{20}Cu_5Y_{10}$ glassy ribbon was produced by melt spinning. Glassy flakes of the as-quenched ribbon were milled together with 5 wt.% multiwall carbon nanotubes and were subsequently exposed to HPT with different numbers of whole turns. We demonstrate the structural and thermal stability of these Mg-based composites against severe shear deformation by HPT and shed light on the effect of CNT addition on their electrochemical H-storage performance. The aim of this series of investigations is to demonstrate whether carbon nanotubes can improve the hydrogen storage performance of Mg-based metallic glass, similar to crystalline alloys.

2. Materials and Methods

2.1. Sample Preparation

An ingot of the master alloy with a nominal composition of $Mg_{65}Ni_{20}Cu_5Y_{10}$ was synthesized by induction melting under a protective argon atmosphere. The purity of the constituents was 99.9%. From the ingot, a fully amorphous ribbon (thickness: ~50 μm) was prepared by rapid quenching of the molten alloy using copper single roller planar flow casting in an inert atmosphere. The tangential velocity of the casting wheel was 40 m/s. Subsequently, the as-quenched ribbon was reduced to small pieces (flakes) in a special attritor device, and then these flakes were milled together with 5 wt.% multiwall carbon nanotubes (Sigma-Aldrich, St. Louis, MO, USA; purity 98%; outer diameter: 6–13 nm; length: 2.5–20 µm) in a stainless-steel vial in an SPEX 8000M mixer mill using ten stainless steel balls (1/4 in.) at 1425 rpm, with a ball-to-powder weight ratio of 10:1 under a protective Ar atmosphere. Based on preliminary experiments [45], the total milling time was selected as 15 min in order to ensure a homogeneous mixture while avoiding unnecessary damage to CNT sections. Thereafter, the $Mg_{65}Ni_{20}Cu_5Y_{10}$ + CNT mixture was pre-compacted into cylindrical disks with a radius of R = 4 mm.

High-pressure torsion of the pre-compacted disks was performed under a 4 GPa applied pressure at room temperature while simultaneously imposing shear strain through N = 1, N = 2 and N = 5 whole revolutions with a torsional speed of 1 rot/min. These revolution numbers are typical in the literature. Hereafter, these samples are denoted as MgNiCuY_CNT_1, MgNiCuY_CNT_2 and MgNiCuY_CNT_5. As a comparison, a disk without torsion was also tested (N = 0), which is denoted as MgNiCuY_CNT_0. As a reference, a CNT-free disk tested with N = 5 whole turns was also prepared (MgNiCuY_5). In the applied setup, stainless steel anvils obey a constrained geometry [51]. Further details of HPT processing are given elsewhere [51]. The final thickness of the strained disks is about L = 0.8 mm. The accumulated shear strain for torsion deformation at a radius r at time t can be represented by

$$\varepsilon(r,t) = \frac{\omega \cdot t \cdot r}{L} = 2\pi \frac{N \cdot r}{L} \quad (1)$$

where ω is the angular velocity [51]. At the perimeter of our disk samples, the shear strain can reach extraordinary values, such as $\varepsilon(R) \approx 1200$.

2.2. Thermal Characterization

A Perkin Elmer power-compensated differential scanning calorimeter (DSC) was applied to explore the thermal stability and crystallization behavior of the rapidly quenched $Mg_{65}Ni_{20}Cu_5Y_{10}$ ribbon and the deformed HPT disk during continuous heating carried out at scan rates of 5, 10, 20, 40 and 80 Kmin^{-1}. The corresponding ΔH crystallization enthalpy values were determined as the area of exothermic peaks. All measurements were carried out under a protective Ar atmosphere. The temperature and the enthalpy were calibrated by using pure In and Al. The activation energy (E_a) of the crystallization processes was determined by the Kissinger analysis [52]. As is known, the dependence of the individual transformation peak temperature (T_i) on the heating rate can be given as

$$\frac{\beta}{T_i^2} = \frac{Z_i R}{E_{a,i}} \exp\left(\frac{-E_{a,i}}{RT_i}\right) \quad (2)$$

where Z and R are the frequency factor and the gas constant, respectively [50]. Plotting $\ln(\beta/T_i^2)$ vs. T_i^{-1} enables the determination of E_a for each thermal event from the slope of the fitted straight line.

2.3. Structural Characterization

The structures of the as-quenched $Mg_{65}Ni_{20}Cu_5Y_{10}$ glass and the torqued disks were examined by X-ray powder diffraction. The measurements were carried out on a Rigaku SmartLab diffractometer using Cu-Kα radiation in θ–2θ geometry. The data were collected from 20° to 100° with a step size of 0.01°.

2.4. Transmission Electron Microscopy

The local structure of the most deformed CNT-containing MgNiCuY_CNT_5 disk was analyzed by high-resolution transmission electron microscopy (HR-TEM). For this purpose, a THEMIS 200 electron microscope was used at an accelerating voltage of 200 kV. Sample preparation was carried out by cutting out a thin lamella from the HPT disk (close to the perimeter) via focused ion beam (FIB) in an FEI QUANTA 3D dual-beam scanning electron microscope. In order to analyze HR-TEM images and to identify the spatial distribution of crystallites, the fast Fourier transform of the images and the corresponding inverse fast Fourier transform (using Fourier filtering) were produced, respectively. For image processing, the Digital Micrograph (Gatan) software (DigitalMicrograph 3.5, Gatan, Inc., Pleasanton, CA, USA) was used. The elemental distribution of the specimen was investigated by energy-dispersive spectroscopy (EDS) built into the TEM device. High-angle annular dark field (HAADF) scanning transmission (STEM) imaging and line EDS scans were also carried out to visualize local elemental concentration variations.

2.5. Electrochemical Hydriding

Electrochemical hydriding/dehydriding experiments were carried out in a three-electrode cell with Hg/HgO as a reference electrode and a counter electrode prepared from Ni mesh. The metal hydride electrode was prepared by mixing the alloy powder (70 mg) with 100 mg of Teflonized carbon black (VULCAN 72 10%PTFE). Pellets with a diameter of 10 mm and thickness of about 1.5 mm were obtained by pressing the as-prepared mixture with a pressure of 150 atm. The electrolyte was a 6 M KOH water solution. Electrochemical charging and discharging were conducted using a galvanostat/potentiostat at a constant current density of 50 mA/g and 20 mA/g, respectively. Room temperature and a cut-off voltage of 500 mV were applied. Two experiments were performed for each sample.

3. Results and Discussion

3.1. Characterization of the As-Spun Ribbon

The XRD pattern of the as-quenched $Mg_{65}Ni_{20}Cu_5Y_{10}$ ribbon is characterized by a significant halo (2 θ~38 deg), which confirms that the alloy is X-ray amorphous (Figure 1). The corresponding linear heating DSC thermogram exhibits typical features of a metallic glass, including the glass transition (T_g = 439K), followed by a three-stage crystallization sequence characterized by T_{x1} = 471 K, T_{x2} = 579 K and T_{x3} = 648 K exothermic transformations (see Figure 2). The width of the supercooled liquid region obtained as $\Delta T_x = T_{x1,onset} - T_g$ corresponds to a remarkably high GFA (ΔT_x = 32 K), in accordance with other Mg-based glasses [24]. The high value of ΔT_x suggests that the undercooled liquid may have a strong resistance to crystallization either by thermal activation or by SPD. The apparent activation energy values of the T_{x1}, T_{x2} and T_{x3} crystallization transformations were determined from the slope of the Kissinger plot (see the inset of Figure 2), and the obtained values ($E_{a,Tx1}$ = 210 kJ/mol, $E_{a,Tx1}$ = 188 kJ/mol and $E_{a,Tx1}$ = 312 kJ/mol) also indicate high thermal stability, with these values being slightly higher than those obtained for $Mg_{54}Cu_{28}Ag_7Y_{11}$ [24]. The total enthalpy release (ΔH) corresponding to the multi-step crystallization process is ΔH = 105 J/g.

The evolution of the structure during the crystallization process can be inferred from the corresponding XRD patterns (Figure 1). As one can notice, linear heating above the glass transition preserves the amorphous nature of the alloy; however, very faint peaks corresponding to Mg_2Ni (JCPDS 35-1225; a = 5.21 Å; c = 13.323 Å) appear on the halo.

The complete disappearance of the amorphous background takes place after the first crystallization event (T_{lin}=525 K), while several crystalline phases develop, such as: Mg_2Cu (JCPDS 02-1315; a = 5.273 Å; b = 9.05 Å; c = 18.21 Å), Ni (JCPDS 04-0850: a = 3.5238 Å), Mg (JCPDS 35-0821; a = 3.2093 Å; c = 5.211 Å), $MgNi_2$ (JCPDS 03-1027; a = 4.815 Å; c = 15.80 Å) and $Mg_{24}Y_5$. As can be seen, the XRD pattern of the T_{lin} = 610 K state is almost identical to that of T_{lin} = 525 K, indicating that no crystalline phase nucleation occurs during the T_{x2} transformation. The pattern taken at T_{lin} = 750 K reveals the formation of additional phases, including Cu_2Mg (JCPDS 03-0987; a = 6.99 Å). However, several low-intensity Bragg peaks cannot be indexed.

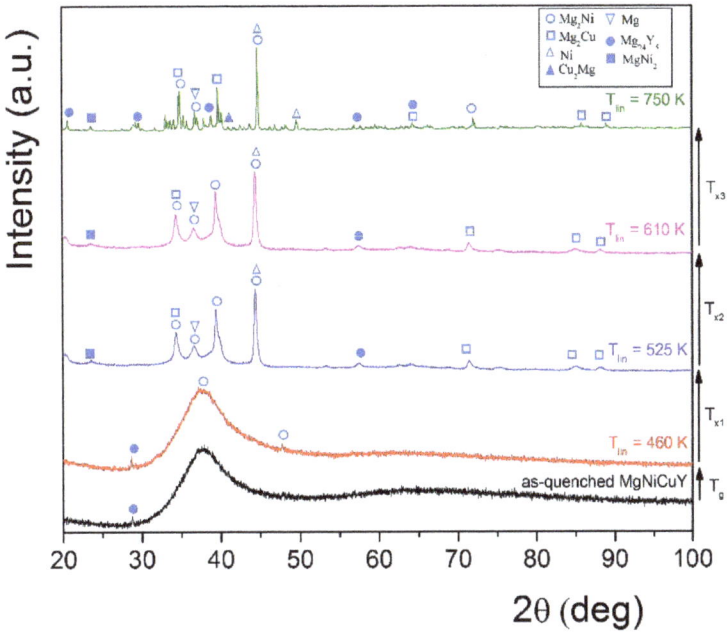

Figure 1. X-ray diffraction patterns for as-quenched $Mg_{65}Ni_{20}Cu_5Y_{10}$ metallic glass and after continuous heating to the indicated temperatures.

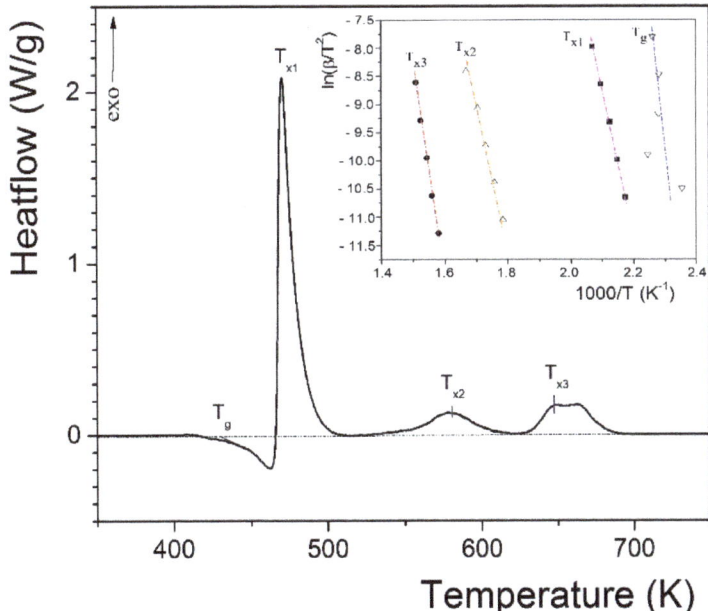

Figure 2. Continuous-heating DSC curve of the as-quenched $Mg_{65}Ni_{20}Cu_5Y_{10}$ metallic glass obtained at 20 Kmin^{-1}. The inset shows Kissinger plots.

3.2. Characterization of HPT-Treated and CNT-Catalyzed $Mg_{65}Ni_{20}Cu_5Y_{10}$

Figure 3 presents an optical micrograph taken on the cross-section of the MgNiCuY_CNT_5 disk. As can be seen, the HPT disk exhibits a constant thickness throughout its whole diameter. However, a small material outflow is recognized at the perimeter. The amorphous alloy shows a featureless, homogeneous microstructure without any detectable cracks, indicating almost full compaction of the glassy flakes into the bulk disk.

Figure 3. Optical micrograph taken on the cross-section of the HPT disk obtained after $N = 5$ rotations.

The XRD patterns recorded on the surface of HPT disks exposed to uniaxial compression (MgNiCuY_CNT_0) and varying rotational strains (MgNiCuY_CNT_1, MgNiCuY_CNT_2, MgNiCuY_CNT_5 and MgNiCuY_5) are reported in Figure 4. Similar to the as-quenched glass, all deformed specimens possess the same amorphous halo centered at $2\theta \approx 37$ deg, while only very faint Bragg peaks of the Mg_2Ni line compound superimpose on the amorphous background. The intensity of these peaks slightly increases with increased torsional straining; however, the amorphous component becomes dominant throughout the whole deformation process. No Bragg peaks corresponding to the CNT additive are recognized in the pattern. In addition, the MgNiCuY_5 disk probably exhibits other crystalline phases, such as Mg_2Cu and $MgCu_2$, as well. As a consequence, the CNT-containing $Mg_{65}Ni_{20}Cu_5Y_{10}$ glass is extremely stable against deformation-induced devitrification and subsequent crystallization.

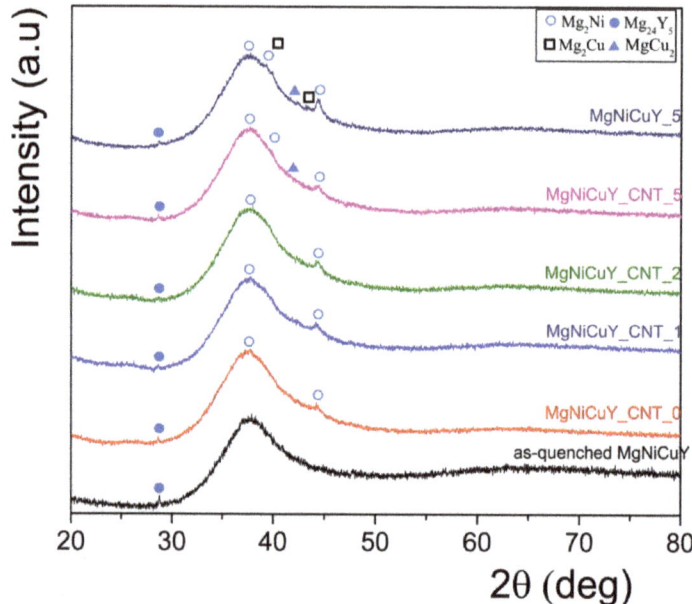

Figure 4. XRD patterns recorded on $Mg_{65}Ni_{20}Cu_5Y_{10}$ HPT disks and the as-quenched ribbon.

In addition to its extreme microstructural stability, the $Mg_{65}Ni_{20}Cu_5Y_{10}$ glassy alloy also exhibits remarkable thermal stability against severe shear deformation during HPT, as confirmed by Figure 5. It is evident from the figure that the continuous-heating DSC thermograms of the HPT-treated disks are very similar to those of the as-quenched ribbon:

i.e., each measurement presents T_g and subsequent T_{x1}, T_{x2} and T_{x3} transformations. At the same time, the total heat release (ΔH_{HPT}) for the HPT disk monotonously decreases with the torsion number, and all values are smaller than ΔH obtained for the rapidly quenched glass (see Table 1). If we suppose that the total heat release ΔH corresponds only to the complete amorphous → crystalline transformation [53], the $\Delta H_{HPT}/\Delta H$ ratio should provide the residual amorphous content of the HPT-deformed disks. Accordingly, the crystalline fraction of the partially crystallized state can be written as

$$\eta = 1 - \frac{\Delta H_{HPT}}{\Delta H} \quad (3)$$

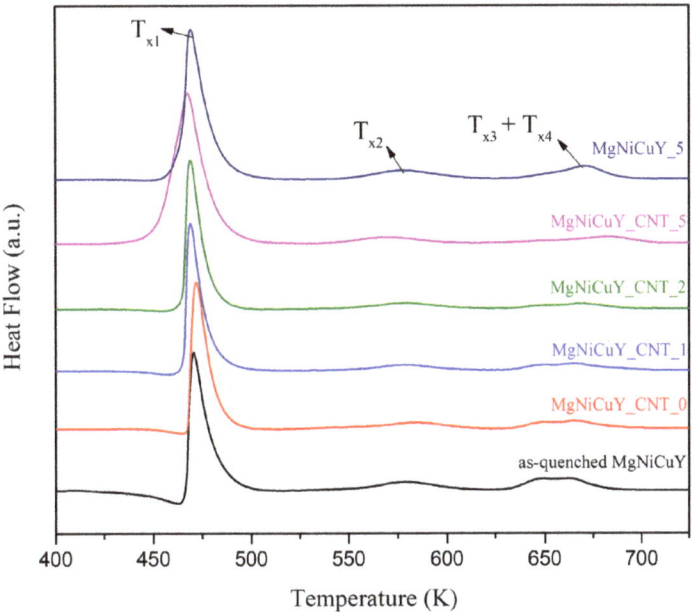

Figure 5. Continuous-heating DSC curves corresponding to the as-quenched $Mg_{65}Ni_{20}Cu_5Y_{10}$ alloy and CNT-containing HPT disks.

Table 1. Total exothermic heat release (ΔH_{HPT}) obtained for the HPT disks and the calculated crystalline fraction (η) values.

Sample	ΔH_{HPT} (J/g)	(η)
MgNiCuY_5	79	0.25
MgNiCuY_CNT_5	74	0.29
MgNiCuY_CNT_2	83	0.2
MgNiCuY_CNT_1	86	0.18
MgNiCuY_CNT_0	103	0.02
As Quenched MgNiCuY	105	

The corresponding η values are also listed in Table 1 and presented in Figure 6 as a function of the torsion number. At first glance, it is evident that severe plastic deformation promotes the intensive nucleation of nanocrystals from the amorphous matrix. As one can recognize, the as-pressed MgNiCuY_CNT_0 disk exhibits only 2% extra crystallinity with respect to the as-quenched state, but η increases significantly with the applied torsion number, reaching a maximum value of η = 0.29 for the MgNiCuY_CNT_5 specimen. At the same time, similar deformation conditions yield slightly smaller crystallinity (η = 0.25) for the CNT-free MgNiCuY_5 alloy. Nonetheless, the massive residual amorphous component undoubtedly confirms the high stability of the $Mg_{65}Ni_{20}Cu_5Y_{10}$ alloy.

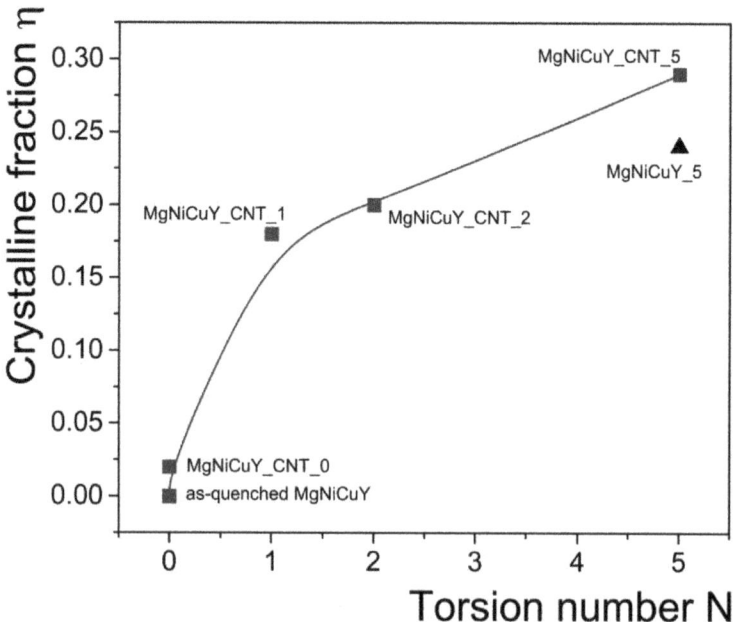

Figure 6. Crystalline fraction obtained for the different HPT disks as a function of the torsion number. The continuous line is a guideline for the eye.

In order to resolve the observed discrepancy between the almost fully amorphous XRD patterns presented in Figure 4 and the reduced amorphous content of the sheared disks, a comprehensive HR-TEM analysis was conducted on the most deformed MgNiCuY_CNT_5 sample.

3.3. TEM Study on the MgNiCuY_CNT_5 HPT Disk

Figure 7a presents a typical HR-TEM micrograph of the MgNiCuY_CNT_5 composite after torsion for $N = 5$ revolutions. It is seen that the material is dominated by an amorphous metal matrix. Lattice fringes of tube-shaped and onion-like carbon structures can also be visualized in the image, which indicates that CNT sections are embedded in this amorphous matrix during SPD processes by HPT. The morphology of the nanotubes suggests severe deformation, since they possess plenty of defects and uneven edges. In the fast Fourier transform of Figure 7a (see the inset), an amorphous halo can be observed, which confirms the disordered structure of the main component of the MgNiCuY_CNT_5 material. Alongside the halo, crystalline Fourier maxima corresponding to the Mg_2Ni phase are also present, in accordance with the XRD study (see Figure 4). The diffraction rings of 2 Å and 1.43 Å can be indexed as (203) and (215), respectively. By using these two Mg_2Ni rings, an inverse Fourier transform image was constructed that displays the spatial distribution of these Mg_2Ni crystals (Figure 7b). Accordingly, crystallites with the size of a few nanometers can be observed in the vicinity of the carbon nanotubes.

The HR-TEM image in Figure 8 shows further evidence of deformation-induced nanocrystallization: i.e., parallel atomic rows within a diameter of ~30 nm are clearly seen within a larger crystal. The lattice distance (see inset) matches reasonably well with the d-value of Mg_2Ni.

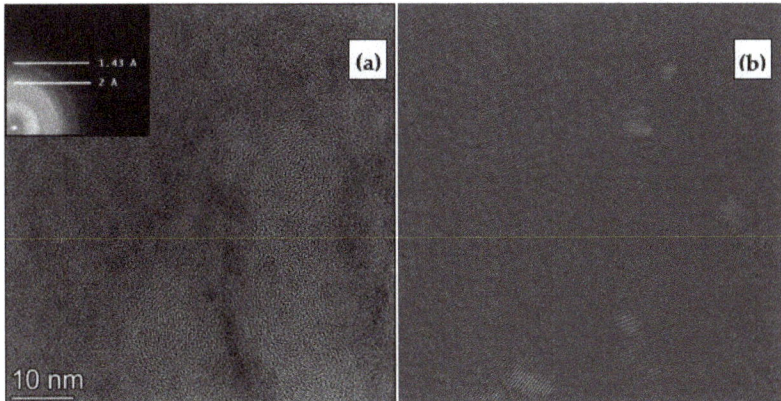

Figure 7. (**a**) HR-TEM micrograph depicting carbon nanotubes embedded in the amorphous matrix of the MgNiCuY_CNT_5 disk. The inset shows the corresponding fast Fourier transform indicating rings of Mg$_2$Ni Fourier maxima. (**b**) Inverse fast Fourier transform constructed using the 1.43 Å and 2 Å rings of the FFT image.

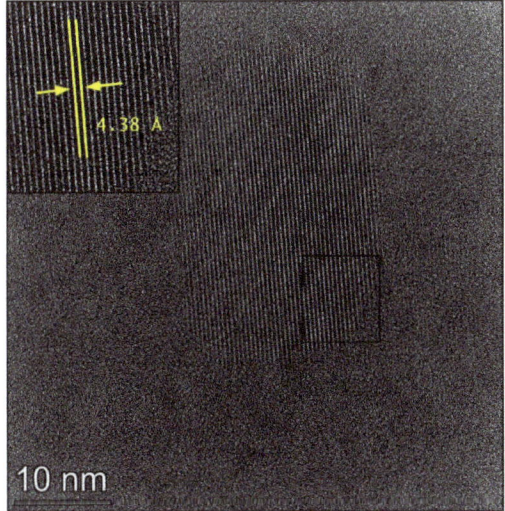

Figure 8. HR-TEM micrograph of a Mg$_2$Ni crystal embedded in the amorphous matrix of the MgNiCuY_CNT_5 composite. The inset shows a magnified view of the lattice fringe.

A supplementary EDS analysis was also carried out on the MgNiCuY_CNT_5 disk to check for elemental inhomogeneities, which would occur as a result of the precipitation of crystalline phases. The HAADF STEM image in Figure 9a depicts the location of the measurement, which was performed along the line marked by a green arrow. The contrast differences observed in the image already indicate local inhomogeneities. The variation in the measured atomic fractions of the constituting elements along the selected line can be observed in Figure 9b. It is seen that the atomic fractions of Mg and Ni vary significantly along the line, and their respective values are lower and higher than their nominal compositions (65 at.% and 20 at.%, respectively). It is noted that the systematically lower concentration of Mg (55 at.%) in positions ranging from 60 to 170 nm might be explained by its unavoidable evaporation during the synthesis of the as-quenched glass. There is also some fluctuation in the atomic fraction of Y, but the average value is close to the nominal one (10 at.%). On the other hand, the atomic fraction of Cu is fairly constant along the marked region; nevertheless, its content is significantly higher than in the as-quenched glass (5 at.%). The increased amount of Cu can be explained by contamination (back sputtering) from the TEM lamella sample holder, which occurs during the thinning process. Changes in the local concentrations of Mg and Ni can be better understood by examining the ratio of their atomic fractions (see Figure 9c). It is clearly seen that the measured area is enriched in Ni relative to Mg as compared to their nominal ratio (marked by the dashed line). The remarkably higher Ni concentration on the left-hand side of the selected area coincides with the brighter area in Figure 9a, which corresponds to a higher average atomic number. These observations also support the formation of deformation-induced Mg_2Ni nanocrystals embedded in the residual amorphous matrix.

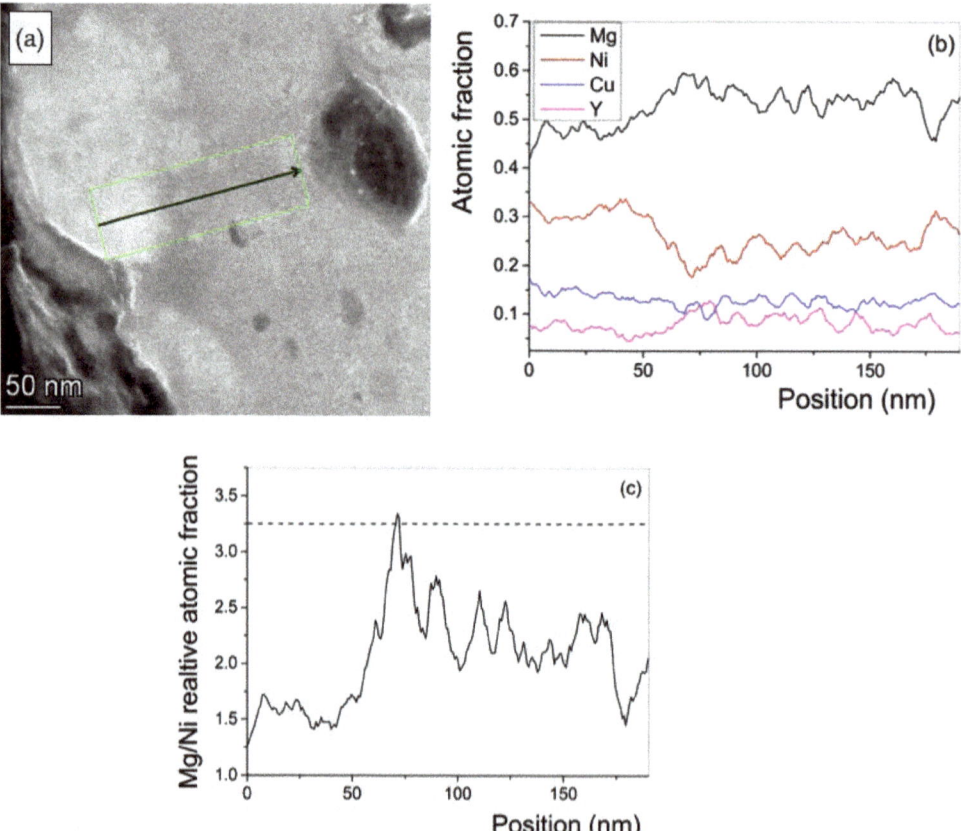

Figure 9. (**a**) HAADF image of the MgNiCuY_CNT_5 disk. The arrow indicates an EDS line measurement. (**b**) Atomic fractions of the constituting elements along the line marked in (**a**). (**c**) Relative atomic fractions of Mg and Ni along the line marked in (**a**); the dashed line indicates the nominal composition.

3.4. Electrochemical Experiments

In order to reveal the hydrogen storage performance of severely deformed amorphous $Mg_{65}Ni_{20}Cu_5Y_{10}$ alloy, electrochemical charge/discharge experiments were carried out at room temperature on the HPT disks. Figure 10 illustrates that all samples containing CNTs exhibit activation after a couple of full charge/discharge cycles, reaching steady-state values of 24–25 mAh/g. These values are slightly lower than those recently obtained for an as-cast Mg-Ni-Cu-La alloy [54]. At the same time, the CNT-free MgNiCuY_5 alloy has a relatively low capacity (<10 mAh/g) over several cycling numbers. Since the capacity of the samples without CNT is substantially lower than that of the samples with CNT, it indicates that CNT plays a key role in the discharge process of Mg-based glasses subjected to severe plastic deformation. As can be seen, MgNiCuY_CNT_0, MgNiCuY_CNT_1 and MgNiCuY_CNT_2 disks reveal a rather similar charge/discharge behavior upon cycling, which might correspond to the similar microstructure obtained in the XRD studies. Interestingly, the most deformed HPT disk (MgNiCuY_CNT_5) shows a slightly lower capacity up to eight cycles, which is followed by a clear breakdown. In our opinion, this phenomenon might be related to the increased density of more damaged CNTs (see Figure 7), which suggests that the best hydrogen absorption performance is associated with the optimal morphology and length of the nanotubes, as was also confirmed for other Mg-based hydrogen absorbing systems [45].

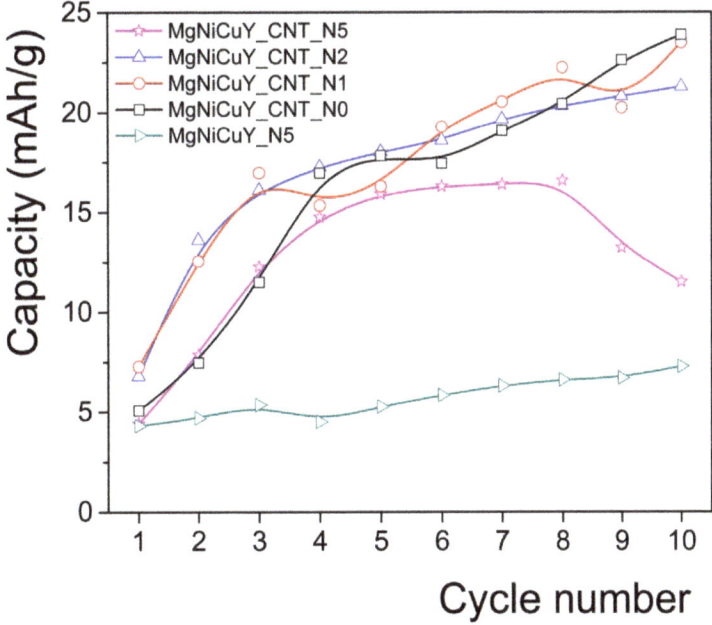

Figure 10. Electrochemical hydriding of MgNiCuY HPT disks.

4. Conclusions

An amorphous $Mg_{65}Ni_{20}Cu_5Y_{10}$ metallic glass with light weight was produced by melt spinning. Amorphous glassy flakes of the as-quenched ribbon were milled together with 5 wt.% multiwall carbon nanotubes and were subsequently exposed to high-pressure torsion with different numbers of whole turns. Complimentary XRD and DSC experiments confirmed the exceptional structural and thermal stability of the glass against the severe plastic deformation generated during HPT. Nevertheless, combined HR-TEM experiments and fast Fourier transform analysis carried out on the most deformed MgNiCuY_CNT_5 disk revealed the formation of deformation-induced Mg_2Ni nanocrystals a few nanometers in size embedded in the amorphous matrix. At the same time, lattice fringes of tube-shaped and onion-like carbon structures confirmed the structural and morphological stability of the CNT sections. The variation in the atomic fractions of the constituting elements determined by HAADF STEM indicated that some regions are enriched in Ni, in accordance with the formation of Mg_2Ni nanocrystals. The electrochemical hydrogen absorption capacity of the HPT-deformed disk containing no nanotubes (MgNiCuY_5) was substantially lower than that of the samples with the CNT additive for several cycles, which confirms that nanotubes play a crucial role in the absorption of hydrogen by Mg-based glasses subjected to severe plastic deformation. It was also established that the most deformed MgNiCuY_CNT_5 disk exhibited a clear breakdown in the electrochemical capacity after eight cycles, which suggests that there exist an optimal morphology and length of the nanotubes.

Author Contributions: Calorimetry measurements, XRD measurements, HPT processing, and writing and editing, Á.R.; pre-compaction of powders, TEM sample preparation and writing, M.G.; ball milling and evaluation of XRD data, M.A.; TEM experiments, V.K.K.; HPT processing, E.S.; preparation of the master alloy, L.K.V.; electrochemical measurements, S.T. and T.S.; melt spinning, M.B. All authors have read and agreed to the published version of the manuscript.

Funding: The electron microscopy facility at the Centre for Energy Research, Budapest, was funded by the European Structural and Investment Funds, grant no. VEKOP-2.3.3–15–2016–0 0 0 02. M.A. is indebted to the Tempus organization for providing a Stipendium Hungaricum scholarship.

Conflicts of Interest: The authors declare no conflict of interest. The funders had no role in the design of the study; in the collection, analyses or interpretation of data; in the writing of the manuscript; or in the decision to publish the results.

References

1. Ren, J.; Musyoka, N.M.; Langmi, H.W.; Mathe, M.; Liao, S. Current Research Trends and Perspectives on Materials-Based Hydrogen Storage Solutions: A Critical Review. *Int. J. Hydrogen Energy* **2017**, *42*, 289–311. [CrossRef]
2. Borgschulte, A. The Hydrogen Grand Challenge. *Front. Energy Res.* **2016**, *4*, 11. [CrossRef]
3. Schlapbach, L.; Züttel, A. Hydrogen-Storage Materials for Mobile Applications. *Nature* **2001**, *414*, 353–358. [CrossRef] [PubMed]
4. Varin, R.A.; Czujko, T.; Wronski, Z.S. *Nanomaterials for Solid State Hydrogen Storage*; Springer Science: New York, NY, USA, 2009.
5. Aguey-Zinsou, K.-F.; Ares-Fernández, J.-R. Hydrogen in Magnesium: New Perspectives toward Functional Stores. *Energy Environ. Sci.* **2010**, *3*, 526–543. [CrossRef]
6. Pasquini, L. The Effects of Nanostructure on the Hydrogen Sorption Properties of Magnesium-Based Metallic Compounds: A Review. *Crystals* **2018**, *8*, 106. [CrossRef]
7. Crivello, J.-C.; Dam, B.; Denys, R.V.; Dornheim, M.; Grant, D.M.; Huot, J.; Jensen, T.R.; de Jongh, P.; Latroche, M.; Milanese, C.; et al. Review of Magnesium Hydride-Based Materials: Development and Optimisation. *Appl. Phys. A* **2016**, *122*, 97. [CrossRef]
8. Révész, Á.; Gajdics, M. Improved H-Storage Performance of Novel Mg-Based Nanocomposites Prepared by High-Energy Ball Milling: A Review. *Energies* **2021**, *14*, 6400. [CrossRef]
9. Li, J.; Li, B.; Shao, H.; Li, W.; Lin, H. Catalysis and Downsizing in Mg-Based Hydrogen Storage Materials. *Catalysts* **2018**, *8*, 89. [CrossRef]
10. Lin, H.-J.; Lu, Y.-S.; Zhang, L.-T.; Liu, H.-Z.; Edalati, K.; Révész, Á. Recent Advances in Metastable Alloys for Hydrogen Storage: A Review. *Rare Met.* **2022**, *41*, 1797–1817. [CrossRef]
11. Hara, M.; Morozumi, S.; Watanabe, K. Effect of a Magnesium Depletion on the Mg–Ni–Y Alloy Hydrogen Absorption Properties. *J. Alloys Compd.* **2006**, *414*, 207–214. [CrossRef]
12. Zhang, Y.; Zhao, D.; Li, B.; Ren, H.; Guo, S.; Wang, X. Electrochemical Hydrogen Storage Characteristics of Nanocrystalline $Mg_{20}Ni_{10-x}Co_x$ (X = 0–4) Alloys Prepared by Melt-Spinning. *J. Alloys Compd.* **2010**, *491*, 589–594. [CrossRef]
13. Si, T.Z.; Liu, Y.F.; Zhang, Q.A. Hydrogen Storage Properties of the Supersaturated $Mg_{12}YNi$ Solid Solution. *J. Alloys Compd.* **2010**, *507*, 489–493. [CrossRef]
14. Kalinichenka, S.; Röntzsch, L.; Baehtz, C.; Kieback, B. Hydrogen Desorption Kinetics of Melt-Spun and Hydrogenated $Mg_{90}Ni_{10}$ and $Mg_{80}Ni_{10}Y_{10}$ Using in Situ Synchrotron, X-Ray Diffraction and Thermogravimetry. *J. Alloys Compd.* **2010**, *496*, 608–613. [CrossRef]
15. Spassov, T.; Köster, U. Thermal Stability and Hydriding Properties of Nanocrystalline Melt-Spun $Mg_{63}Ni_{30}Y_7$ Alloy. *J. Alloys Compd.* **1998**, *279*, 279–286. [CrossRef]
16. Zhang, Q.A.; Jiang, C.J.; Liu, D.D. Comparative Investigations on the Hydrogenation Characteristics and Hydrogen Storage Kinetics of Melt-Spun $Mg_{10}NiR$ (R = La, Nd and Sm) Alloys. *Int. J. Hydrogen Energy* **2012**, *37*, 10709–10714. [CrossRef]
17. Huang, L.J.; Wang, H.; Ouyang, L.Z.; Sun, D.L.; Lin, H.J.; Zhu, M. Achieving Fast Hydrogenation by Hydrogen-Induced Phase Separation in Mg-Based Amorphous Alloys. *J. Alloys Compd.* **2021**, *887*, 161476. [CrossRef]
18. Lin, H.-J.; He, M.; Pan, S.-P.; Gu, L.; Li, H.-W.; Wang, H.; Ouyang, L.-Z.; Liu, J.-W.; Ge, T.-P.; Wang, D.-P.; et al. Towards Easily Tunable Hydrogen Storage via a Hydrogen-Induced Glass-to-Glass Transition in Mg-Based Metallic Glasses. *Acta Mater.* **2016**, *120*, 68–74. [CrossRef]
19. Spassov, T.; Köster, U. Hydrogenation of Amorphous and Nanocrystalline Mg-Based Alloys. *J. Alloys Compd.* **1999**, *287*, 243–250. [CrossRef]
20. Kalinichenka, S.; Röntzsch, L.; Kieback, B. Structural and Hydrogen Storage Properties of Melt-Spun Mg–Ni–Y Alloys. *Int. J. Hydrogen Energy* **2009**, *34*, 7749–7755. [CrossRef]

21. Zhou, H.; Tan, J.; Dong, Q.; Ma, L.F.; Ding, D.Y.; Guo, S.F.; Li, Q.; Chen, Y.A.; Pan, F.S. Nonisothermal Crystallization Behavior of Micron-Sized $Mg_{85}Ni_5Y_{10}$ Amorphous Wires. *J. Non-Cryst. Solids* **2022**, *581*, 121412. [CrossRef]
22. Yang, T.; Wang, P.; Xia, C.; Li, Q.; Liang, C.; Zhang, Y. Characterization of Microstructure, Hydrogen Storage Kinetics and Thermodynamics of a Melt-Spun $Mg_{86}Y_{10}Ni_4$ Alloy. *Int. J. Hydrogen Energy* **2019**, *44*, 6728–6737. [CrossRef]
23. Pang, X.; Ran, L.; Chen, Y.; Luo, Y.; Pan, F. Enhancing Hydrogen Storage Performance via Optimizing Y and Ni Element in Magnesium Alloy. *J. Magnes. Alloys* **2022**, *10*, 821–835. [CrossRef]
24. Révész, Á.; Kis-Tóth, Á.; Varga, L.K.; Lábár, J.L.; Spassov, T. High Glass Forming Ability Correlated with Microstructure and Hydrogen Storage Properties of a Mg–Cu–Ag–Y Glass. *Int. J. Hydrogen Energy* **2014**, *39*, 9230–9240. [CrossRef]
25. Lass, E.A. Hydrogen Storage Measurements in Novel Mg-Based Nanostructured Alloys Produced via Rapid Solidification and Devitrification. *Int. J. Hydrogen Energy* **2011**, *36*, 10787–10796. [CrossRef]
26. Zhang, Q.A.; Liu, D.D.; Wang, Q.Q.; Fang, F.; Sun, D.L.; Ouyang, L.Z.; Zhu, M. Superior Hydrogen Storage Kinetics of $Mg_{12}YNi$ Alloy with a Long-Period Stacking Ordered Phase. *Scr. Mater.* **2011**, *65*, 233–236. [CrossRef]
27. Wu, Y.; Lototskyy, M.V.; Solberg, J.K.; Yartys, V.A. Effect of Microstructure on the Phase Composition and Hydrogen Absorption-Desorption Behaviour of Melt-Spun Mg-20Ni-8Mm Alloys. *Int. J. Hydrogen Energy* **2012**, *37*, 1495–1508. [CrossRef]
28. Wu, Y.; Solberg, J.K.; Yartys, V.A. The Effect of Solidification Rate on Microstructural Evolution of a Melt-Spun Mg–20Ni–8Mm Hydrogen Storage Alloy. *J. Alloys Compd.* **2007**, *446–447*, 178–182. [CrossRef]
29. Zhang, Y.; Li, B.; Ren, H.; Guo, S.; Zhao, D.; Wang, X. Hydrogenation and Dehydrogenation Behaviours of Nanocrystalline $Mg_{20}Ni_{10-x}Cu_x$ (X = 0−4) Alloys Prepared by Melt Spinning. *Int. J. Hydrogen Energy* **2010**, *35*, 2040–2047. [CrossRef]
30. Ding, X.; Chen, R.; Chen, X.; Pu, J.; Su, Y.; Guo, J. Study on the Eutectic Formation and Its Correlation with the Hydrogen Storage Properties of $Mg_{98}Ni_{2-x}La_x$ Alloys. *Int. J. Hydrogen Energy* **2021**, *46*, 17814–17826. [CrossRef]
31. Ding, X.; Chen, R.; Zhang, J.; Chen, X.; Su, Y.; Guo, J. Achieving Superior Hydrogen Storage Properties via In-Situ Formed Nanostructures: A High-Capacity Mg–Ni Alloy with La Microalloying. *Int. J. Hydrogen Energy* **2022**, *47*, 6755–6766. [CrossRef]
32. Denys, R.V.; Poletaev, A.A.; Maehlen, J.P.; Solberg, J.K.; Tarasov, B.P.; Yartys, V.A. Nanostructured Rapidly Solidified $LaMg_{11}Ni$ Alloy. II. In Situ Synchrotron X-Ray Diffraction Studies of Hydrogen Absorption–Desorption Behaviours. *Int. J. Hydrogen Energy* **2012**, *37*, 5710–5722. [CrossRef]
33. Li, H.; Wan, C.; Li, X.; Ju, X. Structural, Hydrogen Storage, and Electrochemical Performance of $LaMgNi_4$ Alloy and Theoretical Investigation of Its Hydrides. *Int. J. Hydrogen Energy* **2022**, *47*, 1723–1734. [CrossRef]
34. Guo, F.; Zhang, T.; Shi, L.; Chen, Y.; Song, L. Mechanisms of Hydrides' Nucleation and the Effect of Hydrogen Pressure Induced Driving Force on de-/Hydrogenation Kinetics of Mg-Based Nanocrystalline Alloys. *Int. J. Hydrogen Energy* **2022**, *47*, 1063–1075. [CrossRef]
35. Han, B.; Yu, S.; Wang, H.; Lu, Y.; Lin, H.-J. Nanosize Effect on the Hydrogen Storage Properties of Mg-Based Amorphous Alloy. *Scr. Mater.* **2022**, *216*, 114736. [CrossRef]
36. Song, F.; Yao, J.; Yong, H.; Wang, S.; Xu, X.; Chen, Y.; Zhang, L.; Hu, J. Investigation of Ball-Milling Process on Microstructure, Thermodynamics and Kinetics of Ce–Mg–Ni-Based Hydrogen Storage Alloy. *Int. J. Hydrogen Energy* **2022**, *in press*. [CrossRef]
37. Kang, H.; Yong, H.; Wang, J.; Xu, S.; Li, L.; Wang, S.; Hu, J.; Zhang, Y. Characterization on the Kinetics and Thermodynamics of Mg-Based Hydrogen Storage Alloy by the Multiple Alloying of Ce, Ni and Y Elements. *Mater. Charact.* **2021**, *182*, 111583. [CrossRef]
38. Cao, W.; Ding, X.; Zhang, Y.; Zhang, J.; Chen, R.; Su, Y.; Guo, J.; Fu, H. Enhanced De-/Hydrogenation Kinetics of a Hyper-Eutectic $Mg_{85}Ni_{15-x}Ag_x$ Alloy Facilitated by Ag Dissolving in Mg_2Ni. *J. Alloys Compd.* **2022**, *917*, 165457. [CrossRef]
39. Guo, F.; Zhang, T.; Shi, L.; Chen, Y.; Song, L. Ameliorated Microstructure and Hydrogen Absorption/Desorption Properties of Novel Mg–Ni–La Alloy Doped with MWCNTs and Co Nanoparticles. *Int. J. Hydrogen Energy* **2022**, *47*, 18044–18057. [CrossRef]
40. Révész, Á.; Gajdics, M. High-Pressure Torsion of Non-Equilibrium Hydrogen Storage Materials: A Review. *Energies* **2021**, *14*, 819. [CrossRef]
41. Révész, Á.; Kovács, Z. Severe Plastic Deformation of Amorphous Alloys. *Mater. Trans.* **2019**, *60*, 1283–1293. [CrossRef]
42. Edalati, K.; Bachmaier, A.; Beloshenko, V.A.; Beygelzimer, Y.; Blank, V.D.; Botta, W.J.; Bryła, K.; Čížek, J.; Divinski, S.; Enikeev, N.A.; et al. Nanomaterials by Severe Plastic Deformation: Review of Historical Developments and Recent Advances. *Mater. Res. Lett.* **2022**, *10*, 163–256. [CrossRef]
43. Strozi, R.B.; Ivanisenko, J.; Koudriachova, N.; Huot, J. Effect of HPT on the First Hydrogenation of $LaNi_5$ Metal Hydride. *Energies* **2021**, *14*, 6710. [CrossRef]
44. Chu, F.; Wu, K.; Meng, Y.; Edalati, K.; Lin, H.-J. Effect of High-Pressure Torsion on the Hydrogen Evolution Performances of a Melt-Spun Amorphous $Fe_{73.5}Si_{13.5}B_9Cu_1Nb_3$ Alloy. *Int. J. Hydrogen Energy* **2021**, *46*, 25029–25038. [CrossRef]
45. Gajdics, M.; Spassov, T.; Kis, V.K.; Schafler, E.; Révész, Á. Microstructural and Morphological Investigations on $Mg-Nb_2O_5$-CNT Nanocomposites Processed by High-Pressure Torsion for Hydrogen Storage Applications. *Int. J. Hydrogen Energy* **2020**, *45*, 7917–7928. [CrossRef]
46. Gajdics, M.; Spassov, T.; Kovács Kis, V.; Béke, F.; Novák, Z.; Schafler, E.; Révész, Á. Microstructural Investigation of Nanocrystalline Hydrogen-Storing Mg-Titanate Nanotube Composites Processed by High-Pressure Torsion. *Energies* **2020**, *13*, 563. [CrossRef]
47. Valiev, R.Z.; Islamgaliev, R.K.; Alexandrov, I.V. Bulk Nanostructured Materials from Severe Plastic Deformation. *Prog. Mater. Sci.* **2000**, *45*, 103–189. [CrossRef]

48. Xu, C.; Lin, H.-J.; Edalati, K.; Li, W.; Li, L.; Zhu, Y. Superior Hydrogenation Properties in a $Mg_{65}Ce_{10}Ni_{20}Cu_5$ Nanoglass Processed by Melt-Spinning Followed by High-Pressure Torsion. *Scr. Mater.* **2018**, *152*, 137–140. [CrossRef]
49. Révész, Á.; Kis-Tóth, Á.; Szommer, P.; Spassov, T. Hydrogen Storage, Microstructure and Mechanical Properties of Strained $Mg_{65}Ni_{20}Cu_5Y_{10}$ Metallic Glass. *Mater. Sci. Forum* **2013**, *729*, 74–79. [CrossRef]
50. Révész, Á.; Kis-Tóth, Á.; Varga, L.K.; Schafler, E.; Bakonyi, I.; Spassov, T. Hydrogen Storage of Melt-Spun Amorphous $Mg_{65}Ni_{20}Cu_5Y_{10}$ Alloy Deformed by High-Pressure Torsion. *Int. J. Hydrogen Energy* **2012**, *37*, 5769–5776. [CrossRef]
51. Zhilyaev, A.; Langdon, T. Using High-Pressure Torsion for Metal Processing: Fundamentals and Applications. *Prog. Mater. Sci.* **2008**, *53*, 893–979. [CrossRef]
52. Kissinger, H.E. Reaction Kinetics in Differential Thermal Analysis. *Anal. Chem.* **1957**, *29*, 1702–1706. [CrossRef]
53. Jiang, J.Z.; Kato, H.; Ohsuna, T.; Saida, J.; Inoue, A.; Saksl, K.; Franz, H.; Ståhl, K. Origin of Nondetectable X-Ray Diffraction Peaks in Nanocomposite CuTiZr Alloys. *Appl. Phys. Lett.* **2003**, *83*, 3299–3301. [CrossRef]
54. Hou, Z.; Zhang, W.; Wei, X.; Yuan, Z.; Ge, Q. Hydrogen Storage Behavior of Nanocrystalline and Amorphous Mg–Ni–Cu–La Alloys. *RSC Adv.* **2020**, *10*, 33103–33111. [CrossRef]

Article

Two Magnetic Orderings and a Spin–Flop Transition in Mixed Valence Compound Mn₃O(SeO₃)₃

Wanwan Zhang [1], Meiyan Cui [2], Jindou Tian [3], Pengfeng Jiang [1], Guoyu Qian [1,*] and Xia Lu [1]

[1] School of Materials, Sun Yat-sen University, Guangzhou 510275, China
[2] State Key Laboratory of Structural Chemistry, Fujian Institute of Research on the Structure of Matter, Chinese Academy of Sciences, Fuzhou 350002, China
[3] Department of Chemistry, University of Science and Technology of China, Hefei 230026, China
* Correspondence: qiangy@mail.sysu.edu.cn

Abstract: A mixed-valence manganese selenite, Mn₃O(SeO₃)₃, was successfully synthesized using a conventional hydrothermal method. The three-dimensional framework of this compound is composed of an MnO₆ octahedra and an SeO₃ trigonal pyramid. The magnetic topological arrangement of manganese ions shows a three-dimensional framework formed by the intersection of octa-kagomé spin sublattices and staircase-kagomé spin sublattices. Susceptibility, magnetization and heat capacity measurements confirm that Mn₃O(SeO₃)₃ exhibits two successive long-range antiferromagnetic orderings with T_{N1}~4.5 K and T_{N2}~45 K and a field-induced spin–flop transition at a critical field of 4.5 T at low temperature.

Keywords: mixed-valence; magnetic properties; topological structures

1. Introduction

Mixed valence transition metal (TM) oxides with three-dimensional electronic configurations are of great significance in the fields of materials chemistry, electrochemical energy and condensed matter physics due to their diverse crystal structures and electronic configurations [1,2]. From the ancient application of Fe₃O₄ in the compass to today's copper-based high temperature superconducting materials, mixed valence TM oxides exhibit exciting and unusual chemical and physical behaviors, including high-temperature superconductors [3], colossal magnetoresistance [4], ion deintercalation [5], metal-insulator transition [6], electrocatalysis/photocatalysis [7,8], etc. More specifically, copper oxides with bidimensional characters, together with the mixed valency of Cu⁺/Cu²⁺ or Cu²⁺/Cu³⁺, are responsible for superconducting properties [9,10]. The ferromagnetic (FM) material La₀.₆₇Sr₀.₃₃MnO₃ exhibits metallic conductivity due to the Zener double exchange mechanism between Mn³⁺ and Mn⁴⁺ ions, but BaFe₁₂O₁₉ (also an FM material) is insulative due to the limitation of the ratio of Fe²⁺ and Fe³⁺ ions [11]. Compound K₂Cr₈O₁₆ (hollandite), with a rare Cr³⁺/Cr⁴⁺ mixed valence state, exhibits a metal-insulator transition in a FM state [12]. The transition metal valence state of cathode material LiMO₂ (M = Mn, Co, Ni) will switch back and forth between M²⁺ and M⁴⁺ during charging and discharging processes in Li-ion batteries [13,14]. X. Yu et al. reported the experimental observation of skyrmionic bubbles with various topological lattices in colossal magnetoresistive manganite La₁₋ₓSrₓMnO₃ [15]. In order to discover new materials with unusual physical/chemical properties, it is necessary to explore new mixed-valence transition metal compounds. The compound Mn₃O(SeO₃)₃ (Mn^II Mn^III ₂O(SeO₃)₃) was first reported by Wildner [16]. Structure analysis confirmed that this compound shows a channel structure with a three-dimensional magnetic topological framework formed by the intersection of octa-kagomé spin sublattices and staircase-kagomé spin sublattices; however, there are few studies regarding its magnetic properties. In this paper, we report the discovery of a mixed valence manganese selenate Mn₃O(SeO₃)₃.

Magnetic measurements indicate that this compound possesses two successive antiferromagnetic (AFM) transitions at low-temperature. Moreover, a spin–flop transition is observed at 2 K with an applied magnetic field of ~4.5 T.

2. Experimental Section

2.1. Synthesis of $Mn_3O(SeO_3)_3$

Single crystals of $Mn_3O(SeO_3)_3$ were obtained using a conventional hydrothermal method. A mixture of 2 mmol $Mn(NO_3)_2 \cdot xH_2O$ (3 N, 0.3943 g), 2 mmol LiI (2 N, 0.2704 g), 1 mmol SeO_2 (4 N, 0.1110 g) and 5 mL deionized water was sealed in an autoclave equipped with a Teflon liner (28 mL). The autoclave was gradually heated to 230 °C at a rate of 1 °C/min, held for 4 days and then naturally cooled to room temperature. The product contained the desired black noodle-like crystals with a 90% yield. The crystals' sizes and morphologies were characterized using a stereomicroscope and field emission scanning electron microscopy (FE-SEM, SU8100, Hitachi, Tokyo, Japan). Figure S1 shows images of the crystals under the stereomicroscope and FE-SEM. It can be observed that the crystal size is approximately 0.3 × 0.08 × 0.05 mm. The product's impurities were manually removed under a microscope. Powdered samples were prepared for physical measurement by crushing small single crystals; purity was confirmed by powder X-ray diffraction (XRD) analysis (Figure 1). Moreover, the reagent LiI acted as a mineralizer, as the quality of crystals was unsatisfactory without it.

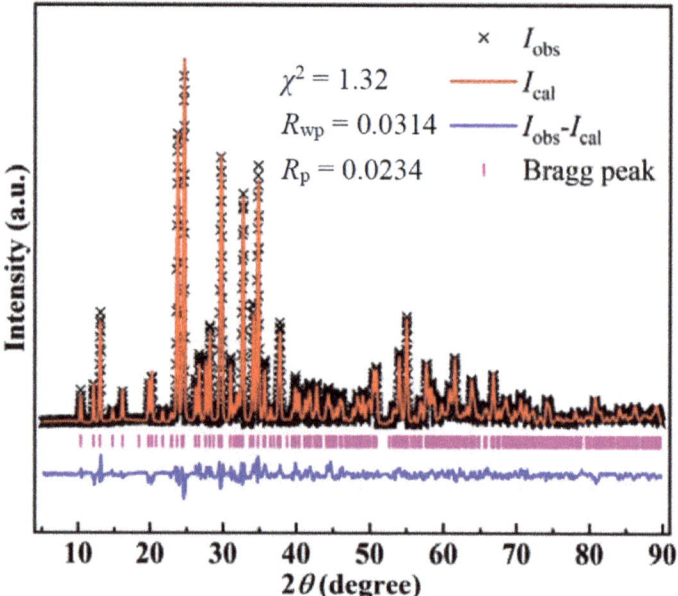

Figure 1. Rietveld refinement of powder X-ray (Cu Kα) diffraction patterns for $Mn_3O(SeO_3)_3$. The refined lattice constants are a = 15.484(9) Å, b = 6.665(8) Å, c = 9.703(1) Å and β = 118.79(4)° with space group $C2/m$, which is consistent with the reported parameters of ref. [16].

2.2. Methods

XRD patterns were collected using a Bruker D8 diffractometer with Cu-K$_\alpha$ radiation (λ~1.5418 Å) at room temperature. Rieltveld refinement was performed using GSAS-EXPGUI software [17]. Refined crystal structures were analyzed using VESTA software [18]. Furthermore, element analysis was observed using FE-SEM with an X-ray energy-dispersive spectrometer (EDS). EDS analysis confirmed the molar ratio of Mn/Se as 3.1/2.0, which is in good agreement with the X-ray structure analysis. Thermogravimetric analysis (TGA) of $Mn_3O(SeO_3)_3$ was collected on NETZSCH STA 449C instruments with an Al_2O_3 crucible from 50 to 900 °C at a rate of 10 °C/min under N_2 atmosphere.

Magnetic measurements of a powdered sample of $Mn_3O(SeO_3)_3$ were performed using a PPMS (Quantum Design, San Diego, CA, USA). The powdered sample (20.6 mg) was placed in a plastic capsule, which was suspended in a copper tube slot. Magnetic susceptibility was measured at 0.1 T from 2 to 300 K. Magnetization was measured at different temperatures at applied field from 0 to 9 T. Heat capacity was measured with the same PPMS system at zero field and determined using a relaxation method on a 5.6 mg sample.

3. Results and Discussion

The structure of compound $Mn_3O(SeO_3)_3$ was first reported by Wildner [16]. $Mn_3O(SeO_3)_3$ crystallizes in the monoclinic system with the space group $C2/m$. As shown in Figure S2, both Mn and Se atoms have three crystallographic sites. The oxidation state is +2 for Mn1 and +3 for Mn2/Mn3. All manganese atoms are coordinated by six oxygen atoms forming MnO_6 distorted octahedra; Mn–O bond lengths range from 2.100(1) to 2.361(8) Å for $Mn1^{2+}O_6$ octahedra and from 1.854(2) to 2.310(6) Å for $Mn2^{3+}O_6$ and $Mn3^{3+}O_6$, respectively. In other words, the degree of distortion for $Mn1^{2+}O_6$ octahedra is smaller than that of $Mn2^{3+}O_6$ and $Mn3^{3+}O_6$. This is due to the Mn^{3+} ($t_{2g}^3 e_g^1$) octahedron with a remarkable Jahn–Teller effect, which may induce a larger structure distortion than Mn^{2+} ($t_{2g}^3 e_g^2$). All selenium atoms are in trigonal pyramid geometry with a stereoactive lone pair of $4s^2$ in Se^{4+} ions; the Se-O bond lengths are approximately 1.70 Å. It should be noted that Se1/Se2/Se3 atoms are surrounded by 4/5/6 manganese atoms with a Se-O-Mn route, respectively. These 4/5/6 manganese atoms contain two Mn^{2+} atoms and 2/3/4 Mn^{3+} atoms, respectively.

As shown in Figure 2, $Mn_3O(SeO_3)_3$ shows a tunnel structure along the b-axis, in which the framework is constituted by MnO_6 octahedra and SeO_3 trigonal pyramids. $Mn1O_6$ octahedra share their edges (O5–O6) to form uniform [-Mn1-] chains along the b-axis. $Mn^{3+}O_6$ octahedra are interconnected via edge-sharing oxygen atoms, forming a two-dimensional [-Mn^{3+}-] layered structure parallel to (001). The detailed linkage mode between manganese ions is shown in Figure 3. Two $Mn2O_6$ octahedra connect to each other by edge-sharing oxygen atoms (O4–O4) to form a [Mn_2O_{10}] dimer along the a-axis. The Mn2-O4-Mn2 angle is 101.34(9)°. $Mn3O_6$ octahedra are interconnected by corner-sharing oxygen atoms (O7) to form uniform [-Mn3-] chains along the b-axis. One $Mn2O_6$ octahedron and two $Mn3O_6$ octahedra are connected in an isosceles triangle configuration. The neighbored [-Mn^{3+}-] layers are separated by [-Mn1-] chains and SeO_3 trigonal pyramids. Furthermore, we noted that $Mn1O_6$ octahedra are interconnected with $Mn2O_6$ octahedra via conner-sharing O5 atoms, but $Mn1O_6$ and $Mn3O_6$ octahedra are connected by SeO_3 groups in the manner of Mn1-O-Se-O-Mn3. After removing the nonmagnetic SeO_3^{2-} groups, the topological arrangement of magnetic Mn ions is a three-dimensional framework (Figure 2b). Mn^{3+} ions form a two-dimensional octa-kagomé lattice parallel to (001) (Figure 2c). The adjacent octa-kagomé layers are connected by [-Mn1-] chains. It is significant that there is a staircase-kagomé lattice composed of Mn^{2+} and Mn^{3+} parallel to (100) in the magnetic topological framework (Figure 2d). The shortest Mn–Mn distance in both [-Mn1-] and [-Mn3-] chains is 3.332(9) Å. However, the detailed connection mode of MnO_6 octahedra in [-Mn1-] chains are edge-sharing, whereas in [-Mn3-] chains it is corner-sharing. The Mn-O-Mn angles in [-Mn1-] and [-Mn3-] chains are 89.77(1)°/102.90(8)° and 127.61(5)°, respectively. The Mn2–Mn2 and Mn2–Mn3 distances in the octa-kagomé lattice are 3.150(0) Å and 3.069(6) Å, respectively, whereas the Mn1–Mn2 distance is 3.902(1) Å.

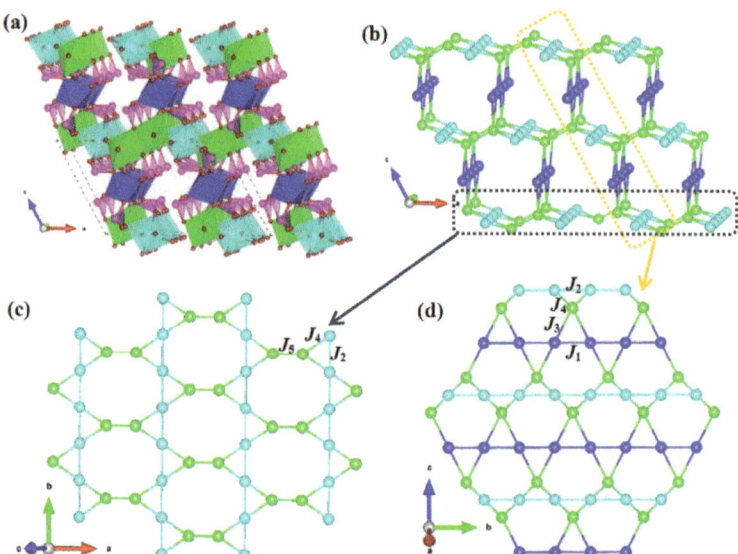

Figure 2. (a) The three-dimensional structure frameworks and (b) topological spin structures of $Mn_3O(SeO_3)_3$; (c,d) show the octa-kagomé and staircase-kagomé spin sublattices, respectively. Here the blue, green, light blue and pink polyhedra represent $Mn1O_6$, $Mn2O_6$, $Mn3O_6$ and SeO_3, respectively. Balls of the above colors represent Mn1, Mn2 and Mn3 ions, respectively.

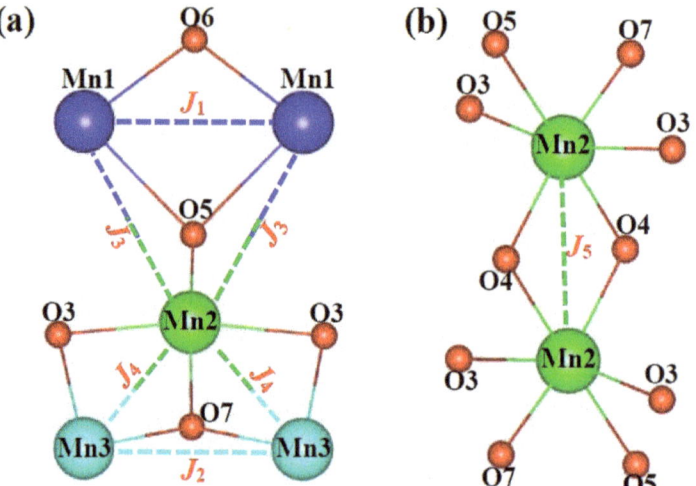

Figure 3. Detailed linkages of five main superexchange pathways in $Mn_3O(SeO_3)_3$. (a) indicates J_1 to J_4 and (b) indicates J_5, respectively.

Figure 4a shows the temperature dependence of magnetic susceptibility $\chi(T)$ of $Mn_3O(SeO_3)_3$ measured at 0.1 T. Magnetic susceptibility increases with decreasing temperature; two peaks can be observed at T_{N1}~4.5 K and T_{N2}~45 K., showing AFM transitions. At high temperature (80–300 K) inverse susceptibility $\chi^{-1}(T)$ follows the Curie–Weiss law with a Weiss temperature of $\theta = -8.89$ K and a Curie constant of $C = 11.03$ emu·mol^{-1}·K. The effective magnetic moment is calculated to be $\mu_{eff} = 5.42(3)\ \mu_B$, obtained by $\mu^2_{eff} = 8C/n$, where $n = 3$. This value of μ_{eff} is slightly smaller than the spin-only value of 5.91(6) μ_B

for Mn^{2+} ($3d^5$, high spin) and larger than the spin-only value of 4.89(9) μ_B for Mn^{3+} ($3d^4$, high spin). As Mn ions are mixed-valent, the theoretical magnetic moment of the titled compound is μ_{theo} = 5.25(9) μ_B obtained by the equation $\mu^2_{eff} = [\mu^2_{eff} (Mn^{2+}) + 2\mu^2_{eff} (Mn^{3+})]/3$. The value of μ_{eff} is quite close to that of μ_{theo}, confirming that Mn ions in the structure are mixed valence. The negative value of θ suggests the presence of dominative AFM interactions between neighboring Mn ions. Figure S3 shows the χT-T curve, in which the value of χT decreases with decreasing temperature, which is characteristic of typical AFM interactions. As shown in Figure 4b, the heat capacity data of $Mn_3O(SeO_3)_3$ show a λ-type peak at T~45 K and a corner-type transition at 4.5 K, providing concrete evidence for the two long-range magnetic orderings observed in the magnetic susceptibility curves.

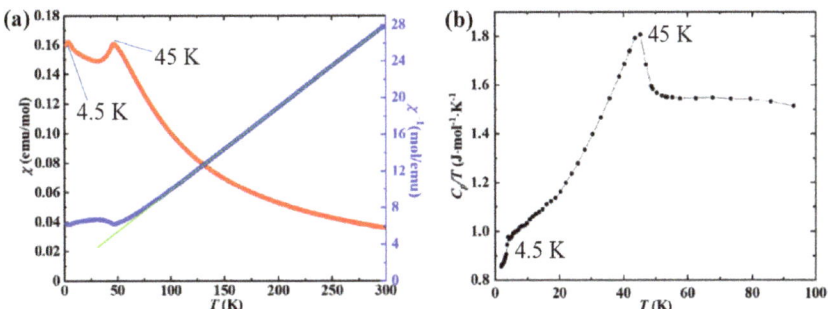

Figure 4. (a) The magnetic susceptibility $\chi(T)$ of $Mn_3O(SeO_3)_3$ and its reciprocal. (b) The heat capacity of $Mn_3O(SeO_3)_3$. The green solid line indicates Curie–Weiss fitting.

To further investigate the magnetic properties of the system, magnetization (M) as a function of applied field (H) was observed at 30 K and 2 K. As shown in Figure 5, at 30 K, magnetization increased linearly with increasing field, and did not saturate at 9 T. Furthermore, no hysteresis or remanent magnetization was observed. These features of the M–H curve suggest that the magnetic anomaly at T~45 K is the onset of an AFM ordering. At 2 K, the magnetization (M) shows a linear increase in magnetization at low field, indicative of a characteristic AFM ground state. A clear change in slope in the magnetization is observed at approximately 4.5 T, indicating field-induced magnetic transition. Furthermore, no hysteresis can be observed on the M–H curve.

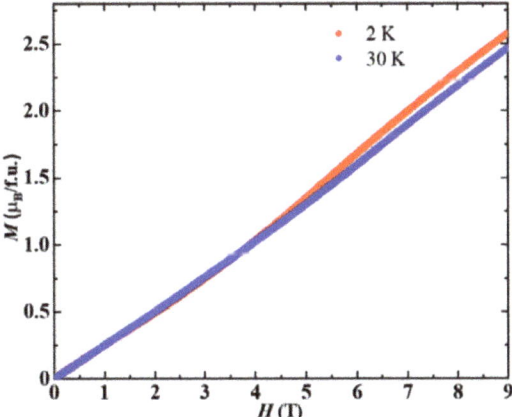

Figure 5. Isothermal magnetization (M) as a function of applied field (H) at 2 K and 30 K.

It is well-known that the magnetic properties of solid magnets are strongly related to their structural features. The three-dimensional manganese topological framework of $Mn_3O(SeO_3)_3$ is formed by the intersection of octa-kagomé lattices and staircase-kagomé lattices. Firstly, we know that the kagomé-like lattices containing equilateral or isosceles triangle sublattices may exhibit strong frustrated magnetic properties [19–21]. We note the value of frustration factor, $f = |\theta|/T_N \sim 0.20$, with Weiss temperature $\theta = -8.89$ K and Neel temperature $T_{N2} \sim 45$ K, ruling out spin frustration in the system. It is well known that primary magnetic interactions originate from the superexchange of Mn-O-Mn. A detailed description of superexchange interactions is shown in Figure 2c,d and itemized in Table 1; there are five main magnetic exchange interactions, numbered J_1–J_5, within the three-dimensional spin-lattice according to the Goodenough–Kanamori–Anderson rules (GKA rules) [22]. There are two isosceles triangular topological configurations composed by J_1–J_4 on the staircase-kagomé spin sublattice. According to the GKA rules, J_2 and J_3 are AFM interactions. J_1 and J_4 both have two Mn-O-Mn superexchange interactions, as the corresponding MnO_6 octahedra share their edges. In general, the spin exchange parameter, J, can be written as $J = J_{FM} + J_{AFM}$, where J_{FM} indicates ferromagnetic exchange and J_{AFM} indicates antiferromagnetic exchange ($J_{FM} > 0$ and $J_{AFM} < 0$) [23]. So, $J_1 = J_{1FM(O5)} + J_{1AFM(O6)}$, and the AFM interaction via O6 in J_1 is negligible; this means that $J_1 \approx J_{1FM(O5)} > 0$. Using the same analytical method, the spin exchange parameter, J_4, is ambiguous, as the magnitude of $J_{4AFM(O7)}$ is difficult to calculate. J_5 should be a weak AFM interaction. This analysis indicates that AFM interactions are dominant in the system, which is consistent with the negative value of θ. It is significant that the neighboring octa-kagomé sublattices are separated by [-Mn1-] chains and that the neighboring staircase-kagomé sublattices are connected by $[Mn_2O_{10}]$ dimers. It is safely said that these two long-range magnetic orders are driven by interlayer magnetic coupling. It is noteworthy that [-Mn1-] chains are composed of Mn^{2+} ions. If Mn^{2+} ions are replaced with nonmagnetic ions with a radius similar to Mn^{2+}, such as Mg^{2+} or Zn^{2+} [24], an octa-kagomé lattice composed of Mn^{3+} ions might form. The expected compounds may exhibit spin-liquid or other quantum physical properties [25,26]. Exploratory synthesis of related compounds is underway.

Table 1. Geometric parameters of dominant magnetic superexchanges in $Mn_3O(SeO_3)_3$ *.

J_1	d(Mn-O) (Å)	d(Mn-Mn) (Å)	Mn-O-Mn Angle (°)	Magnetism
J_1	Mn1-2.130(7)-O6-2.130(7)-Mn1 Mn1-2.361(4)-O5-2.361(4)-Mn1	3.332(9)	Mn1-O6-Mn1 ... 102.90(8)$_{AFM-w}$ Mn1-O5-Mn1 ... 89.77(1)$_{FM-S}$	FM
J_2	Mn3-1.857(1)-O7-1.857(1)-Mn3	3.332(9)	Mn3-O7-Mn3 ... 127.61(5)$_{AFM-S}$	AFM
J_3	Mn1-2.361(4)-O5-2.278(1)-Mn2	3.902(1)	Mn1-O5-Mn2 ... 114.49(4)$_{AFM-S}$	AFM
J_4	Mn2-2.049(5)-O3-2.310(4)-Mn3 Mn2-1.854(2)-O7-1.857(1)-Mn3	3.069(6)	Mn2-O3-Mn3 ... 89.29(7)$_{FM-S}$ Mn2-O7-Mn3 ... 111.59(8)$_{AFM-S}$	?
J_5	Mn2-1.896(5)-O4-2.169(5)-Mn2 Mn2-2.169(5)-O4-1.896(5)-Mn2	3.150(0)	Mn2-O4-Mn2 ... 101.34(9)$_{AFM-w}$ × 2	AFM

* The S or W behind AFM or FM (in the Mn-O-Mn Angle (°) column) refers to the magnitude of the J value; S refers to strong and W refers to weak.

As shown in Figure 6, $Mn_3O(SeO_3)_3$ undergoes a slow weight gain of approximately 0.20% from 100 to 200 °C and a slow weight loss of approximately 0.40% from 300 to 400 °C. As the weight of the sample is only approximately 7 mg, slight weight gain/loss may be caused by instrument error. As the temperature rises, two successive steps of weight loss of approximately 57% occur from 450 to 630 °C, corresponding to the calculated 59% loss of $3SeO_2$. The 2% difference may also be caused by instrument error. Based on this analysis, the final residues of $Mn_3O(SeO_3)_3$ should be Mn_3O_4; however, this was difficult to characterize, as the residues melted in the Al_2O_3 crucible after being heated to 900 °C. We re-selected the sample and sintered it in a smooth quartz crucible at 800 °C in a nitrogen atmosphere for 10 min. We then scraped off the sintered product and performed powder X-ray diffraction analysis. As shown in Figure S4, the residue was confirmed as Mn_3O_4 (PDF #80-0382). This result is consistent with the decomposition characteristics of most manganese-based compounds.

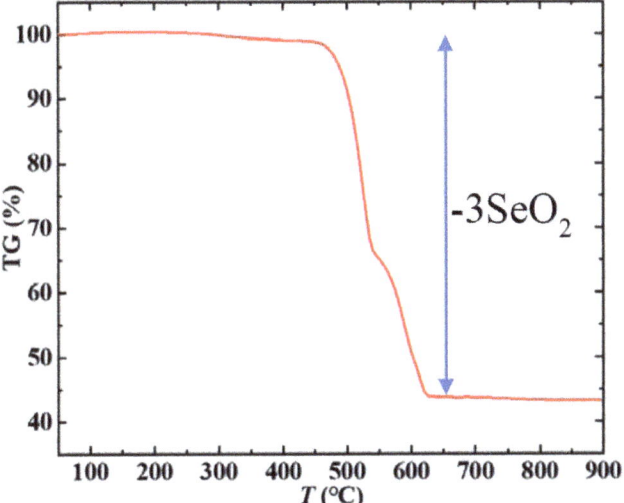

Figure 6. The thermogravimetric curve of $Mn_3O(SeO_3)_3$, which shows that the residual product of decomposition above 630 °C is Mn_3O_4.

4. Conclusions

A mixed-valence compound, $Mn_3O(SeO_3)_3$, was successfully synthesized using a conventional hydrothermal method. The reagent LiI acted as a mineralizer. $Mn_3O(SeO_3)_3$ was shown to have a channel structure with a three-dimensional magnetic topological framework formed by the intersection of octa-kagomé spin sublattices and staircase-kagomé spin sublattices. Magnetic and specific heat data confirmed that $Mn_3O(SeO_3)_3$ exhibits two successive long-range AFM orderings with $T_{N1} \sim 4.5$ K and $T_{N2} \sim 45$ K, and a field-induced spin–flop at a 4.5 T critical field at low temperature. Moreover, magnetic measurements confirmed that the ratio of Mn^{2+}/Mn^{3+} ions in this compound is 1:2, which is consistent with the structural analysis. The exploratory synthesis of Mg^{2+} or Zn^{2+} replaced compounds is in progress.

Supplementary Materials: The following supporting information can be downloaded at: https://www.mdpi.com/article/10.3390/ma15165773/s1, Figure S1. Single crystals of $Mn_3O(SeO_3)_3$ obtained by a conventional hydrothermal method. Figure S2. The oxygen-coordination environments for (a) Mn1, (b) Mn2, (c) Mn3, (d) Se1, (e) Se2 and (f) Se3 atoms in $Mn_3O(SeO_3)_3$. Figure S3. The variation in χT with the temperature of $Mn_3O(SeO_3)_3$. With the decreasing temperature, the value of χT decreases. Figure S4. Powder X-ray diffraction pattern for the final residues of $Mn_3O(SeO_3)_3$ after sintered at 800 °C for 10 min in nitrogen atmosphere.

Author Contributions: W.Z. and G.Q. conceived and designed the experiments. W.Z. prepared the materials and performed the XRD and magnetism testing. J.T. performed the thermogravimetric measurement. P.J. carried out the selection of pure phase crystals. W.Z., M.C. and G.Q. analyzed the data. W.Z. and G.Q. wrote the paper. M.C., J.T., P.J. and X.L. edited the paper. All authors discussed the results and commented on the manuscript. All authors have read and agreed to the published version of the manuscript.

Funding: This research was funded by the Hundreds of Talents program of Sun Yat-sen University and the Shenzhen Science and Technology Program (Grant No. RCBS20200714114820077).

Informed Consent Statement: Informed consent was obtained from all subjects involved in the study.

Data Availability Statement: Data is contained within the article or supplementary material.

Conflicts of Interest: The authors declare that they have no known competing financial interests or personal relationships that could have appeared to influence the work reported in this paper.

References

1. Varma, C.M. Mixed-valence compounds. *Rev. Mod. Phys.* **1976**, *48*, 219–238. [CrossRef]
2. Brown, D.B. *Mixed-Valence Compounds: Theory and Applications in Chemistry, Physics, Geology, and Biology*; Springer Science & Business Media: Berlin/Heidelberg, Germany, 2012.
3. Sleight, A.W. Chemistry of high-temperature superconductors. *Science* **1988**, *242*, 1519–1527. [CrossRef] [PubMed]
4. Uehara, M.; Mori, S.; Chen, C.H.; Cheong, S.W. Percolative phase separation underlies colossal magnetoresistance in mixed-valent manganites. *Nature* **1999**, *399*, 560–563. [CrossRef]
5. Tranquada, J.M. John Goodenough and the many lives of transition-metal oxides. *J. Electrochem. Soc.* **2022**, *169*, 010535. [CrossRef]
6. Saha, R.A.; Bandyopadhyay, A.; Schiesaro, I.; Bera, A.; Mondal, M.; Meneghini, C.; Ray, S. Colossal electroresistance response accompanied by metal-insulator transition in a mixed-valent vanadate. *Phys. Rev. B* **2021**, *104*, 045149. [CrossRef]
7. Wu, T.; Zhao, H.; Zhu, X.; Xing, Z.; Liu, Q.; Liu, T.; Gao, S.; Lu, S.; Chen, G.; Asiri, A.M.; et al. Identifying the origin of Ti^{3+} activity toward enhanced electrocatalytic N_2 reduction over TiO_2 nanoparticles modulated by mixed-valent copper. *Adv. Mater.* **2020**, *32*, 2000299. [CrossRef]
8. Chen, H.; Liu, Y.; Cai, T.; Dong, W.; Tang, L.; Xia, X.; Wang, L.; Li, T. Boosting photocatalytic performance in mixed-valence MIL-53(Fe) by changing Fe^{II}/Fe^{III} ratio. *ACS Appl. Mater. Inter.* **2019**, *11*, 28791–28800. [CrossRef]
9. Michel, C.; Hervieu, M.; Borel, M.M.; Grandin, A.; Deslandes, F.; Provost, J.; Raveau, B. Superconductivity in the Bi-Sr-Cu-O system. *Z. Phys. B* **1987**, *68*, 421–423. [CrossRef]
10. Raveau, B. Copper mixed valence concept: "Cu(I)–Cu(II)" in thermoelectric copper sulfides—an alternative to "Cu(II)–Cu(III)" in superconducting cuprates. *J. Supercond. Nov. Magn.* **2019**, *33*, 259–263. [CrossRef]
11. Zinzuvadiya, S.; Pandya, R.J.; Singh, J.; Joshi, U.S. Low field magnetotransport behavior of barium hexaferrite/ferromagnetic manganite bilayer. *J. Appl. Phys.* **2021**, *130*, 024102. [CrossRef]
12. Hasegawa, K.; Isobe, M.; Yamauchi, T.; Ueda, H.; Yamaura, J.; Gotou, H.; Yagi, T.; Sato, H.; Ueda, Y. Discovery of ferromagnetic-half-metal-to-insulator transition in $K_2Cr_8O_{16}$. *Phys. Rev. Lett.* **2009**, *103*, 146403. [CrossRef] [PubMed]
13. Lin, R.; Bak, S.M.; Shin, Y.; Zhang, R.; Wang, C.; Kisslinger, K.; Ge, M.; Huang, X.; Shadike, Z.; Pattammattel, A.; et al. Hierarchical nickel valence gradient stabilizes high-nickel content layered cathode materials. *Nat. commun.* **2021**, *12*, 2350. [CrossRef] [PubMed]
14. Ran, Q.; Zhao, H.; Hu, Y.; Hao, S.; Liu, J.; Li, H.; Liu, X. Enhancing surface stability of $LiNi_{0.8}Co_{0.1}Mn_{0.1}O_2$ cathode with hybrid core-shell nanostructure induced by high-valent titanium ions for Li-ion batteries at high cut-off voltage. *J. Alloy. Compd.* **2020**, *834*, 155099. [CrossRef]
15. Yu, X.; Tokunaga, Y.; Taguchi, Y.; Tokura, Y. Variation of topology in magnetic bubbles in a colossal magnetoresistive manganite. *Adv. Mater.* **2017**, *29*, 1603958. [CrossRef] [PubMed]
16. Wildner, M. Crystal structure of $Mn(II)Mn(III)_2(SeO_3)_3$. *J. Solid State Chem.* **1994**, *113*, 252–256. [CrossRef]
17. Toby, B.H. EXPGUI, a graphical user interface for GSAS. *J. Appl. Crystallogr.* **2001**, *34*, 210–213. [CrossRef]
18. Momma, K.; Izumi, F. VESTA 3 for three-dimensional visualization of crystal, volumetric and morphology data. *J. Appl. Crystallogr.* **2011**, *44*, 1272–1276. [CrossRef]
19. Balents, L. Spin liquids in frustrated magnets. *Nature* **2010**, *464*, 199–208. [CrossRef]
20. Aidoudi, F.H.; Aldous, D.W.; Goff, R.J.; Slawin, A.M.; Attfield, J.P.; Morris, R.E.; Lightfoot, P. An ionothermally prepared S = 1/2 vanadium oxyfluoride kagome lattice. *Nat. Chem.* **2011**, *3*, 801–806. [CrossRef]
21. Chu, S.; McQueen, T.M.; Chisne, R.; Freedman, D.E.; Muller, P.; Lee, Y.S.; Nocera, D.G. A Cu^{2+} (S = 1/2) kagomé antiferromagnet: $Mg_xCu_{4-x}(OH)_6Cl_2$. *J. Am. Chem. Soc.* **2010**, *132*, 5570–5571. [CrossRef]
22. Goodenough, J.B. *Magnetism and the Chemical Bond*; John Wiley and Sons: New York, NY, USA, 1966.
23. Whangbo, M.H.; Koo, H.J.; Dai, D.; Jung, D. Effect of metal-ligand bond lengths on superexchange interactions in Jahn-Teller d^4 ion systems: Spin dimer analysis of the magnetic structure of Marokite $CaMn_2O_4$. *Inorg. Chem.* **2002**, *41*, 5575–5581. [CrossRef] [PubMed]

24. Shannon, R.D. Revised effective ionic radii and systematic studies of interatomic distances in halides and chalcogenides. *Acta Crystallogr. A* **1976**, *32*, 751–767. [CrossRef]
25. Peng, C.; Ran, S.-J.; Liu, T.; Chen, X.; Su, G. Fermionic algebraic quantum spin liquid in an octa-kagomé frustrated antiferromagnet. *Phys. Rev. B* **2017**, *95*, 075140. [CrossRef]
26. Tang, Y.; Peng, C.; Guo, W.; Wang, J.F.; Su, G.; He, Z. Octa-kagomé lattice compounds showing quantum critical behaviors: Spin gap ground state versus antiferromagnetic ordering. *J. Am. Chem. Soc.* **2017**, *139*, 14057–14060. [CrossRef] [PubMed]

Review

Tug-of-War in the Selection of Materials for Battery Technologies

Wendy Pantoja [1], Jaime Andres Perez-Taborda [2] and Alba Avila [1,*]

[1] Electrical and Electronic Department, Universidad de los Andes, Calle 19A No. 1-82, Bogota 111711, Colombia
[2] Grupo de Nanoestructuras y Fisica Aplicada (NANOUPAR), Universidad Nacional de Colombia Sede De La Paz, Km 9 Via Valledupar, La Paz 202010, Colombia
* Correspondence: a-avila@uniandes.edu.co

Abstract: Batteries are the heart and the bottleneck of portable electronic systems. They power electronics and determine the system run time, with the size and volume determining factors in their design and implementation. Understanding the material properties of the battery components—anode, cathode, electrolyte, and separator—and their interaction is necessary to establish selection criteria based on their correlations with the battery metrics: capacity, current density, and cycle life. This review studies material used in the four battery components from the perspective and the impact of seven ions (Li^+, Na^+, K^+, Zn^{2+}, Ca^{2+}, Mg^{2+}, and Al^{3+}), employed in commercial and research batteries. In addition, critical factors of sustainability of the supply chains—geographical raw materials origins vs. battery manufacturing companies and material properties (Young's modulus vs. electric conductivity)—are mapped. These are key aspects toward identifying the supply chain vulnerabilities and gaps for batteries. In addition, two battery applications, smartphones and electric vehicles, in light of challenges in the current research, commercial fronts, and technical prospects, are discussed. Bringing the next generation of batteries necessitates a transition from advances in material to addressing the technical challenges, which the review has powered.

Keywords: energy storage; ions diffusion; batteries; batteries' components; sustainability

1. Introduction

Batteries are one of the most widely commercialized energy storage systems and have been extensively used for powering portable electronic devices [1–4]. This widespread use of batteries has transformed our daily lives and is leading the future of multifunctional, interconnected, and energy-independent devices [5]. For example, the Internet of Things (IoT) integrates devices that work not just as sensors, but also as transmitters of the sensing signals, and these devices require batteries with a higher level of performance (higher energy density and long cycle life) to power their operations [5,6]. To satisfy the requirements of these applications (size, portability, and flexibility), rapid advances have been made toward exploring new materials. Different materials have been used for the battery components: cathode [7–9], anode [10–12], electrolyte [13–15], and separators [16,17]. Any decision about the next generation of batteries will have to move beyond trial and error toward a material-based selection. This decision has to leverage the best material performance vs. availability.

Historically, the evolution of batteries has been a slow process that combines not only intelligence but also serendipity to integrate the suitable component materials that would enable the development of practical batteries with acceptable parameters: voltage, capacity, and energy density [18]. In 1800, Alessandro Volta discovered that particular liquids allow for the flow of electrical power if they are used as a conductor. Joining silver (Ag) and zinc (Zn) electrodes in an electrolyte, Volta realized that the voltage generated in the terminals could be controlled with stacked voltaic cells [19,20]. Then, in 1802, William

Cruickshank started the mass production of electric batteries (non-rechargeable), changing the Ag electrode to a copper (Cu) one. It was not until 1859 that the rechargeable battery was invented by Gaston Plante, employing an alternative technology that integrates lead (Pb) electrodes and acid as the electrolyte. Afterward, the nickel–cadmium (NiCd) battery was introduced in 1899. The use of Ni and Cd electrodes allowed for a higher energy density than Pb-acid batteries in a smaller and lighter size. The development of NiCd batteries made the use of portable devices possible. Due to safety issues, Cd was replaced with metal–hydride and, quickly, NiMH batteries became the most widely used kind of batteries in 1947 [21]. The 1960s saw the beginnings of lithium (Li) based batteries which had a higher energy density. However, it was only 30 years later that the main difficulties with Li batteries, such as volume expansion, dendrite growth, and side reactions, were acceptably resolved, resulting in the introduction of lithium-ion batteries (LIBs). At that time, these batteries reported the highest energy density by joining a graphite anode and a $LiCoO_2$ cathode [22]. Since then, LIB components have been optimized to increase the energy density.

To manufacture batteries with high energy density, several materials have been used. Metals are the most promising materials for anodes because they can deliver high capacity density. Li is the most studied metal due to its high capacity density and low potential. The concern about using Li is its scarcity. It is estimated that the current Li production cannot meet the demand in the coming decade unless the sustainability of extraction methods could be improved and a recycling process could be effectively developed [23,24]. To alleviate concerns about Li availability, alternative metal anodes have been studied. These alternatives include sodium (Na), potassium (K), zinc (Zn), calcium (Ca), magnesium (Mg), and aluminum (Al), with their respective working ions [15]. Today, the use of metal anodes in practical batteries is still limited by the dendrite growth [22], the large ionic size [25], low-voltage window, and irreversibility [26]. For cathodes, sulfur (S) [7] and oxygen (O) [27] have been studied as an alternative to traditional transition metal oxide electrodes. For electrolytes, switching from carbonates and ionic liquids to polymers and ceramic solid-state is a trend that can address safety concerns and offer additional mechanical properties while fulfilling the functions of separators.

To demonstrate the evolution of batteries, Figure 1 presents a bibliometric review of published articles from 1990 to 2021 and compares seven battery types according to the working ions Li^+, Na^+, K^+, Zn^{2+}, Ca^{2+}, Mg^{2+}, and Al^{3+}. In Figure 1, the difference between estimations and the actual number of publications demonstrates the drop off in research in 2020 and 2021, which could be the result of the pandemic [28].

The pie graphs show that Li-based batteries are the most studied (~70% of the publications), followed by Na-based batteries (~14%) and Zn-based batteries (~7%). This is because LIBs have shown successful practical application in portable electronic devices and electric vehicles, and most research moved to lithium systems for developing efficient battery component materials to achieve higher performance [29]. Current battery technologies based on Li include LIB (~240 W h kg^{-1}) [4], Li-Sulfur (Li−S 2600 W h kg^{-1}) [30,31], Li-Oxygen (Li−O_2 3500 W h kg^{-1}) [32,33], and Li-air [34] technologies.

Na-based batteries have also been widely explored since Li and Na electrode materials have similar structures, and most of the electrode materials discovered for Li batteries have been tested on Na cells [35]. Na-based batteries include Na-ion (SIB) [36], Na-sulfur (~1274 W h kg^{-1}) [37], Na-selenium (Na-Se), and Na-oxygen (Na−O_2) [27]. In the case of Zn-based batteries, Zn-air has been successfully commercialized as non-rechargeable cells, and used in applications such as hearing aids. Today, interest is focused on the possibility of switching to rechargeable cells [38]. Zinc-based batteries include Zn-air (1086 W h kg^{-1}) and Zn-ion (ZIB) technology.

In contrast, the published articles for K, Ca, Mg, and Al represent only ~9% of the total publications. This could be the result of drawbacks that have not yet been resolved and that limit the development of practical batteries. However, there seems to be a growing interest in these new technologies considering that the percentage of publications has increased from 2020 to 2021. Some examples of these technologies are K-ion (KIB) [25], K-S (1023 W h kg^{-1}) [39], Ca-ion (CIB), Mg-ion (MIB), Mg-S, Mg-air [40], Al-ion (AIB) and Al-air batteries.

Previous reviews have addressed battery progress from three perspectives: (1) studying the state of the art of a specific battery part, such as anodes [41], cathodes [42], electrolytes [15], and separators [17], (2) focusing on only one battery technology [39], and (3) comparing two or three battery technologies [43]. The main contribution of this review is that we analyze the materials for anodes, cathodes, electrolytes, and separators from seven battery types according to the working ion: Li$^+$, Na$^+$, K$^+$, Zn^{2+}, Ca^{2+}, Mg^{2+}, and Al^{3+}. For these materials, we studied the parameters (voltage, capacity, current rate, Coulombic efficiency, retention, and cycle number) that determine the battery performance (capacity, energy, and lifetime). This review is organized as follows. Battery specifications are described in Section 2, which includes electrical parameters, battery types according to their mechanical and chemical characteristics, and sustainability factors. Anodes and cathode types are described widely in Sections 3 and 4, respectively. Electrolytes and separators are explained in Section 5. Finally, we added a section on battery applications, and we concluded with the perspectives of materials for manufacturing electrodes, electrolytes, and separators.

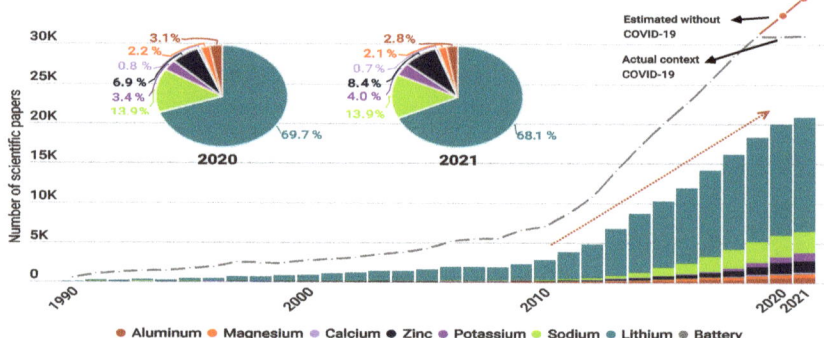

Figure 1. Number of publications per year including words "Aluminum Battery", "Magnesium Battery", "Calcium Battery", "Zinc Battery", "Potassium Battery", "Sodium Battery", and "Lithium Battery". The gray line represents the number of publications per year including "Battery" or "Batteries". The pie charts show the percentage of publications in the years 2020 and 2021. Graph constructed by the authors. Data from Web of science.

2. Battery Specifications

A battery is an electrochemical energy storage system that converts chemical energy into electrical energy. A battery consists of several electrochemical cells which integrate four main components as shown in Figure 2: (1) the anode or negative electrode; (2) the cathode or positive electrode; (3) the electrolyte that is the medium between the anode and cathode; and (4) the separator, a membrane to physically separate the anode and the cathode electrically. During discharge, ions move from the anode to the cathode through the electrolyte, and electrons flow from the anode to cathode through an external circuit. During charge, ions come back to the anode through the electrolyte while an external source forces the electrons to move from the cathode to the anode side. Although Li$^+$ is the most used, other working ions include Na$^+$, K$^+$, Zn^{2+}, Ca^{2+}, Mg^{2+}, and Al^{3+}. It is crucial to keep in mind that the chemistry inside a battery changes according to the working ion used. Therefore, the anodes and cathodes used for Li-based batteries cannot be frequently used in the other battery types.

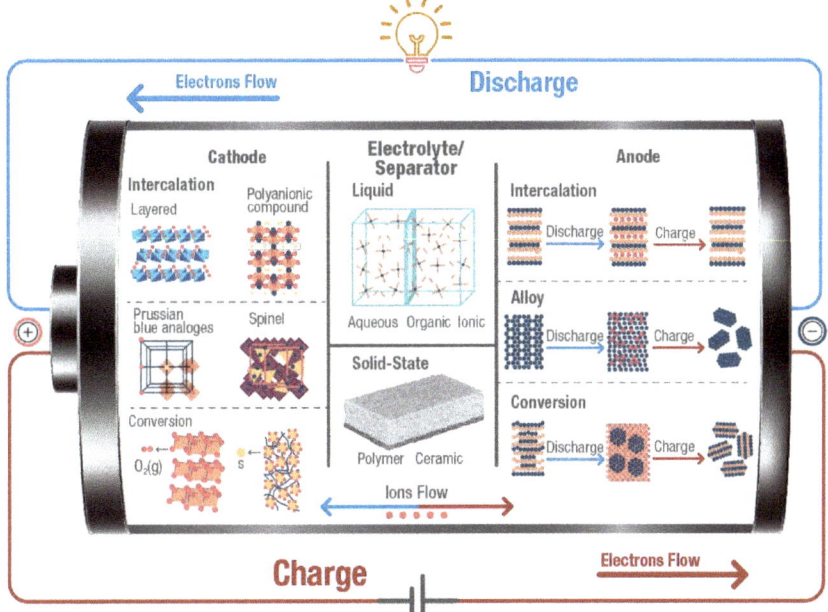

Figure 2. Illustration of the crucial internal components of a battery, showing different types of materials researched for cathodes, anodes, electrolytes, and separators. Arrows indicate the flow of electrons (through the external circuit) and ions (through the electrolyte) during the charging (red) and discharging (blue) process. Graph constructed by the authors.

Figure 2 also presents a general classification of materials used for anodes, cathodes, and electrolytes, and their internal structures according to the operating mechanism. These mechanisms are explained in Sections 3, 4 and 5, respectively. This section provides the terminology, including electrical parameters, types, and sustainability factors, of batteries and their component materials.

2.1. Electrical Parameters

Electrical characteristics are technical operating parameters to assess battery performance. These parameters are used to describe the present condition of a battery, such as state of charge, depth of charge, internal resistance, terminal voltage, and open-circuit voltage, or to compare manufacture specifications, such as capacity, C-rate, nominal voltage, cut-off voltage, energy, power, and cycle life. Electrical parameters are usually presented in graphics to compare different technologies. For example, the Ragone plot is a typical graph that contrasts the energy and power density of different battery chemistries.

The battery parameters used in this review are defined in this section. The first parameter is the capacity. Capacity is the charge that a battery can store and is established by the mass of the active material. Capacity refers to the total amount of Amp-hours (Ah) available when the battery is discharged. To determine the capacity, it is necessary to multiply the discharge current by the discharge time. The second relevant parameter is C-rate, which is defined as the battery discharge current according to battery capacity. C-rate gives a measure of the rate at which a battery is discharged relative to its maximum capacity. For example, 1C rate means that the battery with a capacity of 1000 mA h could be discharged in 1 h at a current of 1000 mA. For this battery, 5 C rate means that the battery is discharged in 12 min at 5000 mA, and a C/2 rate means that the battery is discharged in 2 h at 500 mA. In addition, there is an inversely proportional relationship between the capacity and the C-rate, which means that battery capacity decreases when the C-rate increases.

Another battery parameter is voltage, which indicates the difference between cathode and anode potential. To achieve a high voltage, an ideal cathode should have a high potential, while an ideal anode should have a low potential. Usually, the reported voltage of the battery is called nominal voltage, and the minimum acceptable operational voltage, which defines the "empty" state of the battery, is the cut-off voltage. Power is how many watts are stored, which is calculated by multiplying the voltage by the current density. Commonly, power is given per unit mass, specific power ($W\,kg^{-1}$), or per unit volume, power density ($W\,L^{-1}$). Power density determines the battery size needed to achieve a given performance purpose. Energy is the watt hour that a battery supply at a certain C-rate. Energy is also expressed per unit mass as specific energy ($Wh\,kg^{-1}$) or per unit volume as energy density ($Wh\,L^{-1}$) [44]. Cycle life is the number of discharge–charge cycles the battery performs while maintaining specific performance criteria.

State of charge (SOC%) describes the instant battery capacity as a percentage of maximum capacity. Depth of discharge (DOD%) expresses the battery capacity that has been discharged of maximum capacity as a percentage. Internal resistance is the resistance inside the battery that varies for charging and discharging. If the internal resistance increases, the battery efficiency and thermal stability are reduced since the charging energy is converted into heat.

Table 1 summarizes the standard electrical parameters that are used to evaluate the performance of batteries and their components. The symbol, the unit, and a brief definition of each parameter are included.

Table 1. Selected electrical parameters of batteries [44,45].

Parameter	Symbol	Unit	Description
Voltage	V	V	Cell voltage is cathode potential minus anode potential.
Capacity	C_t	A h	The maximum electrical charge stored in the cell.
Specific Capacity	C_s	$A\,h\,kg^{-1}$	The capacity of electrodes is usually provided per mass of active material.
Energy	E	W h	Maximum energy delivered by a given system with a theoretical voltage and a theoretical capacity: $E_t = C_t \times V$.
Specific Energy	E_s	$W\,h\,kg^{-1}$	Maximum energy of a cell per mass of the whole battery: $E = Em^{-1}$.
Energy Density	E_d	$W\,h\,L^{-1}$	Nominal battery energy per unit volume of the whole battery: $Ed = E/volume$.
Coulombic Efficiency	η_c	%	Ratio of discharging and charging capacity. $\eta c = C_{discharge}/C_{charge}$.
C-rate	C	-	Measure for the charging/discharging current of an electrochemical cell. A C-rate of 1 corresponds to a full charge/discharge within 1 h. $C = i_{applied}/i_{1h}$.
Current Density	J	$A\,m^{-2}$	Electric current per cross-sectional area.
Cycle Life	-	-	Cycle numbers.

2.2. Battery Types

Batteries can be classified in three different ways according to the chemical interaction of their components during the redox reaction. First, batteries are rechargeable if the redox reaction is reversible, and non-rechargeable if the reaction occurs just once. Second, batteries can be categorized as monovalent or multivalent according to the charge carrier ions. Finally, batteries can be organic or inorganic, depending on the material type used for manufacturing.

Rechargeable or Non-Rechargeable

Inside a battery, chemical energy is converted into electrical energy through a redox reaction. The anode is oxidized and delivers electrons to the cathode that is reduced. The electrochemical reaction is irreversible in non-rechargeable systems, also known as primary batteries. In consequence, the battery must be replaced after the discharge. In a rechargeable battery, also called a secondary battery, the chemical reaction is reversible. As a result, the battery can be charged from an external source and restored to the original chemical conditions within the cell [21,46].

Monovalent and Multivalent

A monovalent battery is a mature technology in which each ion generates one electron in the external circuit. On the other hand, in a multivalent battery, one ion generates two or three electrons in the external circuit, depending on the charge carrier ions. Consequently, in multivalent batteries, a higher current density is generated, and also the capacity could be doubled or tripled [47,48]. Common monovalent ions are Li^+, Na^+, and K^+, and the most researched multivalent ions are Zn^{2+}, Ca^{2+}, Mg^{2+}, and Al^{3+}.

Multivalent batteries are in the research stage, and technical challenges, such as instability and short cycle life, must be addressed to manufacture commercial applications successfully. Instability and short cycle life could be ascribed to volume expansion, interface degradation and active losing. For example, the electrode volume expansion, generated by the extra electrons, causes electrode breaks. In addition, Al and Ca electrodes reversibility was recently demonstrated [49,50], and it is still necessary to improve stability and the cycle life of these systems [47]. Finally, the cell assembly of multivalent batteries requires strong atmosphere control procedures to avoid contaminants such as water or oxygen, which could generate the formation of the passive film in an anode electrode [47].

Organic and Inorganic

Commercial batteries are built with inorganic materials since they have a higher specific capacity than organic materials. These inorganic materials include heavy metals, such as Co, Pb, and Ni, and alkaline metals, such as Li. The concern about using these materials is because of the negative environmental impact from their toxicity and danger to human safety. Conversely, organic batteries are built using organic battery materials composed of C, H, O, N, or S. Some of the most common organic materials are based on metal-organic frameworks (MOFs) and covalent organic frameworks (COFs), that are crystalline porous materials with large surface areas, well-defined crystalline structures and highly ordered pores [51]. Interest in these organic materials surged due to their low cost and high availability. They are studied as active material in electrodes as well as electrolytes and separators [52]. Organic batteries exhibit a high rate of performance and a longer cycle life than inorganic batteries. This is to due to the fact that the redox process in organic batteries is fast, and it does not imply changes in the layer structure of intercalation materials used in inorganic batteries [52–55]. However, the low conductivity limits their practical application, and, therefore, it is necessary to continue researching solutions for this challenge [51].

Flow Batteries

Flow batteries are an energy storage system based on electrochemical technology in which at least one electrode should be a solution. The difference between a traditional and flow battery is that the charge-discharge process occurs directly in a conventional battery since there is no spatial parting between the energy conversion unit and active material. On the other hand, in a flow battery, the energy conversion unit and active material are physically separated from each other [2]. The flow battery promises to be an alternative for large battery systems by pumping fluids from external tanks through a membrane that resembles a battery. This operation mechanism limits their application in wearable and portable devices, generating issues due to corrosion, high cost, and adverse environmental and safety impact [56].

Batteries can also be rigid or flexible according to the fabrication processes, material mechanical properties, and internal configuration. Rigid batteries have hard packing, and they are manufactured based on the slurry-casting method and also by dry electrode technology [57]. In contrast, flexible batteries (FB) are based on multilayers of a separator sandwiched by two electrodes, with a versatile packing [58]. The advantages and disadvantages of rigid and flexible batteries are described in the following paragraphs.

Rigid Batteries or Flexible Batteries

Rigid batteries are the largest commercial battery market, which provides a wide range of capacities. The rigid packages offer mechanical stability and protection to the internal components. Although today there are a large number of battery sizes and shapes, rigid batteries can be classified into four types of shape: coin, cylindrical, prismatic, and pouch [1] (Figure 3).

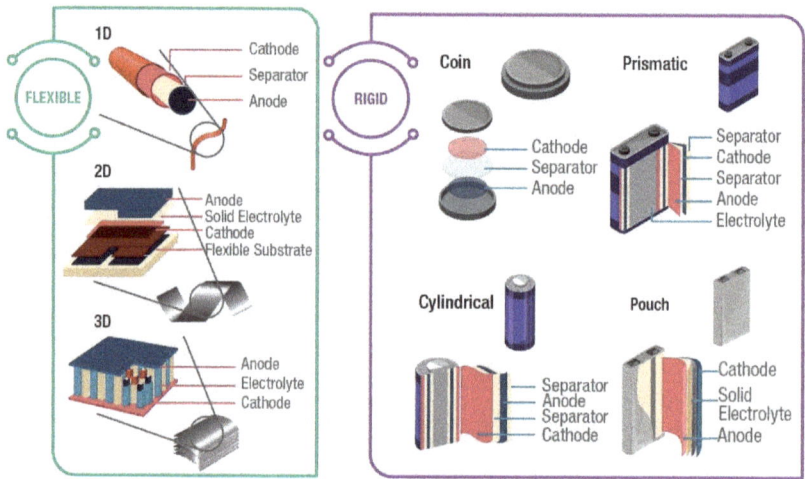

Figure 3. Typical battery configurations. Flexible: 1D, 2D, and 3D. Rigid: coin, cylindrical, prismatic, and pouch. Graph constructed by the authors.

Coin cells, also known as button cells, are small disk batteries that consist of a single cell encased in stainless steel, as presented in Figure 3. Typically, these cells have a diameter of 5 mm to 25 mm and a height of 1 mm to 6 mm. Voltage in coin cells is between 1.4 V and 3 V, and the capacity is between 1 mA h and 2000 mA h. Applications of coin cells include powering small portable electronic devices, such as wristwatches and hearing aids. Therefore, these cells should exhibit a long service life, at least one year, since they are frequently non-rechargeable cells. Commercial coin cells include chemistries such as Lithium-manganese dioxide, and Zinc-air [59].

Cylindrical cells consist of layered electrodes and separators rolled and encased in a metal casing as shown in Figure 3. These cells have different sizes, varying in the range of 8 mm to 20 mm in diameter and 25 mm to 70 mm in height. Standard size references are A, AA, or AAA for alkaline and Ni-metal–hydride, which have a voltage of 1.5 V and a capacity between 700 mA h to 3000 mA h. For LIB, the most common size is 18,650, which has a voltage of 3.7 V and a capacity of 3900 mA h. Usually, cylindrical cells are used in portable devices, power tools, medical instruments, laptops, e-bikes, and electric vehicles due to their high specific energy, good mechanical stability, and their ability to be rechargeable or non-rechargeable. It is also easy to implement automatic manufacturing for this battery [59].

In prismatic cells, the electrodes are usually manufactured in a flattened spiral to have a very thin profile. As presented in Figure 3, the cell is contained in a rectangular package. Currently, no standard size exists; each manufacturer can design prismatic cell batteries to satisfy specific requirements of different applications. Voltage in prismatic cells is between 3 V and 3.7 V, and the capacity is between 800 mA h to 400 mA h. These cells offer better space usage with a thin profile design, increasing their manufacturing cost. Additionally, they exhibit less efficiency in thermal management, producing swelling and shorter cycle life than the cylindrical design. Applications of these cells include mobile phones, tablets, and low-profile laptops [60].

Pouch cells are a soft and lightweight battery design. These cells were created using conductive foil welded to the electrodes and eliminating the metal enclosure to support expansion during battery operation. Similar to prismatic cells, pouch cells do not have a standard form, giving freedom to manufacturers to design customized cells. Commonly, pouch packs are used by Li-polymer batteries for portable applications that demand high load currents, such as drones and hobby gadgets. However, cell expansion is a hazard since pouch packs can grow from 10 % over 500 cycles, and the pressure created can crack the battery cover, generating ignition [60].

FB are highly interesting since they can satisfy the superior flexibility and durability required for wearable and portable electronic devices. The market for flexible, printed, and thin-film batteries is expected to be $109.4 million by 2025 [61]. To supply the emergent demand for bendable and stretchable devices, battery components and packaging materials should be flexible in tolerating stress [62]. Therefore, alternative fabrication techniques, such as 3D printing, should be developed [63].

Currently, there are two approaches to manufacturing FB: (1) developing new flexible materials, and (2) designing innovative structures [62,64]. Flexible materials include carbon-based (carbon nanotubes CNT and graphene), metal-based, hybrid nanocomposite, and conducting polymers. Metal-based materials require particular structure manufacturing to exhibit flexible behavior, such as serpentine layouts and buckled structures, or using a flexible substrate [65–67]. Hybrid nanocomposites integrate the electrical properties of nanostructured rigid filler in a flexible way.

On the other hand, suggested structural designs to achieve mechanical deformations can be organized inside one of three groups: (1) one-dimensional (1D) cells; (2) two-dimensional (2D) cells; and (3) three-dimensional (3D) cells (Figure 3). One-dimensional cells include wire and ribbon shapes, which allow a deformation with different degrees of freedom. Wire structures can be a coaxial or non-coaxial design, and the device performance is influenced by the geometry of the materials used. Two-dimensional cells integrate thin-film and planar shapes. These cells are based on thin (1–10 mm thickness) film or a single-layer material. Furthermore, some 3D architectures, such as kirigami and origami, have been designed to achieve several bending modes. Three-dimensional cells are commonly used in batteries with solid electrolytes. Their design consists of interpenetrating electrodes or multi-layered devices, which are highly stretchable in the direction perpendicular or parallel to that of the electrodes [58,68].

2.3. Sustainability Factors

Battery cost is determined by different elements, including the availability of materials, cell chemistry, and the manufacturing process [23]. For example, in the last 20 years, different chemistries and materials have been tested to improve LIB performance [24]. This has increased energy stored in LIBs from around 200 W h L^{-1} to more than 700 W h L^{-1}, and reduced the costs by 30 times, to around $ 100/kWh [23].

The availability of materials to supply increasing energy demand has generated debate since commercial LIBs are manufactured with lithium and cobalt, which are scarce raw materials (Figure 4). It has been forecast that this demand for electric energy could reach up to 1000 GW h by 2025, and it will at least double by 2030. As a result, it has been estimated that the demand for these materials cannot be met, thus, increasing the cost [24]. To address raw material availability concerns, novel battery technologies based on abundant materials have been researched.

Figure 4 illustrates the abundance of the Earth's crust, costs, Young's modulus, and the electrical conductivity of raw materials used in commercial and prototype batteries. Na, K, Zn, Ca, Mg, and Al are promising materials for batteries since they are more abundant than Li. In addition, the costs of these abundant materials are lower than the costs of Li. Although in most batteries, the materials are used in the form of compounds instead of elements as presented in the Figure, identifying all compounds is challenging. Therefore, the elements that are used to create the compounds were selected. These results suggest that battery costs can be reduced using abundant and cheaper materials for manufacturing. Young's modulus is a mechanical property that refers to the ratio of stress to a strain of a material. Figure of merit (f_{FoM}) of materials flexibility exposes that a small Young's modulus provides a high flexibility, being a critical parameter in guiding an appropriate selection of materials for the design of flexible batteries [69,70]. Finally, high electrical and ionic conductivity are crucial to promote electron transference, determining the rate performance in batteries [62].

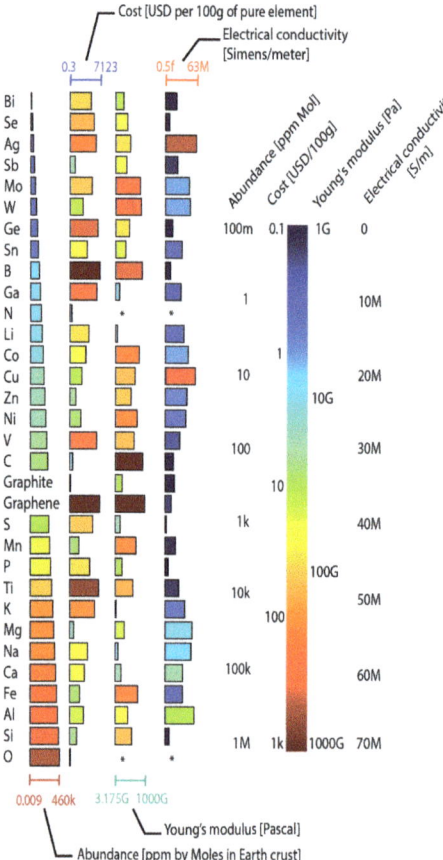

Figure 4. Elements used in battery manufacturing organized from low to high availability. The first column represents the abundance in the earth's crust. The second shows the cost per 100 g of the pure element. The third represents the Young modulus. The fourth shows the electrical conductivity. The length of the bars is normalized between the lowest and highest value for each parameter. The * symbol indicates that there is no value. Graph constructed by the authors. Data from Chemicool.com.

The manufacturing process is another critical point that affects battery cost. Commercial batteries integrate raw materials that are distributed around the world, as shown in Figure 5, while the countries that fabricate batteries are not the ones that produce raw materials. For instance, the raw materials for LIBs, such as lithium and cobalt, are distributed in South America and Africa, while the significant battery manufacturing companies are in China, Germany, Japan, the Republic of Korea, and the USA. As a result, battery companies must import raw materials to manufacture batteries. Another challenge in manufacturing is to achieve sustainable production through a reliable provision of raw materials and appropriate management of materials at the end of battery life. Some proposed solutions include reusing waste battery materials, resource conservation, useful creation, and reliable mining policies [71].

Figure 5. Geographical distribution of mineral resources vs. battery manufacturing companies. (A) Countries with mineral resources, (B) countries with mineral resources and manufacturing companies, and (C) manufacturing countries. Note: the length of the bar indicates the relative fraction of the total production. Graph constructed by the authors. Data from US Geological Survey, Mineral Commodity Summaries 2020.

Batteries manufactured annually will grow as the population and demand for portable electronic devices increase. Although batteries can help reduce the negative impacts of fossil fuels, it is necessary to address environmental pollutants that batteries generate during manufacturing, use, transportation, collection, storage, treatment, disposal, and recycling [72].

3. Anodes

The anode is the negative electrode of a battery that oxidizes (loss of electrons) during the discharge process [22], and it plays different roles according to the work ion, critical for the operation of rechargeable and non-rechargeable batteries. An ideal anode for rechargeable batteries should have a high capacity, low potential against cathode material, low volume expansion, long cycle life, low cost, and environmental compatibility [73]. Metals, such as Li, Na, K, Mg, Ca, Zn, and Al, have been explored as potential anodes due to their high energy density and low potential.

One of the main limitations of using metals as anodes is that they are susceptible to dendrite growth. Dendrite growth is a phenomenon that consists of the growth of branches on the anode surface, reducing the energy density and cycle life. Additionally, dendrites cause safety concerns since they can break the separator, generating a short circuit and battery explosion. The mechanism and behavior of dendrite growth are still being researched, and some models to describe them are (1) the thermodynamic model, (2) the space–charge model, (3) the stress and inelastic deformation model, (4) the film growth model, and (5) the phase field kinetics model [74]. To address these metal anode issues, alternative materials have been researched. These materials have been classified into three types according to the electrochemical mechanisms of operation: (1) intercalation, (2) alloys, and (3) conversion. The advantages and disadvantages of each mechanism are explained in the next paragraphs.

The intercalation mechanism consists of the intercalation and deintercalation of ions into the crystal lattice of the host material [75]. Common intercalation anodes materials are based on carbon, e.g., graphite, graphene sheets, hard carbon, and soft carbon, and titanium oxides, such as $Li_4Ti_5O_{12}$ (LTO) and TiO_2. Carbon-based materials have good working potential, low cost, and safety. The issues with using carbon anodes are high voltage hysteresis and high irreversibility capacity. Titanium oxide-based materials also have low cost, long cycle life, and high power capacity, but they are limited by a low energy density [73].

In the alloy mechanism, two or three elements are combined in a well-controlled process to produce desirable properties. The operation is governed by the chemical reaction of $xA^+ + xe^- + M \rightarrow A_xM$. Usually, A and M represent metals or metalloids such as Li, Na, K, Zn, Ca, Mg, Al, Si, Sn, Ge, and P. Alloys can be in a liquid or solid state at room temperature, and both metals can act as electrochemically active materials. Alloy anodes are promising materials since they have a high specific capacity, low potential, and these anodes have been shown to avoid dendrite growth [11]. Some challenges that alloys face are the volume expansion and the secondary reactions during the charge and discharge cycle. This volume expansion causes mechanical fractures, instability of the SEI, and swelling at the electrode level [73].

Conversion anodes are compounds that include oxides, fluorides, phosphides, and sulfides. The conversion redox reactions result in the formation of the metallic phase, which involves the breakdown of a single-crystalline parent material to polycrystalline metallic particles dispersed in an amorphous alkali oxide matrix [12]. Conversion reaction is determinate by the chemical reaction of $M_xR_y + (y*n)A \rightarrow xM + yA_nR$ where M is a transition metal, R: O, S, Se, P, H, and A represent Li, Na, K, Zn, Mg, Ca, or Al. The main advantage is the high specific and volumetric capacity. The practical application of conversion anodes is limited by the following points: (1) the large volume change, (2) the large voltage hysteresis between charge/discharge profiles, (3) the cycle instability, (4) the sloping region in the charge-discharge profile; (5) the low Coulombic efficiency, near 75%, and (6) the low diffusion coefficient of alkali ions in the material and the diffusion path length. Strategies studied for improving the performance of conversion materials are (1) size control of particles, (2) morphological control, (3) composition control of the material (sulfides, selenides, phosphides, hydrides, polymers, and carbon materials), and (4) architectural control (heterostructures based on patterned electrodes and thin-film deposition techniques) [12,73].

Figure 6 presents the increasing research on "negative electrodes". Although anodes for Li-based batteries are the most studied, since 2010, there has been an increasing interest in abundant elements.

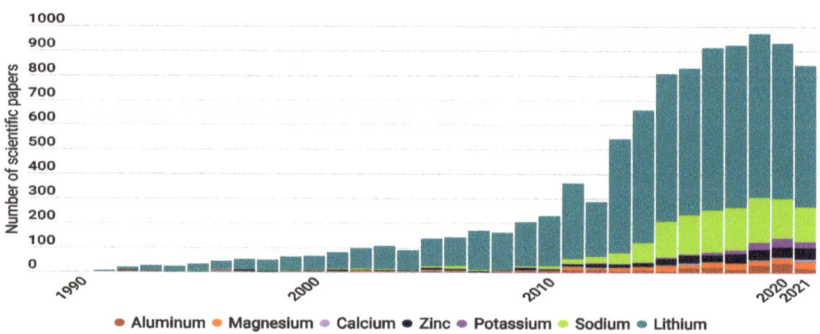

Figure 6. Publications per year about negative electrodes, comparing seven battery types: Li, Na, K, Zn, Ca, Mg, and Al. Graph constructed by the authors. Data from Web of Science.

In this section, we present the anode materials commercialized and researched for each battery type, describing their main properties and showing the strategies explored to address the challenges.

3.1. Anodes for Lithium-Based Batteries

The anodes discussed for Li batteries include pure Li metal, intercalation (carbon and titanium-based), alloys (Si, Ge, B, Al), and conversion (transition metal oxides (TMO), transition metal sulfides (TMS), phosphides, and nitrides) materials.

Li is the alkali metal with the highest theoretical specific capacity (3860 mA h g^{-1}) and the most negative potential -3.04 V vs. the standard, hydrogen electrode (SHE) [22]. Li anodes have been studied since the 1970s when Stanley Whittingham introduced Li anodes in room-temperature batteries [4]. These anodes were not commercialized due to safety issues, such as explosions caused by the reactive nature of Li and dendrite growth [76]. There have been three directions to tackle the Li anode issues: (1) designing structured anodes, (2) assessing organic or solid-state electrolytes (Section 5), and (3) replacing Li anodes. As an example of structured anodes, a stable Li-metal anode composed of 2D arrays of NbN nanocrystals was explored as a Li host [77], achieving high metallic conductivity, high ion transport channels, and high capacity (2340 mA h g^{-1}). The synthesized material showed some properties that could suppress Li dendrite growth, such as thermodynamic stability against Li, high Li affinity, fast Li-ion migration, and Li-ion transport through the porous 2D nanosheets. Consequently, the anode achieves a higher 99 % Coulombic efficiency after 500 cycles. Structured Li anodes continue in a research field with some proof-of-concept for potential practical applications [22,34].

Carbon-based materials have been used by the battery industry to overcome Li anode obstacles [22,73]. LIBs are currently the most commercialized battery, and they use a graphite anode. Graphite has a theoretical specific capacity of 372 mA h g^{-1}, and a low working potential compared to Li, allowing a good cycle life [78]. A limitation that this material faces is a low power density due to the diffusion rate of Li into graphite, which is between 1×10^9 cm^2 S^{-1} and 1×10^7 cm^2 S^{-1}. Alternative carbon materials explored to develop high-capacity anodes are hard carbon, especially porous hard carbon. Yang et al. synthesized two typical porous carbons, achieving a reversible experimental capacity of 433 mA h g^{-1} and 503 mA h g^{-1} [79]. Ultrafine layered graphite has also been explored as anode. Chen et al. showed that reducing the size of the layered graphite particles ($\sim$10 times) improves the Li-ions intercalation in the graphite crystals. As a result, the anode can deliver >500 mA h g^{-1} in the first cycle at 40 mA g^{-1}, and 393 mA h g^{-1} in the second [80].

Silicon has been studied as an alternative to Li and carbon anodes. Silicon is an alloy with a theoretical specific capacity of 4200 mA h g^{-1} and working potential of 0.4 V. The Si properties that limit its application in practical batteries include a low electrical conductivity (1.6×10^{-3} S m^{-1}) and Li-ion diffusivity (1.9×10^{-14} cm^2 S^{-1}) [81]. Moreover, Si exhibit a large volume change during charge/discharge ($\sim$300 %) and unstable solid–electrolyte interface (SEI), resulting in a rapid capacity drop, large irreversible capacity, and poor rate performance [82]. Strategies to overcome these issues include nanostructured Si, porous Si, and polymer binders (poly(vinyli-dene fluoride) (PVDF), carboxymethylcellulose sodium(CMC), poly(acrylic acid)).

To develop flexible Si anodes, MXenes have been researched due to their metallic conductivity, good hydrophilicity, and excellent mechanical properties. Reducing dimensions to change properties has been studied by Zhang et al. The authors developed an MXene nanosheet structure to confine Si-C nanoparticles [82]. The MXene framework provides high conductivity and reduces the volume change during the lithiation/delithiation process. As a result, the MXene-bonded Si-C film electrode shows flexibility, stability, high capacity, and superior rate performance. Si-C film exhibits a capacity of 2276.3 mA h g^{-1} in the first cycle and is capable of delivering 1040.7 mA h g^{-1} after 150 cycles at a current density of 420 mA g^{-1}, and it remains a capacity of 553 mA g^{-1} at a current density of 8.4 A g^{-1}.

Germanium (Ge) has a high theoretical specific capacity (1600 mA h g^{-1}), high electrical conductivity (2.20 S m^{-1}), high Li-ion diffusivity (6.25 × 10^{-12} cm^2 S^{-1}), and an isotropic lithiation which minimizes fracturing in the anodes [83]. The main drawback of Ge alloy materials is their huge volume expansion (~230%) that causes anode pulverization and cracking, reducing their cycle life. Another issue is their high cost and scarcity. Some strategies that have been researched to overcome these issues include low-dimensional nanomaterials (nanowires, nanobelts, nanoparticles, and nanotubes) [84], coating design, and porous structures. As an example, a peroxide route to produce peroxogermanate GeO$_2$ thin films was demonstrated for the first time, which could be an alternative process to germanium and germanium oxide coatings [85]. The GeO$_2$ thin film deposited on graphene oxide shows an initial discharge capacity of 2067 mA h g^{-1} at a current of 100 mA g^{-1}, and it is capable to deliver a capacity of 740 mA h g^{-1} at a current rate of 2000 mA g^{-1}. The concern using this material is its low Coulombic efficiency (69%). In another study, micro-sized porous Ge powders were synthesized and tested as anode material [83]. The researchers suggest that reducing diffusion lengths for the lithium-ion allows for a rapid charge and discharge compared to the bulk material. Additionally, it is possible to improve the electrode's mechanical integrity through porous materials capable of hosting the volume change within their pores. As a result, the anode delivers a capacity of 1300 mA h g^{-1} at 1000 mA g^{-1} after 340 cycles, and a 469 mA h g^{-1} at 8000 mA g^{-1} after 1800 cycles. However, a low Coulombic efficiency is exhibited in the first cycle.

Alloy anodes with two metal components are studied as an effective way to protect alkali metal anodes, such as Li-B, Li-Al, Li-Mg, Li-In, Li-Zn, and Li-Sn Li-Na [11]. For example, Zhong et al. showed that the Li-B alloy is capable of maintaining its structural stability during repeated cycles [86]. Li-B anodes exhibit a capacity of 213 mA h g^{-1} with a retention capacity of 74.5% after 200 cycles. They also found that a Li-Al layer coating on the Li-B anode is capable of improving the retention capacity to 86.4% after 200 cycles. It is suggested that this improved performance can be associated with the effective suppression of Li-dendrite growth and the reduction of side reactions induced by the mixed ion/electronic conductor of the Li-Al layer on the Li-B electrode. Li-Mg alloy was explored since it can reach a capacity of 2950.3 mA h g^{-1}, and it shows a discharge capacity of 606.5 mA h g^{-1} after 200 cycles working in a Li-S battery [87].

Anode materials for Li-based batteries are summarized in Table 2. We include the highest reported voltage range, capacity, current density, number of cycles, and Coulombic efficiency.

Table 2. Anode materials for Li-based batteries.

Anode Material	Voltage Range (V)	Specific Capacity (mA h g^{-1})	Current (mA g^{-1})	Cycles	η_c
Intercalation					
Graphite [88]		449	0.1 [a]	100	
Porous Hard Carbon [79]		503	0.2 [a]		
UF layerd Graphite [80]	0.1–2.0	393	40	800	
Li$_4$Ti$_5$O$_{12}$ NP/CNTs [89]	1.2–2.0	173	0.1 [a]	1000	98.5
TiO$_2$ [90]	0.8	330			
TiO$_2$/CNT [91]		316	66	100	
TiO$_2$/G [92]		272	168	100	
Alloy					
Ge [93]	0.02–1.2	1300	250	340	60
GeO/GO [85]	0–2.5	1000	250	50	69
Li−B [86]		213.2		200	

Table 2. Cont.

Anode Material	Voltage Range (V)	Specific Capacity (mA h g^{-1})	Current (mA g^{-1})	Cycles	η_c
Li−B−Al [86]		211.9	100	200	83.9
Li−Mg [94]		607	50	200	
Si MW [81]		3038	400	200	90
Si/C film [82]	0.1–2.5	2276	8400	150	73
Si/G [95]		3500	12,600	300	
Si NP/CNTs [96]		1629	200	200	
Sn NP/G [97]		1022	2000	1000	
SnO$_2$ NC/G [98]		1865	500	500	
Sn/SnO$_2$/G	0.01–3.0	2970	100	75	44
ZnCo$_2$O$_4$/rGO		1613	500	400	
Conversion					
Co$_3$O$_4$/rGO	0.01–3.0	2313	100	500	74
Co$_{0.85}$Se NS/G		680	50	300	
CuMn$_2$O$_4$/G	0.01–3.0	1491	50	150	75
GaS$_x$ NF/CNTs [99]		2118	120	100	

^{a}: C-rate. CNT: carbon nanotubes, G: graphene, GO: graphene Oxide, NC: nanocrystals, NF: nanofilms, NP: nanoparticles, NS: nanosheets, NW: nanowires, UF: ultrafine.

3.2. Anodes for Sodium-Based Batteries

The anodes discussed for Na batteries include pure Na metal, intercalation (hard carbon- and titanium-based), alloys (Si, Ge), and conversion (TMO and TMS) [100].

Na has been used as an anode since it has a specific capacity of 1166 mA h g^{-1}, a voltage of −2.21 V vs. SHE, and it is also abundant in the Earth's crust [100]. The development of Na-metal anodes started in the 1960s, using liquid Na in batteries that worked at high temperatures (573.15 K). These batteries had an efficiency of 87%, and they required an external energy source to control the high operating temperature, causing high manufacturing costs and safety issues [41]. Current research focuses on developing room-temperature Na batteries, while maintaining a low-cost.

Similar to Li, graphite was explored as anode material for Na, but the intercalation was not favorable. On the other hand, hard carbon based materials are a promising anode which posses a low Na storage voltage and also they are low cost and non-toxic. However, to use hard carbon anodes it is necessary to solve the low initial Columbic efficiency, the insufficient long cycle stability and the poor rate performance [101]. An optimization strategy of hard carbon anodes is structure control. For instance, Arie et al. prepared sheet-like structures with sufficient mesopores and micropores (larger than that of graphite) to facilitate the insertion and extraction of Na$^+$ during the charge-discharge process [102]. Authors developed a hard carbon derived from the ground leaves of used tea bags and showed a stable cycle profile, maintaining a specific capacity of 193 mA h g^{-1} after 100 cycles at a current density of 100 mA g^{-1} and capacity of 127 mA h g^{-1} after 200 cycles at 1000 mA g^{-1}. Alternatively, Ding et al. fabricated an interconnected spiral nanofibrous hard carbon that was able to recover after mechanical deformation [103]. The material possessed a highly disordered carbonaceous structure with an interlayer spacing of ~0.48 nm, which was able to store Na$^+$ and had a gravimetric capacity of 200 mA h g^{-1} at a current of 1000 mA g^{-1} after 1200 cycles.

Ti-based materials have also been an alternative anode for Na batteries. Titanium dioxide TiO$_2$ has a high theoretical capacity of 335 mA h g^{-1}, high rate performance, and good cyclability [104]. However, experimental results showed low electronic conductivity and capacities in the range from 100 mA h g^{-1} to 150 mA h g^{-1}. To increase the conductivity and kinetics for Na$^+$ storage, Bayhan et al. prepared 2D TiO$_2$/TiS$_2$ hybrid nanosheets as the anode [105]. The hybrid nanosheet showed a capacity of 245.89 mA h g^{-1} in the first cycle and then it was stabilized at 329.63 mA h g^{-1} at a current density of 1000 mA g^{-1} during 140 cycles. In addition, TiO$_2$/TiS$_2$ hybrid nanosheets showed a good cycling performance with capacities of 171.63 mA h g^{-1} and 134.05 mA h g^{-1} at current densities of 10 and 20 A g^{-1}.

Silicon also has been studied as an alloy anode for Na batteries. In 2009, the phase diagram between Na and Si by Morito et al [106] was demonstrated, suggesting that bulk Si is not a promising anode because it exhibits poor Na diffusion kinetics. In 2015, Xu et al. experimentally proved reversible electrochemical Na$^+$ ion uptake in Si [107]. The anode shows a reversible capacity of 279 mA h g^{-1} at a current rate of 10 mA g^{-1} and a capacity retention of 248 mA h g^{-1} after 100 cycles at 20 mA g^{-1}.

Germanium as an alloy anode presents some challenges in storing Na in its crystalline structure due to the large ionic size of Na$^+$ (1.02 Å) compared to Li$^+$ (0.76 Å) and presents a high volume expansion (500%) [108]. The reduction of dimensions has been studied as an alternative to alloying Ge electrochemically with Na to form Na$_x$Ge phases [109]. For example, mesoporous germanium phosphide (MGeP$_x$) microspheres were tested [108], showing a specific capacity of 1268 mA h g^{-1} in the first cycle, and a Coulombic efficiency of 65.28%. The loss of capacity in the first cycle is associated with the electrolyte decomposition on the electrode surface for SEI formation, generating that the anode delivers a capacity of 704 mA h g^{-1} after 100 cycles at a current density of 240 mA g^{-1}. Another alternative studied to improve the anode electrochemical performance was the synthesis of GeTe nanocomposite modified by amorphous carbon (GeTe/C) [109], which exhibits good Na storage characteristics of 335 mA h g^{-1} at 300 mA g^{-1} and 300 mA h g^{-1} at 900 mA g^{-1}.

Phosphorus (P) has a high theoretical capacity of 2596 mA h g^{-1}, and it exists in three allotropic forms: white P, red P, and black P [110]. Due to the low stability and toxicity, white P is not used as an anode. Red P is more stable than white P, but it has lower electronic conductivity ($\sim 10^{-14}$ S cm^{-1}), working as an electronic insulator. On the other hand, black P is the most thermodynamically stable allotropic, and it is a semiconductor useful for energy storage. Black P is an anisotropic layered material and the bulk electrical conductivity is $\sim 10^2$ S cm^{-1} [110]. The challenges in P-based materials are low conductivity, volume swelling, and unstable SEI. To address these issues, some strategies include designing nanostructures, using conductive agents (carbon materials), and manufacturing 3D nanostructures of P. For example, to achieve red P anodes, red P nanoparticles were homogeneously embedded in porous nitrogen-doped carbon nanofibers (P/C) [111]. This material exhibits a good rate capability of 1308 mA h g^{-1} at a current rate of 200 mA g^{-1}, and 343 mA h g^{-1} at 10 A g^{-1}. Additionally, it is capable of maintaining 81% after 1000 cycles. A hybrid phosphorene graphene (P/G) composite has been tested as anode through computational calculations [112]. The calculated specific capacity is 372 mA h g^{-1}, and the average open circuit voltage is 0.53 V. To achieve flexibility, red P was encapsulated in porous multichannel Carbon nanofibers (Phosphorus/PMCNFs) as flexible anodes for Na batteries [113]. The material shows a rate capability capacity of 500 mA h g^{-1} at 10 A g^{-1} and 700 mA h g^{-1} at 2 A g^{-1} after 920 cycles. The authors suggest that the improved Na storage performance is due to the special core/shell structure of P/PMCNFs.

Zhu et al. studied copper phosphide nanocrystals as anode material for Na-ion batteries due to their high specific capacity [114]. To improve the performance, the authors reported a 3D nanoarchitecture consisting of a heterostructured assembly of Cu$_3$P-C nanosheets. The thin Carbon shell serves as an electron conductor and accommodates the volume change of the Cu$_3$P single-crystalline nanosheet. The 3D Cu$_3$P−C shows a capacity retention of 286 mA h g^{-1} after 300 cycles at 100 mA g^{-1} and 156 mA h g^{-1} after 1000 cycles at 1000 mA g^{-1}. Furthermore, a vanadium phosphide–phosphorus composite V$_4$P$_{7/5}$P was investigated as an anode [115]. This composite delivers a high reversible discharge capacity of 738 mA h g^{-1}, with an initial Coulombic efficiency of 85.9% at 363 K. Moreover, a carbon nanotube-backboned mesoporous carbon (TBMC) material was designed and synthesized for the impregnation of red P [116]. Multi-walled carbon nanotubes facilitate the electron transfer, while the mesoporous carbon layers offer voids to load appropriate amounts of P but leave enough space to alleviate the huge volume change of the P upon sodiation. The P/TBMC composite shows a capacity of 1000 mA h g^{-1} at 50 mA g^{-1} and 430 mA h g^{-1}

retained at 8 A g^{-1}. In addition, this material is capable of maintaining a capacity over 800 cycles at 2.5 A g^{-1}.

TMS are promising materials for anodes due to their high theoretical capacity, good cycling stability, easily controlled structure, and modifiable chemical composition [117]. Common TMS explored as anodes are copper sulfide (CuS), vanadium sulfide (VS$_2$ and VS$_4$), molybdenum sulfide (MoS$_2$), iron sulfide (FeS$_2$), and cobalt sulfide (CoS$_2$). However, the practical application of TMS is limited by low electronic conductivity and large volume expansion. A strategy to reduce the negative impact of volume expansion is to manufacture porous structures. Zhang et al. fabricated single-layered mesoporous MoS$_2$/carbon composites with fast kinetics and long durability which obtained good Na$^+$ absorption on the surface of the MoS$_2$ [118]. As a result, the anode was able to show a capacity of 570 mA h g^{-1} after 150 cycles at 50 mA g^{-1}, and it also reached 385 mA h g^{-1} after 1000 cycles at 1 A g^{-1}.

Anode materials for Na-based batteries are summarized in Table 3. We include the highest reported voltage range, capacity, current density, number of cycles, and Coulombic efficiency.

Table 3. Anode materials for Na-based batteries.

Anode Material	Voltage Range (V)	Capacity (mA h g^{-1})	Current (mA g^{-1})	Cycles	η_c
Alloy					
Si		279	10	100	
Intercalation					
Hard carbon [103]	~0.0–3.0	250	1000	1200	
crumpled G [119]	~0.0–2.5	125	1000	500	
porous G/SbO$_x$ [120]	~0.0–3.0	350	50	100	
porous multilayered G [121]	~0.0–3.0	392	100	100	
G/Co$_{0.85}$Se nanosheets [122]	~0.0–2.5	180.7	500	100	
G nanosheets/Fe$_2$O$_3$ [123]	~0.0–3.0	400	100	200	
G/P stacks [124]	0.0–2.0	1706	260	60	
G/SnS$_2$ stacks [125]	~0.0–2.5	618.9	200	100	
G/TiNb$_2$O$_7$ [126]	~0.0–3.0	200	200	70	
N-doped G sheets [127]	~0.0–3.0	115.5	50	260	
N-doped G/NaTi$_2$(PO$_4$)$_3$ [128]	~1.5–3.0	75	20 [a]	200	
N-rich G [129]	~0.0–3.0	250	50	250	
N-/S-doped G sheets [130]	~0.0–3.0	289	100	100	
2D TiO$_2$/TiS$_2$ [105]	0.1–3.0	329.63	1000	140	
Phosphorene					
Black P/C [131]	~0.0–3.0 [132]	958 [132]	2000	500	58.5
Cu$_3$P−C [114]	~0.0–3.0	378.9	1000	1000	89
P/C-amorphous [133]	~0.0–2.0	1764	250	140	87
P-layered black [134]	0.0–2.0	1500	125	25	
P/C composite layers [135]	~0.0–1.5	1500	100	100	
P/G hybrid [136]	0.0–1.5	2400	0.02 C	100	
P-PMCNFs [113]	~0.0–2.0	2260	100	90	70
P-C [111]		1308	200		
P-TBMC [116]	~0.0–2.0	1544	8000	800	69.8
V$_4$P$_7$/$_5$P [115]	~0.0–2.0	738	8000	100	85.9

Table 3. Cont.

Anode Material	Voltage Range (V)	Capacity (mA h g^{-1})	Current (mA g^{-1})	Cycles	η_c
TMS					
MoS$_2$ nanosheets [137]	~0.0–3.0	386	40	100	
MoS$_2$/C nanosheets [138]	~0.0–2.9	280	1 C	300	
MoS$_2$/G sheets [139]	0.1–2.3	218	25	20	
MoSe$_2$ nanoplates [140]	0.1–3.0	369	0.1 C	50	
WSe$_2$ [141]	0.1–2.5	117	0.1 C	30	
WSe$_2$/C [142]	~0.0–3.0	270	0.2 C	50	
MoS$_2$/G Carbon [118]	0.5–3.2	310	5000	2500	

^{a}: C-rate. C: carbon, G: graphene, NF: nanofibers, P: phosphorus PMCNFs: porous multichannel flexible freestanding, carbon nanofibers.

3.3. Anodes for Potassium-Based Batteries

The anodes discussed for K-based batteries include K metal, carbon materials, organic materials, alloys, and metal-based compounds.

K metal has a theoretical capacity of (685 mA h g^{-1}), a low potential (−2.92 V vs. SHE), and it is also abundant in the Earth's crust [143]. Challenges to developing K as an anode are (1) dendrite growth that generates safety issues, and (2) severe side reactions that limit the capacity and cause poor kinetics.

Common carbon materials studied for K-based batteries are graphite, expanded graphite, graphene, hard carbon, soft carbon, heteroatom-doped carbon, and biomass-derived carbon [25]. The main challenge for carbon anodes is the large size of K$^+$ (2.72 Å). The interaction between K and carbon (KC$_8$) was observed and studied in 1932, demonstrating a theoretical capacity of 279 mA h g^{-1}. The electrochemical K$^+$ insertion in graphite was reported for the first time in 2015, in a nonaqueous electrolyte [144]. The anode showed a reversible capacity of 273 mA h g^{-1} at 27.9 mA g^{-1}, but a low electrochemical performance. It only maintains 80 mA h g^{-1} at 279 mA g^{-1} and capacity drops from 197 mA h g^{-1} to 100 mA h g^{-1} after 50 cycles at 139.5 mA g^{-1}. To improve the performance of carbon anodes, a nongraphitic soft carbon was synthesized, exhibiting a capacity of 273 mA h g^{-1} at 6 mA g^{-1}, and a high capacity of 210 mA h g^{-1} and 185 mA h g^{-1} at 279 mA g^{-1} and 139 mA g^{-1}, respectively. This soft carbon shows improved cyclability with a capacity retention of 81.4% after 50 cycles at 558 mA g^{-1}. To enhance the K$^+$ diffusion, low-cost and commercial expanded graphite has been studied due to its good conductivity and enlarged interlayer spaces [145]. This material exhibits a capacity of 263 mA h g^{-1} at 10 mA g^{-1} and maintains a capacity of 174 mA h g^{-1} after 500 cycles at 200 mA g^{-1}. Hard carbon microspheres (HCSs) are an alternative to improve the anode performance by dimension reduction. HCSs show a reversible capacity of 262 mA h g^{-1}, and they are capable of delivering 190 mA h g^{-1} at a rate of 558 mA g^{-1} and 136 mA h g^{-1} at a rate of 1395 mA g^{-1} [146].

Tetratitanate K$_2$Ti$_4$O$_9$, a titanium-based material, has been studied as an intercalation anode material [147]. The material was tested for the first time in 2016, delivering a discharge capacity of 80 mA h g^{-1} at a current density of 100 mA g^{-1} and 97 mA h g^{-1} at 30 mA g^{-1}. Another alternative studied is K$_2$Ti$_8$O$_{17}$, which showed a first capacity of 275 mA h g^{-1} at a current of 20 mA g^{-1} and 44.2 mA h g^{-1} at current density of 500 mA g^{-1}, but the capacity dropped to 110.7 mA h g^{-1} after 50 cycles [148].

Alloy materials, such as Sn, Sb, and Bi, are alternative anodes for K batteries since they have a high theoretical capacity. The main concern of alloys is the large volume expansion during the reaction due to the larger ionic radius of K$^+$ [149]. Some strategies to overcome this issue are morphology optimization and surface engineering, which allow anodes to obtain better electrochemical performance.

Organic anode materials have some advantages: (1) the precursor is renewable, which exactly fits in the requirements of being low cost, (2) the synthesis of organic electrodes is usually conducted by a low-temperature process, enabling low-energy consumption, (3) the organic materials are composed of elements with low atomic weight (C, H, O, N, S, etc.), giving rise to high theoretical gravimetric capacities, (4) the flexible molecular structure of organic materials is expected to favorably accommodate large-size K ions without much spatial hindrance, and (5) the satisfactory electrochemical performance by modifying the structure and functional groups [25].

On the other hand, transition metal oxides and transition metal sulfides have been reported as conversion anodes, with high theoretical capacities and redox reversibility. For example, Co_3O_4-Fe_2O_3 nanoparticles in a super P carbon matrix (Co_3O_4-Fe_2O_3/C) were fabricated to improve the conductivity and to reduce the impact of volume change [150]. The anodes deliver a reversible capacity of 220 mA h g^{-1} at 50 mA g^{-1}.

Anode materials for K-based batteries are summarized in Table 4. We include the highest reported voltage range, capacity, current density, number of cycles, and Coulombic efficiency.

Table 4. Anode materials for K-based batteries.

Anode Material	Voltage Range (V)	Capacity (mA h g^{-1})	Current (mA g^{-1})	Cycles	η_c
Alloy					
Co_3O_4–Fe_2O_3/C [150]	0.01–3.0	770	1000	50	54
CoS–G [151]	0.01–2.9	434.5	4 [a]		64.4
KTiO [147,152]	0.01–3.0	151	500	900	65.4 [148]
$KTi_2(PO_4)_3$/C [153]	1.2–2.8	75.6	5 [a]		75
MoS_2 [154]	0.5–2.0	98	2.86 [a]		74.4
MoS_2–RGO [155]	0.01–3.0	679	500		30
Sb_2S_3–S [156]	0.1–3.0	548	1000		69.7
chSb NP/3D C [157]	0.01–2.0	478	1000		68.2
SnO_2–G–C [158]	0–2.5	519	1000	100	44
Sn_4P_3/C fiber [159]	0.1–2.0	514	2000	200	64
SnP_3/C [160]	0.01–2.0	697	1200	50	58
SnS_2/RGO [161]	0.01–2.0	355	2000		56
Ti_3C_2 [162]	0.01–3.0	136	300		27
Ti_3CNT [163]	0.01–3.0	710		28.40	
$TiSe_2$ [164]	1.0–3.0	92.7	1000		67.1
VSe_2 [165]	0.01–2.6	366	2000		69.10
Carbon					
Graphite [144]	0.01–1.5	273	200 [166]	200 [167]	87 [168]
expanded Graphite [145]	0.01–3.0	267	200	500	81
hard Carbon [169]	0.01–1.5 [146]	300	1395	100	87
soft Carbon [144]		273		50	50
Organic					
K_2PC [170]	0.1–2.0	245	2 [a]		44
K_2TP [171]	0.1–2.0	305.8	1000		76.1

[a]: C-rate. C: carbon, CNT: carbon nanotubes, G: graphene, K_2PC: potassium 2,5-pyridinedicarboxylate, K_2TP: dipotassium terephthalate, P: phosphorous, RGO: reduce graphene oxide

3.4. Anodes for Zinc-Based Batteries

The anode discussed for Zn batteries is Zn metal. Zn metal has been thought of as an ideal anode material used in both non-rechargeable and rechargeable Zn-based cells. This material has many attractive properties, such as high capacity (820 mA h g^{-1}), nontoxicity, relatively low redox potential (−0.76 V vs. SHE), high safety, and low cost [172]. The Zn anodes explored in recent years focus on modifying the basic concepts, and these anodes can be organized into foil-, paste-, slurry-, and structure types [173].

The main concerns of Zn anodes are passivation, irreversibility, corrosion, and the growth of dendrites during the plating/stripping process [172]. Passivation reduces the surface contact between the electrolyte and Zn anode, generating low conductivity. The dendrite growth increases the surface area of the Zn anode, corrosion, and other surface-dependent reactions, causing low Coulombic efficiency, poor capacity, and limited cycle life. Proposed solutions to overcome these concerns include designing a nanostructure Zn metal anode, adding additives in the electrolyte, or changing Zn salt concentrations in the electrolyte. Electrolytes for Zn-based batteries are studied in Section 5.4. In this section, we focus on novel anodes for Zn batteries.

The nanostructured Zn anode is proposed as an alternative to overcome passivation and dissolution issues [174]. This anode was fabricated with ZnO nanoparticles wrapped with graphene oxide nanosheets, and the test after 150 cycles showed a retention capacity of 86%. An ion-sieving Carbon nanoshell coated ZnO nanoparticle anode was also studied as an anode material to address the same problems [175]. Results showed that cyclability improves in comparison with Zn foil. Zn sponge anodes for higher stability were explored by Stock et al. [176]. These sponge anodes were approached from two factors: (1) using an energy-saving and low-temperature preparation method; and (2) stabilizing the pore system with a lightweight anion-exchange ionomer.

Anode materials for Zn-based batteries are summarized in Table 5. We include the highest reported voltage range, capacity, current density, number of cycles, and Coulombic efficiency.

Table 5. Anode materials for Zn-based batteries.

Anode Material	Voltage Range (V)	Capacity (mA h g^{-1})	Current (mA g^{-1})	Cycles	η_c
Carbon-coated ZnO [175]	1.5–2.0	155.5	1	42	48.8
hyper dendritic Zn [177]		232.6	0.2	100	77
Zn sponge [176]	0.9–2.2	164	0.03	36	
Zn on Cu foam [178]		690	25.4	9000	31
Zn foil with IL membrane [179]	0.8–2.3			107	
ZnO−Ag−polypyrrole [180]	1.2–1.9	437	1	300	87.5
Zn foil [181]		0.4	0.06	147	100
ZnO [182]		269.8	0.5	1000	100
ZnO/C [183]	1.2–2.0	266.7	0.17	400	
ZnO with ionomer layer [184]	1.0–1.9	124.5	0.78	67	75
ZnO in Carbon matrix [174]	1.4–2.0	241.9	4.96	150	82.2
Zn sponge advanced [185]	1.4–1.9	310	3.1	141	96.6

3.5. Anodes for Calcium-Based Batteries

The anodes discussed for Ca-based batteries include metal Ca, carbon-based, alloys, and organic materials.

Calcium possesses multivalent charge carrier ions (Ca^{+2}), a low potential (−2.87 V vs. SHE), and a high capacity (1337 mA h g^{-1}). To achieve successful Ca anodes, it is necessary to produce a reversible plating and stripping of Ca [186]. The development of Ca anodes started in the 1960s, using Ca anodes in batteries that work at high temperature levels (>723.15 K). The main challenge reported for Ca anodes was the failure of Ca electrodeposition due to the passivation layer formed in the anode by the electrolyte decomposition. Some attempts to study Ca anodes focused on non-rechargeable cells until 2016, when the feasibility of Ca electrodeposition was demonstrated, enabling operation at a lower temperature between >323.15 K and 373.15 K [50]. According to Ponrouch et al., the Ca electrodeposition is possible if the following four requisites are satisfied: (1) migration of solvated Ca^{2+} ion inside the electrolyte; (2) low desolvation energy barrier at the electrolyte/passivation layer interface; (3) migration of the desolvated Ca^{2+} ions through the passivation layer; and (4) low energy barrier for nucleation and low growth of Ca at the electrode substrate interface. Achieving reversible Ca anodes requires more efficient electrolytes (Section 5).

Carbon-based anodes have been explored to overcome the difficulties caused by Ca plating and stripping. The use of carbon anodes requires Ca intercalation, which is a challenge due to the large ionic radius (1 Å) that hinders smooth intercalation into the host lattice. To achieve successful intercalation at room temperature, Wu et al. employed an isotropic graphitic layered structure called mesocarbon microbead (MCMB) as an anode [187], showing a reversible discharge capacity of 66 mA h g^{-1} at a current rate of 2C and 62 mA h g^{-1} at 1 C after 300 cycles with 94 % retention.

Alloy anodes focus on using Si and Sn. The electrochemical decalciation of CaSi$_2$ was tested with experimental and computational analysis, showing a capacity of 240 mA h g^{-1} at moderate temperatures (373.15 K) [188], an average voltage of 0.37 V, and a volume expansion of 306 %. On the other hand, the alloying/de-alloying process of Sn anode (Ca$_7$Sn$_6$) exhibited a high capacity of 526 mA h g^{-1} with a volume expansion of 136.8 % [189]. The Sn was proved in a full cell, exhibiting a discharge capacity of 50 mA h g^{-1} in the first cycle that increases to 85 mA h g^{-1} at the 200th cycle and then is reduced to 80 mA h g^{-1} after 350 cycles. The main challenges of alloy materials are large voltage hysteresis, high volume expansion, and low Coulombic efficiency.

Organic anodes explored include polyaniline (PANI) and polyimide poly (PNDIE). PANI has been explored as an anode because it has a lower specific weight than inorganic materials. For example, PANI was deposited over carbon cloth by in situ polymerization, showing a discharge capacity of 123 mA h g^{-1} at a current of 150 mA g^{-1} with a Coulombic efficiency of 99.7 % and a retention of 84 % after 200 cycles [190]. On the other hand, PNDIE has reported a specific capacity of $\sim$160 mA h g^{-1} at -0.45 V [191] with a capacity retention of 80 %, 105 mA h g^{-1} after 4000 cycles at 925 mA g^{-1} and a Coulombic efficiency >99 %. Alternative organic anode materials investigated include PTCDA.

Anode materials for Ca-based batteries are summarized in Table 6. We include the highest reported voltage range, capacity, current density, number of cycles, and Coulombic efficiency.

Table 6. Anode materials for Ca-based batteries.

Anode Material	Voltage Range (V)	Capacity (mA h g^{-1})	Current (mA g^{-1})	Cycles	η_c
Alloy					
Ca [50]				30	85
Ca-Si [188]	0.37	240		1	
Ca-Sn [192]	0.8	526		350 [189]	80
Carbon					
Graphene [193]		225			
MCMB [187]	4.6	62	1 [a]	300	82
Organic					
PANI [190]	0.4	114	150	200	99
PNDIE [191]	-0.45	160	915	4000	99
PTCDA [194]		80			

[a]: C-rate. CC: carbon cloth, MCMB: mesocarbon microbead, PANI: polyaniline, PNDIE: polyimide, polynaphthalenetetracarboxiimide, PTCDA: perylenetetracarboxylic dianhydride.

3.6. Anodes for Magnesium-Based Batteries

The anodes discussed for Mg batteries include pure Mg metal, intercalation (carbon-based), alloys (In, Sn, Sb, Pb and Bi), and conversion (transition metal oxides (TMO), transition metal sulfides (TMS), phosphides, and nitrides) materials.

Mg metal has been studied as an anode since it possesses multivalent charge carrier ions (Mg^{+2}), a low potential (-2.37 V vs. SHE), a high capacity (2205 mA h g^{-1}), and it does not form dendrite. The application of Mg metal as the anode in commercial batteries is restricted due to the electrochemically inactive layer that is generated on the anode surface. This layer is an electronic and ionic insulating surface film that obstructs any electrochemical reaction, affecting the battery efficiency. It is recommended to prevent the formation of the passive surface film to achieve practical Mg anodes [195].

Lithium titanate Li$_4$Ti$_5$O$_{12}$ (LTO) has been explored as an intercalation anode for Mg batteries since it exhibits low volume changes during Mg ion intercalation–deintercalation.

LTO showed a specific capacity of 175 mA h g^{-1} at current density of 15 mA g^{-1} and 55 mA h g^{-1} at 300 mA g^{-1}, and a high cycling stability with 100% Coulombic efficiency and capacity retention of 99.9% after 500 cycles [196].

Alloy materials which are reversibly electrochemical with Mg include some p-block elements, such as In, Sn, Sb, Pb and Bi, which form MgIn, Mg$_2$Sn, Mg$_3$Sb$_2$, Mg$_2$Pb and Mg$_3$Bi$_2$ at low voltage [197]. Early studies on alloys have focused on Bi because it has a theoretical gravimetric capacity of 385 mA h g^{-1} with Mg and a rhombohedral crystalline structure. In 2012, the electrodeposition of Bi as anode showed a maximum specific capacity of 257 mA h g^{-1} and of 222 mA h g^{-1} after 100 cycles [198]. In addition, superionic conductivity was described but only in phases >976.15 K. Larger capacities are achieved by replacing the Bi layers with nanotubes. A specific capacity of 350 mA h g^{-1} and a Coulombic efficiency initial of 95% was reported for Bi nanotubes [199]. Although the electrochemical behavior exhibited with this nanostructure is remarkable, it is difficult to achieve commercial systems due to its exotic nature and manufacturing costs. Other composites tested as anodes include Bi$_{1-x}$Sb$_x$ and Sb [198]. A maximum specific capacity of 298 mA h g^{-1} was exhibited in the first cycle and 215 mA h g^{-1} after 100 cycles by Bi$_{0.88}$Sb$_{0.12}$, while Sn showed a poor cycling performance after 16 cycles, ~16 mA h g^{-1}. Tin (Sn) as an alloy anode has been tested because it is more abundant than Bi and tin has a high theoretical capacity over 890 mA h g^{-1} [200]. The use of Sn faces a big challenge since experimental tests show non reversible reactions in the Sn anode. SnSb alloys have been explored as an alternative to overcoming the challenges that Sn and Sb have individually. The theoretical capacity of SnSb is 768 mA h g^{-1} and experimental results show a high reversible capacity of 420 mA h g^{-1} and cyclic stability, delivering 350 mA h g^{-1} after 100 cycles [200].

Anode materials for Mg-based batteries are summarized in Table 7. We include the highest reported voltage range, capacity, current density, number of cycles, and Coulombic efficiency.

Table 7. Anode materials for Mg-based batteries.

Anode Material	Voltage Range (V)	Capacity (mA h g^{-1})	Current (mA g^{-1})	Cycles	η_c
Mg$_3$Bi$_2$ [198]	0.23	257	385	100	86
Mg$_3$Bi$_2$ NT [199]		350	19	200	95
Bi$_{0.88}$Sb$_{0.12}$ [198]		298	1 [a]	100	75
In [201]	0.09	425	0.01 [a]		
LTO [196]		175	15	500	
Sb [198]		16	1 [a]	50	
SnSb [200]		420	50	200	

[a]: C-rate. LTO: Li$_4$Ti$_5$O$_{12}$, NT: nanotubes.

3.7. Anodes for Aluminum-Based Batteries

Metal Al is a promising anode since it has multivalent charge carrier ions (Al$^+$), high capacity (2980 mA h g^{-1}), a relatively low potential (−1.66 V vs. SHE), low cost, and it is the most abundant metal on the Earth's crust [202]. The main challenge of using Al as an anode is the highly stable passivation layer, causing an electrochemical inertness [7,47,203]. Al anodes are usually foils, which work as both active material and current collector. However, this limits the active area. Three-dimensional thin film has been studied to increase the active surface area [54]. The 3D thin-film fabricated exhibited a capacity of 165 mA h g^{-1} and a retention of 86% after 500 cycles. To address passivation layer problems, researchers have focused on testing different electrolytes chemistries (Section 5).

Figure 7 compares the current research in anode materials for each battery type, presenting the theoretical capacity, experimental capacity, current density, and the number of cycles for each material. Figure 8 shows the progress per year of the specific capacity in anode materials in the research field, highlighting the most relevant studied materials per year. The line color represents the battery type. For Li- and Na-based batteries, two curves are presented. The continuous lines represent traditional anode materials, which operate through an intercalation mechanism. Dot lines represent conversion materials studied in the next generation of batteries.

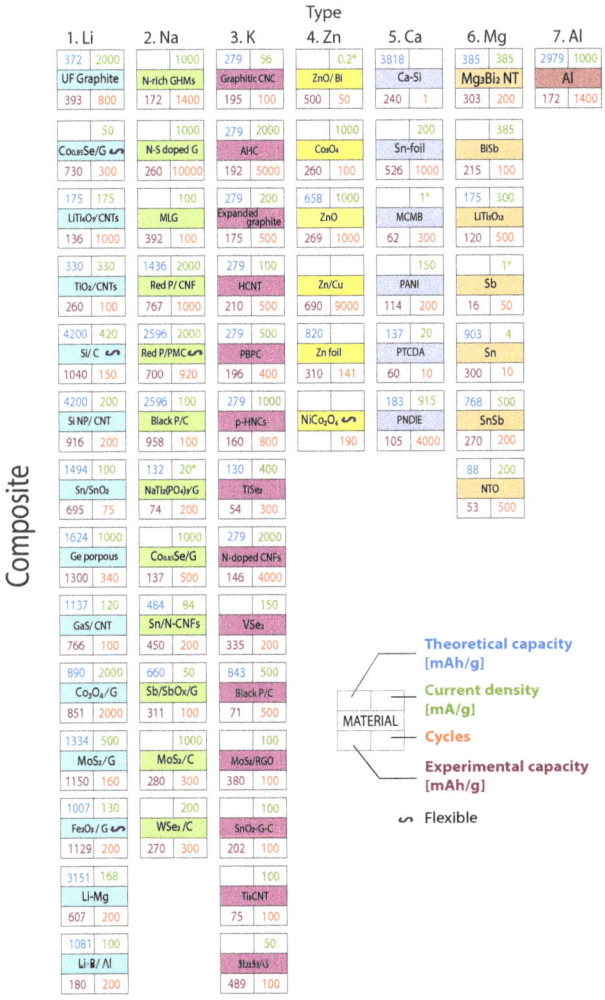

Figure 7. Current research in anode materials. Columns organize the seven battery types. Rows represent the composite material group. Theoretical capacity, experimental capacity, current density, and the number of cycles are also included. A special symbol highlights flexible anode materials. Graph constructed by the authors from references in Tables 2–7.

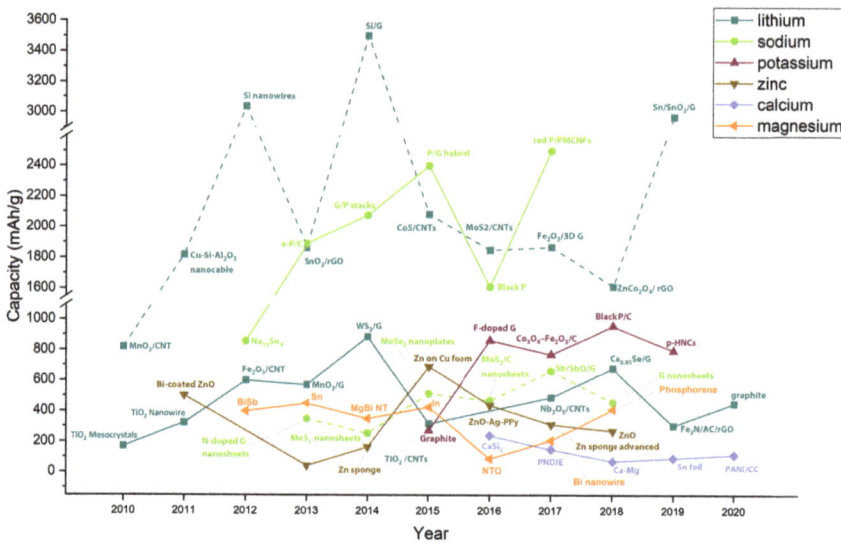

Figure 8. The advance of the anodes' experimental specific capacity, comparing the seven battery types: Li, Na, K, Zn, Ca, Mg, and Al. Continuous line: traditional intercalation materials. Dot lines conversion materials. Graph constructed by the authors from references in Tables 2–7. NW, nanowires; NR, nanorods; NB, nanobelts.

4. Cathodes

The cathode is the positive electrode of a battery that is reduced (gain of electrons) during the discharge process [46]. Similar to the anode, the cathode plays different roles according to the work ion, establishing the operation of rechargeable and non-rechargeable batteries. Generally, cathodes have a lower capacity than anodes, limiting the battery performance [9]. Therefore, there is an interest in improving the capacity of cathodes by optimizing the chemical, electrochemical, and physical properties of materials [204]. An ideal cathode should have high performance, high potential, low cost, and low environmental impact [204]. A high cathode performance implies that the cathode offers a large reversible storage capacity at the desired electrochemical potential. A cathode with high potential allows the development of high energy density batteries [9]. Although the materials' intrinsic nature determines the electrochemical properties of electrodes, it is possible to vary their microstructures with different synthesis methods and conditions. For example, dopants can be introduced to modify the crystal parameters to improve cyclic stability and specific capacity. The storage of ions in cathodes occurs via two mechanisms: intercalation and conversion.

In the intercalation, the electrode material must contain space to store and release working ions reversibly [204]. Intercalation cathode materials can be classified into three kinds according to their chemical composition: (1) transition metal compounds, (2) polyanionic compounds, and (3) Prussian Blue [9].

- Transition metal compounds, oxides, or complex oxides have olivine (1D), layered (2D), or spinel (3D) crystal structures [8,204]. Olivine crystal structures have 1D tunnels to allow ions to flow, causing lower rate capability. Reducing the size of the active material is a strategy to address this issue. Layered oxides have a general formula A_xBO_2, where A represents the ion carrier such as Li, Na, K, Zn, Ca, Al, and Mg, and B represents one or more metal ions such as Ni, Co, Fe, Mn, and Cu. Spinel oxides have a general formula AB_2O_4, where A represents the ion carrier such as Li, Na, K, Zn, Ca, Al, and Mg, and B can be Ti, V, and Mn [9,205]. The layered and spinel oxides offer good electronic conductivity and high densities.

- Polyanionic compounds have a general formula $A_xBB'(XO_4)_3$, where A represents one ion carrier, Li, Na, K, Zn, Ca, Al, or Mg; B could be V, Ti, Fe, Tr, Al, or Nb; and X is P or S. Polyanionic compounds offer higher thermal stability and safety than the layered and spinel oxide cathodes due to the covalent bond between the oxygen and the P, S, or Si. Moreover, polyanionic cathodes include abundant transition metals, such as Fe, which contributed to their applications in storage devices for renewable energy sources. The use of polyanionic compounds requires synthesized small particles with coated conductive carbon due to the poor electronic conductivity, increasing the cost, reducing the volumetric energy density, and leading to low performance [9].
- Prussian blue analogues (PBA) have a general formula $A_xB_1B_2(CN)_6$. A is usually Li, Na, K, Zn, Ca, Mg, or Al, while B_1 and B_2 can be Fe, Mn, Ni, Co, or Cu. The use of PBA as an electrode is due to two structural characteristics: (1) large 3D diffusion channels that facilitate its inward and outward transport by the weak interaction with the diffusing ion, and (2) control of the $[B_2(CN)_6]^{-4}$ vacancies that improve the crystallinity by changing the stoichiometry and the preparation conditions. Moreover, PBA has a high theoretical specific capacity, a simple synthesis, and a low cost [206].

Opposite to intercalation, in conversion, the material does not have active intercalation sites, but the material reacts electrochemically during discharge, breaking chemical bonds and creating new ones [8]. For this reason, the bulk material may react electrochemically during discharge. Conversion mechanism occurs in metal–air and metal–S technologies. During metal–S battery operation, S is reduced electrochemically to produce metal sulfide. The reaction is expressed as $S_2 + 2nM^+ + 2ne^- \rightarrow nM_2S$, where M is Li, Na, K, Zn, Mg, Ca, or Al, and $1 \leq n \leq 8$ [39]. In theory, sulfur can be combined with any metal anode to form metal sulfide. Although S has a low electrochemical potential (0.4 V), it exhibits extremely high theoretical specific capacity (1675 mA h g^{-1}). The main limitation of using S is the volumetric changes due to the density changes during cycling. In addition, the use of conversion materials is limited by their irreversibility. Some strategies to improve the reversibility are the development of small particle sizes (<20 nm in diameter), and the combination with alloying materials [7,204].

Metal–air batteries follow the reaction $O_2 + 4e^- \rightarrow 2H_2O + 4OH^-$ with aqueous electrolytes and $xM^+ + O_2 + xe^- \rightarrow M_xO_2$ ($x = 1$ or 2) with aprotic electrolytes [207]. These technologies offer high theoretical energy densities. However, rechargeable metal–air batteries suffer slow kinetics and overpotential, which limit their practical application.

The selection criteria to choose cathode materials include the abundance in the Earth's crust, eco-friendly nature for processing, usage and recycling, and cost. Figure 9 presents the increasing interest in the research about "positive electrodes". In this section, we present the cathode materials commercialized and researched for each battery type, describing their main properties and showing the strategies explored to address the challenges.

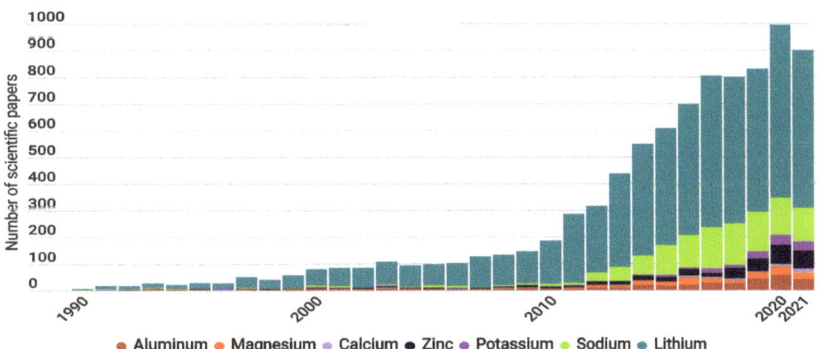

Figure 9. Publications per year about positive electrodes, comparing seven battery types: Li, Na, K, Zn, Ca, Mg, and Al. Graph constructed by the authors. Data from Web of Science.

4.1. Cathodes for Lithium-Based Batteries

The cathodes discussed for Li batteries include layer (lithium cobalt oxide LCO ($LiCoO_2$), lithium nickel cobalt aluminum oxide NCA ($LiNiCoAlO_2$), and lithium nickel cobalt manganese oxide NMC ($LiNiMnCoO_2$), spinel (lithium manganese oxide LMO ($LiMn_2O_4$), and lithium nickel manganese spinel LNMO ($LiNiMnO_4$), olivine (LFP—lithium iron phosphate ($LiFePO_4$), and S, and O_2. The introduction of cathodes for batteries is not straightforward. For example, LCO began to be successfully used in commercial batteries in 1991, 11 years after its discovery [9].

LCO is a layered material that has a theoretical capacity of 273 mA h g^{-1}. LCO has a rhombohedral structure and can achieve open-circuit voltages of 4 V to 5 V and stable operating voltages at 3.7 V. The main issues of LCO are a short life span, low thermal stability, and limited load capabilities [9]. Therefore, in practical application, LCO only achieves ∼140 mA h g^{-1}. To develop a high voltage and high energy density cathode, multi-functional material coatings have been studied [208]. Xing has demonstrated that cover LCO electrode with metal fluoride, metal phosphate, Li metal oxide and Li metal phosphate, increase the specific capacity to 220 mA h g^{-1}, and the voltage cycle stability (3 V–4.6 V). The use of LCO as a cathode requires that it be charged and discharged at a current equal to its C-rating. Forcing a fast charge or applying a load higher than 1 C causes overheating and undue stress. Finally, LCO cathodes raise some concerns about their toxicity, high price, and limited production. For example, Co represents up to 60% of the materials cost for battery manufacturers and, to be profitable, these industries require a continual supply of economic Co. To enhance longevity, loading capabilities, and cost, new materials such as nickel, manganese, and aluminum, have been integrated into LCO cathodes. In addition, to improve the rate capability, researchers focus on the control of particle morphology, while to achieve high capacity, they focus on increased charge voltage.

NMC has a specific capacity and operating voltage similar to LCO, as well as a lower cost since the Co content is lowered [9]. Different Li:M ratios have been studied to determine the optimal value to produce this material to favor physical and electrochemical properties. NMC-111, NMC-442, and NMC-532 are currently the state of the art of these cathode materials. To achieve higher specific energy and lower costs, Ni-rich NMC cathode materials, such as NMC-811 and NMC-622, will be developed in the coming years.

$LiFePO_4$ (LFP) was identified as a polyanionic compound in 1987, but it was suggested as a potential cathode for rechargeable lithium batteries by JB Goodenough et al. in 1997 [9]. LFP has good electrochemical performance with low resistance, high current rating, long cycle life, good thermal stability, enhanced safety, and tolerance if abused. Therefore, it seems to be an alternative to replace the LCO cathode. Although LFP is more tolerant to full charge conditions when kept at high voltage for a long time, it shows a low electronic conductivity (∼10^{-9} S cm^{-1}) and low theoretical capacity (170 mA h g^{-1}). These values are related to its olivine structure, which has a one-dimensional diffusion channel and limits the diffusion of ions. Strategies to increase the capacity include morphology and particle size control, doping, surface coating, and the addition of prelithiation additives [209]. Wang et al. proposed a new morphology joining porous LFP microspheres with carbon and CNTs [210]. As a result, the cathode material showed a discharge capacity of 115 mA h g^{-1} at 1700 mA g^{-1}, maintaining 113 mA h g^{-1} after 1000 cycles. The reduction in dimensions, using nanoscale materials, is another option explored to improve the electronic and ionic transport lengths [3].

$LiMn_2O_4$ (LMO) was first reported by M. Thackeray et al. in 1983, and it was commercialized as a cathode material by Moli Energy. LMO has a spinel structure consisting of a three-dimensional structure (usually composed of diamond shapes connected into a lattice) that improves ion flow on the electrode generating lower internal resistance and a more stable structure than LCO. In addition, LMO possesses a competitive cost advantage with the lowest price ($10 kg^{-1}), non-toxicity, three-dimensional Li^+ diffusion pathways, high thermal stability, and enhanced safety. Some disadvantages of using LMO are a limited cycle life, low capacity (theoretical capacity: 148 mA h g^{-1}), and poor high-temperature performance due to its instability in the electrolyte and capacity loss [3].

NCA has been proposed since 1999 for specific applications, showing high specific energy, good specific power, and a long life span. NCA is a compound that integrates Co and Al in a LiNiO$_2$. The integration of Al helps to minimize negative phase transitions, improve the thermal stability of LiNiO$_2$, and keeps the crystal structure stable. On the other hand, Co helps to reduce cation mixing and also stabilize the layered structure. In addition, Co is electrochemically active, improving the performance of NCA with higher specific capacity (200 mA h g^{-1}), high energy density (200 W h kg^{-1}), and long cycle life [3]. To increase NCA cathode performance, some researchers have explored the use of CNTs to modify NCA surfaces. For example, LiNi$_{0.8}$Co$_{0.15}$Al$_{0.05}$O$_2$/CNT (NCA/CNT) composite has been studied, exhibiting a reversible capacity of 181 mA h g^{-1} with a discharge retention of 96% after 60 cycles at a current of 50 mA g^{-1}. The result suggests that NCA/CNT material enhances the capacity of pristine NCA by 18%, and at a high current rate of 1000 mA g^{-1}, it can deliver a reversible capacity of 160 mA h g^{-1} [211].

To address the Co concern as a critical material, novel research focuses on a new class of cobalt-free materials. Muralidharan et al. have developed a material called lithium iron aluminum nickelate with the formula LiNi$_x$Fe$_y$Al$_z$O$_2$ (x + y + z = 1). The results showed a good performance with a specific capacity of 200 mA h g^{-1}, and 80% capacity retention after 100 cycles at a rate capability of C/3 [212]. The cobalt-free material was synthesized through the sol-gel method that allows varying aluminum and iron composition amounts. Li-rich manganese-based cathode materials have also been explored, such as the cobalt-free material Li$_{1.2}$Ni$_{0.2}$Mn$_{0.6}$O$_2$. However, their practical application is limited by a low Coulombic efficiency during the first cycle, a low rate capability, and a pronounced capacity and voltage fading during cycling. The influence of synthesis conditions have been studied as a critical factor in the characteristics and electrochemical performance of electrode materials [213]. Zhao et al. showed that the increase in calcination temperature (1073.15 K) could improve the layered structure of Li[Li$_{1/5}$Fe$_{1/10}$Ni$_{3/20}$Mn$_{11/20}$]O$_2$, delivering a discharge capacity of 251.9 mA h g^{-1}.

The next generation of batteries focuses on a conversion mechanism with higher specific capacities than intercalation materials. For example, sulfur has a theoretical capacity of 1675 mA h g^{-1} [214]. The use of sulfur requires addressing some challenges: (1) the polysulfides formed during the discharge and the low electrical conductivity, causing high internal resistance of the batteries and limiting the active material utilization efficiency and rate capability; (2) the migration of polysulfides onto the Li anode by a shuttle effect, generating an electrochemically inactive layer that reduces the battery efficiency; (3) the volume change of sulfur during the cycling process, resulting in a volumetric increase in ~79 % after discharge. This significant volume change makes Li$_2$S lose its electrical contact with the conductive substrate or the current collector, causing a fast capacity fading and fast degradation of the cell due to mechanical stress [7,31,215–217]. The strategies to overcome these challenges include using coatings (polymeric species, ceramic membranes, and carbon materials), and encapsulating S into porous carbon matrices [218].

A 3D interconnected porous carbon nanosheets/CNT (PC/CNT) has been explored as the host for sulfur loading [219]. The S-PC/CNT composite showed a high specific capacity of 1485.4 mA h g^{-1} at a current of 836 mA g^{-1}, and 1138 mA h g^{-1} at 3344 mA g^{-1} with 40% of retention after 400 cycles. In addition, this material was tested at high current, delivering a capacity of 749 mA h g^{-1} at 6688 mA g^{-1}.

Other research has studied a hierarchical pore-structured CNT particle host containing spherical macropores to overcome the issues of uniformly impregnating highly active S [220]. This spherical macropore structure (SM-CNTPs) improves the penetration of S into the carbon host in S melt diffusion. The S-SM-CNTP cathode delivers a high specific capacity of 1343 mA h g^{-1} at a current of 334 mA g^{-1}, and 1138 mA h g^{-1} at 3344 mA g^{-1}. In addition, in the latter, capacity retention of 70% was observed after 100 cycles.

To address the flexibility requirements of emerging applications, a carbon nanotube foam (CNTF) has been suggested as a flexible cathode [221]. This cathode is free standing, mechanically flexible, and binder-free 3D interconnected. CNTF has an initial specific capacity of 1378 mA h g^{-1} at a current rate of 334 mA g^{-1} and shows retention of 53.1% after 1000 cycles. At 1672 mA g^{-1}, the cathode delivers a capacity of 1004 mA h g^{-1}. Additionally, Amin et al. explored a flexible organic S-based cathode. Sulfur-linked carbonyl-based poly ((2.5-dihydroxyl-1.4-benzoquinonyl sulfide) (PDHBQS) was synthesized and embedded in single-wall carbon nanotubes (SWCNTs) [222]. The PDHBQS-SWCNTs cathode showed a specific capacity of 182 mA h g^{-1} at a current rate of 50 mA g^{-1} and of 75 mA h g^{-1} at a current rate of 5000 mA g^{-1}. It was tested in a potential window of 1.5 V to 3.5 V, showing retention of 89 % with 250 mA h g^{-1} after 500 cycles.

Another alternative studied to improve the performance of S cathodes is the use of conductive ZrB$_2$ nanoflakes with only 2 wt % conductive Carbon [223]. B has a metallic nature, a suitably high tap density (4.2 g cm^{-3}), robust chemical adsorption on LiPS, and the exposed B sites of the crystal facet (001) in the ZrB$_2$ skeleton kinetically facilitates the fragmentation of high-order LiPS into shorter chains. Also, ZrB$_2$ nanoflakes exhibit a lower barrier for redox reactions from solid Li$_2$S to S since the intense binding strength on exposed Zr sites alters the reaction pathway of delithiation for Li$_2$S. As a result, the ZrB$_2$−S electrode shows a specific discharge capacity of 1243 mA h g^{-1} at a current of 0.2 C, and it can deliver 356 mA h g^{-1} at 5 C. Moreover, the electrode retains a discharge capacity of 831 mA h g^{-1} after 250 cycles at 0.5 C and a capacity of 586 mA h g^{-1} after 600 cycles.

Li−air and Li−O$_2$ are alternative technologies, which have the highest energy density (3500 W h kg^{-1}) among battery technologies. The challenges of O$_2$ electrochemistry have been addressed by developing different catalysts, porous electrode materials, and stable electrolyte solutions [33,224].

The next stage of batteries demands replacing pure O$_2$ gas with air from Earth's atmosphere. These Li-air batteries require the selective filtration of O$_2$ gas from air and avoiding undesired reactions with other air components, such as N$_2$, water vapor (H$_2$O), and carbon dioxide (CO$_2$) [34]. For example, the Li anode reacts with N$_2$ gas to produce lithium nitride (Li$_3$N). In addition, the water vapor from moist air forms LiOH and CO$_2$ impurities from Li$_2$CO$_3$. The accumulation of these materials reduces cyclability, and it causes high overpotential. Therefore, the development of Li-air batteries requires an O$_2$ selective membrane with a high O$_2$ permeability but also rejects other gas molecules [34].

Cathode materials for Li-based batteries are summarized in Table 8. We include the highest reported voltage range, capacity, current density, number of cycles, and retention.

Table 8. Cathode materials for Li-based batteries.

Cathode Material	Voltage Range (V)	Capacity (mA h g^{-1})	Current (mA g^{-1})	Cycles	Retention (%)
LiCoO$_2$ [208]	3–4.6	220	C/4 *	100	
LiFePO$_4$/C/V$_2$O$_3$ [225]	3.4	140	750	30	100
LiFePO$_4$/CNTs [210]		115	1700	1000	98
LiFePO$_4$/G [226]		123	1700	1000	89
CNT/LiNi$_{0.5}$Mn$_{1.5}$O$_4$ [227]		140	70	100	96
LiMn$_2$O$_4$		80			
CNT/LiNi$_{0.8}$Co$_{0.15}$Al$_{0.05}$O$_2$ [211]		189	50	60	96
LiCo$_{1/3}$Ni$_{1/3}$Mn$_{1/3}$O$_2$ [228]	4.6	158		30	99
LiNi$_x$Fe$_y$Al$_z$O$_2$ [212]	3–4.5	200	1 *	100	80
V$_2$O$_5$/CNTs [229]		298	150	200	71
PDHBQS-SWCNTs [222]	1.5–3.5	182	5000	500	89
Sulfur					
S coating on hydroxylated CNTs [230]		1274		100	57

Table 8. Cont.

Cathode Material	Voltage Range (V)	Capacity (mA h g^{-1})	Current (mA g^{-1})	Cycles	Retention (%)
S encapsulated in spherical CNTs particles [220]	1.5–2.8	1343	3344	100	70
S encapsulated in CNTs [219]	1.5–3	1485	3344	400	60
S embedded in CNT foam [221]		1379	1672	200	76
S wrapped on CNT array [231]		1092		50	64
S wrapped on CNTs [232]		1065		300	77
ZrB$_2$–S [223]		1243	5 *	600	89

*: C-rate, CNT: carbon nanotubes, G: graphene, PDHBQS: poly-dihydroxyl-benzoquinonyl sulfide, SWCNTs: single-wall carbon nanotubes, S: sulfur.

4.2. Cathodes for Sodium-Based Batteries

The cathodes discussed for sodium batteries include layered transition metals, oxides, sulfides or fluorides, spinel structure, polyanionic compounds, Prussian blue analogues, polymers, organics [100,233], sulfur, and oxygen.

In the 1980s, layer transition metal oxides were studied by Delmas and Hagenmuller. These sodiated transition metal oxides (Na$_{1-x}$MO$_2$, M: transition metal) were classified into two groups depending on both the alkali metal position and the number of alkali metal layers in the structure perpendicular to the layering. The first group is the P2 type, where P represents a prismatic structure, and the second group is the O3 type, where O is an octahedral environment. Common transition metals used are Fe and Co.

Fe-based materials, such as Na$_{1-x}$FeO$_2$ and derivatives, have advantages such as non-toxicity and low cost due to the abundance of Fe. The use of these materials is limited by a low capacity and an irreversible structural change. Martinez De Ilarduya et al, have demonstrated that add Mn increase the capacity. They studied Na$_{2/3}$[Fe$_{1/2}$Mn$_{1/2}$]O$_2$, which delivered a specific capacity of 190 mA h g^{-1} and an average voltage of 2.75V. However, it had a poor cycling stability [234]. Ni was also explored to improve the performance of layered cathodes. Na[Ni$_{1/3}$Fe$_{1/3}$Mn$_{1/3}$]O$_2$ showed a specific capacity of 123 mA h g^{-1} and 80% of retention capacity after 100 cycles.

Co-based materials have also been studied in layered cathodes. For example, NaCoO$_2$ is a layered material that has excellent reversibility with a capacity of 125 mA h g^{-1} and a retention of 86% after 300 cycles. The main concern of the integration of cobalt in cathode materials is the increase in their cost and Co toxicity.

Ni-based materials are another alternative to cathodes. Na[Ni$_{0.5}$Mn$_{0.5}$]O$_2$ has been proposed, showing a specific capacity of 141 mA h g^{-1}, and a retention of 90% after 100 cycles [235]. Moreover, the authors demonstrate that using conductive CNTs, the apparent diffusion coefficient of Na ions in the layered composite electrode can be increased, with better rate capability of 80 mA h g^{-1} at a current of 480 mA g^{-1}.

In P2-type Na layered oxides, the Na$^+$ kinetics and cycling stability at high rates depend on superstructures (single-phase domains characterized by different Na+/vacancy-ordered) generated by the Na concentration and the voltage range used to test the cathode, which causes low performance. To address the low performance that these materials exhibit, P2-Na$_{2/3}$Ni$_{1/3}$Mn$_{1/3}$Ti$_{1/3}$O$_2$ (NaNMT) has been studied. (NaNMT) shows that structure modulation to construct a completely disordered arrangement of Na-vacancy within Na layers can be better than ordered structures [236]. Disordered P2-NaNMT maintains a 83.9% capacity retention after 500 cycles at 1 C, and delivers a reversible capacity of 88 mA h g^{-1} with an average voltage of 3.5 V.

Mn-based compounds have also been researched in sodium batteries due to the low cost of Mn. These materials have been studied since 1970. Na$_x$MnO$_2$ (x = 0.44–1) has a three-dimensional structure at lower x values (x = 0–0.44) and two-dimensional structure at higher x values (x > 0.5) [237]. However, the stability of α−NaMnO$_2$ and β−NaMnO$_2$ dependent on temperature, the former is stable at low temperature and the latter is stable at high temperature. The electrochemical analysis shows a charge capacity of 208 mA h g^{-1} (α−NaMnO$_2$) and 191 mA h g^{-1} (β−NaMnO$_2$) at a current of 10 mA g^{-1}, but low Coulombic efficiencies of 84 and 70%, respectively.

On the other hand, transition metal polyanion compounds (Na$_x$MM'(XO$_4$)$_3$, X = P, S) have displayed significant thermal stability and a high voltage due to strongly covalent bonds. Commonly, polyanion materials with P are combined with V, Co, Ni, Fe, and Mn. For example, Na$_3$V$_2$(PO$_4$)$_3$ is the most common with a voltage of 3.4 V and a capacity of 107 mA h g^{-1}. To increase the voltage of Na$_3$V$_2$(PO$_4$)$_3$, Na–V fluorophosphate family has been studied, exploring several stoichiometries such as NaVPO$_4$F and Na$_3$V$_2$O$_{2x}$(PO$_4$)$_2$F$_{3-2x}$. Furthermore, Co-based materials have been investigated in polyanionic compounds. For instance, NaCoPO$_4$ has an olivine structure with a calculated voltage of 4.19 V, but this material needs to be tested experimentally. Ni-based materials also have a theoretical voltage over 4.5 V, but it is necessary for experimental results. Other polyanionic compounds for Sodium-based batteries are NaFePO$_4$, Na[Fe$_{0.5}$Mn$_{0.5}$]PO$_4$. Among the polyanionic compounds using sulfates have also been explored, such as Na$_2$Fe$_2$(SO$_4$)$_3$, demonstrating a voltage of 3.8 V. The main challenge of polyanion compounds is that they exhibit lower electrical conductivity than Oxides, low gravimetric capacity because of heavy polyanion groups, and low volumetric energy densities.

Another alternative studied for cathodes is the use of mixed polyanionic compounds. Na$_2$Fe(C$_2$O$_4$)SO$_4$·H$_2$O was explored in both experimental and simulation methods [238]. This cathode showed average voltages of 3.5 V and 3.1 V and a capacity of 75 mA h g^{-1} at 44 mA g^{-1} after 500 cycles with 99 % of Coulombic efficiency. The low cost and environmental friendliness of this material suggest it is a promising material in practical applications.

Prussian blue and Prussian blue analogues are organic materials with the chemical formula Fe$_4^{III}$[Fe^{II}(CN)$_6$]$_3$. They have been studied as cathodes for Na batteries since they have a 3D open framework and it is possible adjust their structure and chemical composition [239]. However, the structure defects, crystal water and interface instability restrict the achievement of high capacity, high rate capability and long cyclability. To reduce the mechanical degradation and improve the electrochemical cyclability, Hu et al. have proposed a concentration–gradient composition method [240]. The purpose was to gradually increase Ni while Mn decreases from the interior of the particle surface in Na$_x$Ni$_y$Mn$_{1-y}$Fe$^-$(CN)$_6$·nH$_2$O. Although the results showed that the capacity decreased with the increasing Ni:Mn ratio (from 120 mA h g^{-1} when x = 0 to 110, 98, and 82 mA h g^{-1} when x = 0.1, 0.3, and 0.5, respectively at 0.2 C), the cycling performance improved. The capacity retention for g−(Ni$_{0.1}$Mn$_{0.9}$)HCF was ∼95 % after 100 cycles, while for MnHCF was 52.5 %. The material is able to deliver ∼80 mA h g^{-1} at 5 C with a retention of 93 % after 1000 cycles.

To improve the volumetric performance of a cathode material, a recent strategy focused on the fabrication process was explored. Compact highly dense metal oxide quantum dots-anchored nitrogen-rich reduced graphene oxide (HD−TiO$_2^-$N−RGO) hybrid monoliths were designed through a large number of ultrasmall TiO$_2$−QDs (∼4.0 nm) which were homogeneously anchored onto N- RGO. The HD−TiO$_2$−N−RGO compact monolith exhibited a high gravimetric capacity of 203.4 mA h g^{-1} without degradation after 100 cycles at 100 mA g^{-1}. Furthermore, a high stability lifespan with over 91% capacity retention after 1000 cycles at 2000 mA g^{-1} was demonstrated [241].

Cathode materials for sodium-based batteries are summarized in Table 9. We include the highest reported voltage range, capacity, current density, number of cycles, and retention.

Table 9. Cathode materials for Na-based batteries.

Cathode Material	Voltage Range (V)	Capacity (mA h g^{-1})	Current (mA g^{-1})	Cycles	Retention (%)
Layer metal oxides					
O$_3$-NaNi$_{0.5}$Mn$_{0.5}$O$_2$ [242]	2.0–4.0	133	468	500	70
O$_3$-NaNi$_{0.12}$Cu$_{0.12}$Mg$_{0.12}$Fe$_{0.15}$Co$_{0.15}$Mn$_{0.1}$Ti$_{0.1}$Sn$_{0.1}$Sb$_{0.04}$O$_2$ [243]	2.0–3.9	110	360	500	83
α−NaMnO$_2$ [237]	2.0–3.8	175	10	50	
β−NaMnO$_2$ [237]	2.0–3.8	130	10	50	
P2−Na$_{2/3}$Ni$_{1/3}$Mn$_{1/3}$Ti$_{1/3}$O$_2$ [236]	2.5–4.1	88	1 [a]	500	83.9
NaNi$_{0.5}$Mn$_{0.2}$Ti$_{0.3}$O$_2$ [244]	2.8	135	240	200	85
Na$_3$Ni$_{1.5}$Cu$_{0.5}$BiO$_6$ [245]	3.2	94	10.8	200	62
NaNi$_{0.5}$Mn$_{0.5}$O$_2$ CNT [235]	2.5	141	12	100	90
Na$_{0.67}$[Mn$_{0.6}$Ni$_{0.1}$Fe$_{0.3}$]O [246]	4.3	200	13	25	75
Na$_{0.7}$CoO$_2$ Microspheres [247]	2.9	125	5	300	86
Na$_{0.67}$[Fe$_{0.5}$Mn$_{0.5}$]O$_2$ [234]	2.7	183	15	10	90
Na$_{0.9}$Cu$_{0.22}$Fe$_{0.3}$Mn$_{0.48}$O$_2$ [248]	3.2	100	10	100	
NaNi$_{1/3}$Fe$_{1/3}$Mn$_{1/3}$O$_2$ [249]		123	130	100	80
Na$_{0.67}$Mn$_{0.67}$Ni$_{0.28}$Mg$_{0.05}$O$_2$ [250]		123	0.1 [a]		
Na$_{2/3}$Ni$_{1/3}$Mn$_{1/3}$Ti$_{1/3}$O$_2$ [236]		88	17.3	500	83
HD−TiO$_2$−N−RGO [241]		203.4	100	100	91
Prussian blue					
Na$_x$Ni$_{0.1}$Mn$_{0.9}$Fe$^-$(CN)$_6$·nH$_2$O [240]		110		1000	95
Prussian white					
Na$_{1.92}$FeFe(CN)$_6$ [251]	3	150	600	1000	75
Polyanionic Compoound					
Na$_2$Fe(C$_2$O$_4$)SO$_4$·H$_2$O [238]	1.7–4.2	~75	44	500	85
Na$_3$V$_2$(PO$_4$)$_2$F$_3$ [252]	3.7	120	0.05 [a]		
Na$_2$Fe(C$_2$O$_4$)SO$_4$ [238]	3.1	170	44	500	85
Sulfur					
S sugar derived [253]	0.8–2.6	700	1675	1500	81
S/Fe–HC [254]	0.8–2.7	1023	100	1000	38

[a]: C-Rate. HD: highly dense, RGO: reduced graphene oxide.

4.3. Cathodes for Potassium-Based Batteries

The cathodes discussed for potassium batteries include layered TMO (Co-, Fe-, Mn-, and V-based), Prussian blue analogues, polymers, organics, S, and O.

Cobalt-based materials have been studied as cathode since 2017. For instance, K$_{0.6}$CoO$_2$ has an hexagonal symmetry crystal structure, and it is able to deliver a capacity of 74 mA h g^{-1} working in a voltage range of 1.7–4.0 V, maintaining 64 mA g^{-1} after 300 cycles [255]. The main challenge of using Co-based materials is their high costs and toxicity, causing an increasing interest in searching for cheaper and eco-friendly alternative materials.

Mn-based materials have been proposed as cathodes for potassium batteries since 2016. For example, K$_{0.3}$MnO$_2$ is a layered material that allows the intercalation of K ions with an experimental specific capacity of 136 mA h g^{-1} at 27.0 mA g^{-1} in a voltage range of 1.5–4.0 V, but a low capacity retention (58% after 50 cycles) [256]. The results also show that the cathode performance has a high dependence on the window potential, exhibiting a high capacity retention (91% after 50 cycles and 57% after 685 cycles) between 1.5 and 3.5 V with a specific capacity of 65 mA h g^{-1} at 27 mA g^{-1}. This loss of capacity is also reported for K$_{0.5}$MnO$_2$ that delivers a specific capacity of 140 mA h g^{-1} between 1.5 and 4.2 V with a low retention (~35 % after 20 cycles), and 93 mA h g^{-1} between 1.5 and 3.9 V at a current of current rate of 20 mA g^{-1} with a capacity retention of 70% after 50 cycles [257]. The loss of capacity of these materials at a higher potential is associated with irreversible phase transitions at higher potentials.

Similar to layered oxide studies for Na-based batteries, the K$^+$ transport kinetics and storage sites are limited in these materials, causing a low performance and capacity. Disordered structures instead of ordered structures have been also studied in K-based batteries to address these issues. For instance, layered oxide K$_x$Mn$_{0.7}$Ni$_{0.3}$O$_2$, which has a high redox potential and highly symmetric crystalline structure, has been explored as K$^+$/vacancy disordering [258]. The results show that K$_{0.7}$Mn$_{0.7}$Ni$_{0.3}$O$_2$ delivers the best performance, with high discharge capacity of 125.4 mA h g^{-1} at 100 mA g^{-1} and 83.8 mA h g^{-1} at 1000 mA g^{-1}, and an average discharge voltage of 3.0. Additionally, K$_{0.7}$Mn$_{0.7}$Ni$_{0.3}$O$_2$ exhibits a good cyclic stability, retaining a capacity of 78.4 mA h g^{-1} after 800 cycles at 1000 mA g^{-1} (capacity retention of 88.5%).

V-based materials deliver high voltage plateaus in Potassium batteries. Some examples include KVPO$_4$F and KVOPO$_4$ that can maintain an average working voltage of 4.02 V and 3.95 V, respectively [259]. These materials were tested in the voltage range of 4.8–2.0 V at 6.65 mA g^{-1}, showing a capacity of 70 mA h g^{-1}. Moreover, KVPO$_4$F and KVOPO$_4$ show reversible capacities of 92 mA h g^{-1} and 84 mA g^{-1} in the voltage range of 5–2.0 V with a working voltage of ~4.0 V. Another option of V-based material is KVP$_2$O$_7$, which shows an average discharge potential of 4.2 V and a capacity of 60 mA h g^{-1} [260]. To improve the material capacity of the cathode, some researchers have focused on developing compounds analogous to the Li and Na equivalent. For example, KV$_2$(PO$_4$)$_2$F$_3$ has been proposed [261]. This material is capable of delivering a capacity of 104 mA h g^{-1} at lower voltage of 3.7 V.

Prussian blue analogue (PBA) materials are common multi-metal redox couples that have the formula K$_x$M^{III}[Fe^{II}(CN)$_6$] where M^{III} (trivalent transition ion) is replaced with M^{II} (bivalent transition ion) to obtain higher theoretical capacities. An example of PBA studied as a cathode is K$_2$Fe[Fe(CN)$_6$] which deliver more than one redox couple, at 3.55–3.26 V and 4.1–3.91 V, and it exhibit a specific capacity of 110 mA h g^{-1}. Another example is K$_{1.6}$Mn[Fe(CN)$_6$]$_{0.96}$, which shows 4.12–3.69 V and presents a higher initial capacity of 125 mA h g^{-1}. Another approach explored is increasing the content of K in these PBA materials to raise the capacity. For example, K$_{1.75}$Mn[Fe(CN)$_6$]$_{0.93}$ exhibited a discharge capacity of 130 mA h g^{-1} at 30 mA g^{-1} after 100 cycles [262], and K$_{1.89}$Mn[Fe(CN)$_6$]$_{0.92}$ delivered a reversible capacity of 146.2 mA h g^{-1} at 0.2 C [263].

Organic materials have been studied as the cathode of potassium batteries since 2015. They have impressive electrochemical performances, and they also are inexpensive and eco-friendly. Organic materials that have a crystal structure usually possess larger interlayer spacings since they are held together by van der Waals forces instead of ionic or covalent bonding. Some common organic cathodes are anthraquinone-1,5-disulfonic acid sodium salt (AQDS), oxocarbon salts, perylene tetracarboxylic dianhydride (PTCDA), and PAQS. AQDS exhibited a first discharged capacity of 114.9 mA h g^{-1}, and 78 mA h g^{-1} after 100 cycles [264]. OxoCarbon salts K$_2$C$_6$O$_6$ have good performance, showing 212 mA h g^{-1} at 40 mA g^{-1} and 164 mA h g^{-1} at 2000 mA g^{-1} [265]. PTCDA showed a specific capacity of 131 mA h g^{-1} and two discharge plateaus at around 2.4 and 2.2 V [266]. PAQS delivered a high capacity of 200 mA h g^{-1} with a capacity retention of 75% after 50 cycles at a rate of C/10, and it has two slopes averaged at 2.1 and 1.6 V [267].

In addition, S has been explored as an alternative cathode for K-based batteries. The main challenges using S are as follows: (1) cycle stability due to capacity fade and shuttle effect; (2) the discharge voltage not being in a flat region but lying on the sloping part of the curve; and (3) high operating temperature. To address these issues, some research has been focused on testing the cathode materials studied for Li- and Na-based batteries. For instance, a pyrolyzed polyacrylonitrile/sulfur nanocomposite (SPAN) has been studied to assess their performance in a room temperature battery. SPAN shows a high reversible capacity of 270 mA h g^{-1} at a current rate of 125 mA g^{-1} working in a voltage range of 0.8 V to 2.9 V [268]. To reduce and avoid the shuttle effect, a confined and covalent sulfur cathode has been explored, operating at room temperature. This cathode can deliver an energy density as high as ~445 W h kg^{-1}, a Coulombic efficiency close to 100%, and superior cycle stability with a capacity retention of 86.3% over 300 cycles at a voltage 3.0 V [269].

Another option explored is the use of carbon materials to avoid the formation of soluble polysulfides. For example, a microporous carbon-confined small-molecule sulfur composite has been tested, showing a reversible capacity of 1198.3 mA h g^{-1} with retention of 72.5% after 150 cycles and Coulombic efficiency of ~97% [270].

Cathode materials for potassium-based batteries are summarized in Table 10. We include the highest reported voltage range, capacity, current density, number of cycles, and retention.

Table 10. Cathode materials for K-based batteries.

Cathode Material	Voltage (V)	Capacity (mA h g^{-1})	Current (mA g^{-1})	Cycles	Retention (%)
Transition metal Oxide					
NaCoO$_2$ [271]	2.9	80	0.05 [a]	50	80
NaCrO [272]	2.95	88	0.05 [a]	200	71
K$_{0.6}$CoO$_2$ [273]	3	82	11.8	300	87
K$_{0-3}$CrF$_6$	5.43	284			
KFMO [274]	2.45 [275]	178	1000	300	87
K$_{0.3}$MnO$_2$ [256]	1.5–3.5	136	27.9	685	91
K$_{0.5}$MnO$_2$ [257]	3.6	140	20	50	76
KNiCoMnO [276]	3.1	76.5	20	100	87
K$_{0.5}$V$_2$O$_5$ [277]	2.5	90	10	250	81
Prussian blue analogue					
KCuFe(CN)$_6$ [278]		60	83 [a]		83
K$_4$Fe(CN)$_6$ [279]	3.6	65.9	20	400	68
KFeFe(CN)$_6$	3.75 [280]	122 [262]	100 [281]	1000 [281]	90 [280]
KMnFe(CN)$_6$ [263]	3.9 [282]	142	10 [a] [262]	100	96
KNiFe(CN)$_6$ [283]		59	41 [a]		95
SWCNT−PB [284]				1000	80
MWCNT−PB [284]				1000	60
RGO−PB−SSM [285]		90	10 [a]		87
Polyanionic compounds					
FePO$_4$	2.1	160	4	50	
FeSO$_4$F	3.5		0.05 [a]		
KVPO$_4$F [259]	4	92	665	50	97
KVP$_2$O$_7$ [260]	4.4	60	20 [a]	100	83
K$_3$V$_2$(PO$_4$)$_2$F$_3$ [261]	3.7	104	250	100	97
K$_3$V$_2$(PO$_4$)$_3$/C [286]	3.6	54	20	100	
Organic					
Anthraquinone [264]	1.7	114	13	100	
K$_2$C$_6$O$_6$ [265]	1.7	213	1 [a]	100	
PAQS [267]	2.4	200	50	300	80
PTCDA [287]	2.1	130	13	50	
Sulfur					
Catholyte: S + K$_2$S$_x$ [288]		402	1	1000	100
Confined and covalent S [269]		873.9	100	300	86.3
CMK-3/S [289]		606	10	10	40
S-CNT [290]		1140	167.5	50	52.6
Microporous C/S [270]		1198.3	20	150	72.5
Pyrolyzed poly acrylonitrile/S [268]	0.8–2.9	710	125	100	54
Sulfurized Carbonized polyacrylonitrile [291]		1050	837.5	100	95

[a]: C-rate. CNT: carbon nanotubes.

4.4. Cathodes for Zinc-Based Batteries

The cathodes discussed for Zn batteries include layered materials (Mn Oxides, and V-based), Prussian blue, and organic materials.

Early research in cathode materials for Zn-based batteries was focused on Mn oxides since they possess a high valence state and phases that facilitate the intercalation of Zn ions, and also they are eco-friendly and offer a low cost. Current Mn-based materials studied include the following phases: $\gamma-MnO_2$, $\alpha-MnO_2$, and $ZnMn_2O_4$ [172]. To research the electrochemical mechanism of the former as cathode material, a mesoporous $\gamma-MnO_2$ was synthesized and characterized, showing a structural transformation from tunnel type to spinel type [292]. The cathode delivered an initial discharge capacity of 285 mA h cm^{-1} in a voltage range of 0.8–1.8 V. $\alpha-MnO_2$ has been studied since 2009, but its electrochemical reaction mechanism is still a topic of discussion [172]. Currently, there are three mechanisms proposed: (1) zinc intercalation/deintercalation; (2) conversion reaction; and (3) H$^+$ and Zn^{2+} co-insertion. Another Mn-based material explored is $ZnMn_2O_4$, which is inspired by the success of $LiMn_2O_4$. $ZnMn_2O_4$ has a spinel structure, and it was tested as the host material for intercalation of Zn^{2+} cations, showing a specific capacity of 150 mA h g^{-1} at a high current of 500 mA g^{-1} with a retention of 94 % for 500 cycles.

V-based cathode materials have been widely studied as a cathode for Zn-based batteries. For instance, V_2O_5 is a layered material capable of storing Zn ions in the interlayers. Additionally, the role of structural H_2O on Zn^{2+} intercalation has been studied in bilayer $V_2O_5 \cdot nH_2O$, suggesting that the H_2O improves the Zn^{2+} diffusion due to the water functions as a lubricant that reduced electrostatic interactions with the V_2O_5 framework [293]. This cathode was capable of delivering a high initial capacity of 381 mA h g^{-1} at a current density of 60 mA g^{-1}, and 372 mA h g^{-1} at 300 mA g^{-1}. Moreover, V_2O_5 was tested at 5000 mA g^{-1} for over 4000 cycles, showing a long cycle life [294]. Other V material explored are as follows: vanadium dioxide VO_6, which has a special tunnel-like framework; VO_2 that has a capacity of 274 mA h g^{-1} and an ultra-long lifespan of 10,000 cycles with a capacity retention of 79 % [295]; and $VO_2(B)$ nanofibers that have ultrafast kinetics of Zn^{2+} due to the tunnels into the material and little structural change on Zn^{2+} intercalation (capacity of 357 mA h g^{-1} at 100 mA g^{-1}) [296]. V materials also can be improved with a pre-intercalation of cations. For instance, $H_2V_3O_8$ nanowire cathode exhibits a high capacity of 423.8 mA h g^{-1} at 100 mA g^{-1} with capacity retention of 94.3 % for over 1000 cycles [297]. Recently, layered $Mg_{0.1}V_2O_5 \cdot H_2O$(MgVO) nanobelts were proposed for practical Zinc battery systems [298]. Additionally, a concentrated 3 M $Zn(CF_3SO_3)_2$ polyacrylamide was used as a gel electrolyte. As a result, the cathode showed a capacity of 470 mA h g^{-1}, and it is capable of delivering 345 mA h g^{-1} at 500 mA g^{-1}. The assembled system has 95 % capacity retention over 3000 cycles operating in a temperature range from 243.15 K to 353.15 K.

Prussian blue materials suggested for Zinc batteries include copper hexacyanoferrate (CuHCF) and zinc hexacyanoferrate (ZnHCF). CuHCF showed a specific capacity of $\sim$50 mA h g^{-1} at a current rate of 60 mA h g^{-1} with an average discharge voltage of 1.73 V and a retention of 96.3 % after 100 cycles [299]. ZnHCF, which has a rhombohedral crystal structure, showed a first cycle capacity of 65.4 mA h g^{-1} at a current rate of 60 mA g^{-1} with an average voltage of 1.7 V [300].

Polyanionic compounds as cathode in zinc batteries are $Li_3V_2(PO_4)_3$ [301], which shows a capacity of 113.5 mA h g^{-1}, $Na_3V_2(PO_4)_3/C$ [302] that a capacity of 97 mA h g^{-1}, and $Na_3V_2(PO_4)_2F_3$ [303] a capacity of 50 mA h g^{-1}.

Cathode materials for zinc-based batteries are summarized in Table 11. We include the highest reported voltage range, capacity, current density, number of cycles, and retention.

Table 11. Cathode materials for Zn-based batteries.

Cathode Material	Voltage Range (V)	Capacity (mA h g^{-1})	Current (mA g^{-1})	Cycles	Retention (%)
Layered Oxides					
Manganese					
α−MnO$_2$ nanorods [304]	0.8 – 1.8	115.9	5000	4000	97.7
α−MnO$_2$ nanofibers [305]	1.0 – 1.85	285	1520	5000	92
α−MnO$_2$/rGO [306]	1.0 – 1.9	382	300	3000	94
MnO$_2$ [307]	1.0 – 1.8	70	1885	10,000	
MnO$_2$−PANI [308]	1.0 – 1.8	125	2000	5000	
Vanadium					
H$_2$V$_3$O$_8$ nanowires [309]	0.2–1.6	173.6	5000	1000	94.3
H$_2$V$_3$O$_8$ nanowires/ GO [297]	0.2–1.6	394	300	2000	87
Li−V$_2$O$_5$·nH$_2$O [310]	0.4–1.4	192	10,000	1000	
LiV$_3$O$_8$ [311]	0.6–1.2	140	133	65	
Mg$_{0.1}$V$_2$O$_5$·H$_2$O(MgVO)	0.1–1.6	470	5000	3000	95
Bilayer V$_2$O$_5$·nH$_2$O [293]	0.2–1.6	200	6000	900	71
VO$_2$ [295]	0.7–1.7	133	10,000	10,000	79
VO$_2$ (B) [296]	0.3–1.5	357	100	50	
V$_2$O$_5$ [312]	0.2–1.6	372	5000	4000	91.1
V$_6$O$_{13}$ [313]	0.2–1.5	240	4000	2000	92
V$_3$O$_7$·H$_2$O/rGO [314]	0.3–1.5	245	1500	1000	79
VO$_2$/rGO [315]	0.3–1.3	240	4000	1000	99
VS$_2$ flake [316]	0.4–1.0	125	200	250	99.7
Zn$_{0.25}$V$_2$O$_5$·nH$_2$O nanobelts [317]	0.5–1.4	260	2400	1000	80
others					
Ag$_{0.4}$V$_2$O$_5$ [318]	0.4–1.4	144	20	4000	
Ca$_{0.25}$V$_2$O$_5$·nH$_2$O [319]	0.6–1.6	70	20	3000	96
K$_2$V$_6$O$_{16\cdot 2.7}$H$_2$O nanorod [320]	0.4–1.4	188	6	500	82
Na$_{0.33}$V$_2$O$_5$ [321]	0.2–1.6	218.4	1	1000	93
NaV$_3$O$_8$ [322]	0.3–1.25	165	4	1000	82
Na$_2$V$_6$O$_{16\cdot 1.63}$H$_2$O [323]	0.2–1.6	158	5	6000	90
Na$_{1.1}$V$_3$O$_{7.9}$/rGO [324]	0.4–1.4	171	300	100	
NH$_4$V$_4$O$_{10}$ [325]	0.4–1.4	255.5	10	1000	
MoO$_2$/Mo$_2$N nanobelts [326]	0.25–1.35	113	1	1000	78.8
MoS$_2$ [327]	0.3–1.5	161.7	1	1000	97.7
Prussian blue analogues					
CuHCF [299]	0.45–1.4	50	10,000	1000 [328]	80
ZnHCF [329]	0.8–1.9	68	300	200	85
Polyanionic compound					
Li$_3$V$_2$(PO$_4$)$_3$ [301]	0.7–2.1	113.5	1500 [294]		
Na$_3$V$_2$(PO$_4$)$_3$/C [302]	0.8–1.7	97	50	200 [330]	74
Na$_3$V$_2$(PO$_4$)$_2$F$_3$ [303]	0.8–1.9	50	1000		
Organic					
Polyaniline [53]	0.5–1.5	82	5000		
Quinones [331]	0.2–1.8	120	500	3000	92

4.5. Cathodes for Calcium-Based Batteries

The design of cathodes for Ca-based batteries suffer several technical bottlenecks that have limited the electrochemical calcium intercalation in known materials [47,186,332]. For example, the large size of Ca-ions limit a rapid insertion and de-insertion of Ca^{+2}, requiring materials with sufficient crystallographic space to insert the Ca-ions. A few successful cathodes have been tested electrochemically, and the research has been focused on demonstrating the reversible capacity of Ca in layered (Co-based and V-based), Prussian blue, and organic materials.

In 2016, the intercalation of Ca^{+2} was successfully proved in a calcium cobaltate cathode (CaCo$_2$O$_4$), showing that the reversible capacity change in the range of 30 mA h g^{-1}

to 100 mA h g^{-1} depending on the experimental conditions (current density and voltage range) [333]. Moreover, Ca extraction was achieved for the first time in a 1D framework for Ca$_3$Co$_2$O$_6$ [334]. This compound was widely studied for its magnetic properties and its crystal structure, and the result opens new routes toward the research of 1D structures as electrodes in calcium batteries. In addition, the electrochemical intercalation of Ca$_{2+}$ in layered TiS$_2$ using alkyl carbonate-based electrolytes was proved [335], showing a capacity of 520 mA h g^{-1} at C/100 and 210 mA h g^{-1} at C/50, with a working average voltage of 1.5 V. This material requires further research to evaluate cyclability. Graphite has been explored in Ca batteries that can work stably at room temperature and high voltage. Graphite cathodes have shown a capacity retention of 95% after 350 cycles with a voltage of 4.45 V [189]. The Ca^{+2} intercalation in the MnFe(CN)$_6$ has been researched, showing a first cycle capacity of ~80 mA h g^{-1}, and after 30 cycles it is capable of delivering a capacity of ~50 mA h g^{-1} [336].

On the other hand, molybdite Ca$_x$MoO$_3$ was shown to be electrochemically active in Ca batteries with nonaqueous electrolytes, and molybdenum and molybdenum oxide has a low toxicity [337]. Although the perovskite-type CaMoO$_3$ was found to be unsatisfactory due to the low mobility of Ca in its framework, the structure of the orthorhombic α−MoO$_3$ phase (nonplanar double-layers of MoO$_6$-octahedra separated by a van der Waals gap) is suitable for intercalation reactions of monovalent and divalent cations in both aqueous (protic) and nonaqueous (aprotic) media. The results showed an experimental reversible first cycle capacity of about 180 mA h g^{-1} with an average voltage of 1.3 V, but after 12 cycles the capacity is ~100 mA h g^{-1}.

Early research in V-based material for Ca batteries showed that vanadium oxides are reversible as Ca hosts, estimating discharge capacities of 400 mA h g^{-1} [338]. Recently, bilayered Mg$_{0.25}$V$_2$O$_5$H$_2$O has been explored as a stable cathode [339]. This material exhibited a capacity of 80 mA h g^{-1} in the first cycle and 120 mA h g^{-1} in the second cycle at a current rate of 20 mA g^{-1} and a capacity of 70.2 mA h g^{-1} at a current of 100 mA g^{-1} with a capacity retention of 86.9% after 500 cycles. VOPO$_4$·2H$_2$O has also been studied because it has a higher working potential than layered Vanadium Oxide. VOPO$_4$·2H$_2$O cathode delivered a discharge capacity of 100.6 mA h g^{-1} at 20 mA g^{-1}, and 42.7 mA h g^{-1} at 200 mA g^{-1} with an average voltage of 2.8 V and cycling during 200 cycles [340].

Organic material explored as cathodes include a structured potassium copper hexacyanoferrate (CuHCF) with high redox-potential and a sufficiently large open channel structure that accommodates storage and diffusion of a Ca^{+2}. CuHCF nanoparticles showed a crystal lattice structure and delivered a capacity of 50 mA h g^{-1} at 300 mA g^{-1} with 94% capacity retention after 1000 cycles [190].

Furthermore, sulfur and air cathodes have been proposed for Ca batteries. Early work in Ca/S batteries focused on non-rechargeable systems due to irreversible processes; these systems can achieve a capacity of 600 mA h g^{-1} [341].

Cathode materials for calcium-based batteries are summarized in Table 12. We include the highest reported voltage range, capacity, current density, number of cycles, and retention.

Table 12. Cathode materials for Ca-based batteries.

Cathode Material	Voltage Range (V)	Capacity (mA h g^{-1})	Current (mA g^{-1})	Cycles	Retention (%)
Carbon					
Graphite [189]	4.45	70		350	95
Layered Oxide					
CaCoO [334]	3	150		30	
MoO$_3$ [337]	1.3	180	2	12	~50
V$_2$O$_5$ [338]	3.2	465	200 [340]	200 [340]	
Layered sulfide					
TiS$_2$ [335]	1.5	520	1/50 [a]	1	
S/meso -C [341]	0.75	600			
Polyanionic compound					
VOPO$_4$·2H$_2$O [340]	2.8	100	200	200	
Prussian blue analogue					
K$_2$BaFe(CN)$_6$	0–0.8	55.8		100	
Na$_{0.2}$MnFe(CN)$_6$ [336]	0–3.5	70		35	
Organic					
CaCuHCF [190]		50	300	1000	95

[a]: C-rate. PANI: polyaniline, PAQS: polyanthraquinonyl sulfide, PTCDA: perylene-tetracarboxylicacid-dianhydride, NB: nanobelts, NF: nanofibers, NR: nanorod, NW: nanowires.

4.6. Cathodes for Magnesium-Based Batteries

The cathodes discussed for magnesium batteries include Chevrel phase, spinel, layered materials (Mn oxides, and V-based), polyanion compounds, Prussian blue, and organic materials. Finding a stable cathode material for Mg batteries has resulted in some challenges due to the difficulty in entering and diffusing the inorganic materials by the divalent cation (Mg^{+2}), and the necessity to research compatible electrolytes with these cathodes [8,47].

Mo$_6$S$_8$ is a Chevrel phase (CP) intercalation material that showed for the first time the reversibility of Mg^{+2} ions. Its structure has a quasi-simple-cubic packing of the Mo$_6$S$_8$, which forms 3D channels available for Mg^{+2} transfer. With the high mobility of Mg^{+2} and fast interfacial charge transfer, Mo$_6$S$_8$ is the most successful cathode material at room temperature to date, exhibiting excellent intercalation kinetics and reversibility with a capacity of 120 mA h g^{-1} at 1.2 V [342]. The electrochemical performance of nanosized and microsized Mo$_6$S$_8$ has also been researched to improve the voltage and capacity.

Spinel materials as cathodes for Mg-based batteries have the general formula MgT$_2$X$_4$, where X can be O, S, or Se, and T is a transitional metal such as Ti, V, Cr, Mn, Fe, Co, or Ni. The challenge of using spinel cathodes is the low ion mobility and difficult intercalation reversibly at room temperature.

Layered materials are an alternative for cathodes. For example, layered TiS$_2$ consists of stacking sequences of TiS$_2$ slabs, composed of stacking of close-packed two-dimensional triangular lattices of sulfur. Additional layered oxide materials studied include V-based, such as V$_2$O$_5$, and Mo- based, such as MoO$_3$. Some researchers suggest increasing the distance between layers in layered TiS$_2$ and the crystal volume of spinel TiS$_2$ to benefit Mg^{2+} mobility since layered and spinel TiS$_2$ are sensitive to the size of octahedral and tetrahedral sites.

Polyanionic compounds are an alternative for cathode materials that have a 1D diffusion channel. The olivine structure consists of a distorted hexagonal close-packed (hcp) framework of oxygen with tetrahedral sites occupied by P or Si and two distinct octahedral sites: 4a occupied by Mg and 4c occupied by M ions (M: Fe, Mn, or Co) [8]. These polyanionic compounds include phosphates, such as FePO$_4$, and silicates, such as MgMSiO$_4$.

Conversion cathodes for Mg-batteries are a promising option to achieve higher energy density than intercalation materials. In the conversion mechanism, chemical bonds are broken, and new ones are created during the insertion and extraction of Mg. Conversion materials include many transition metal oxides, sulfides, chloride, and organic compounds.

Although different Mg electrolyte systems have been developed, including producing high reduction–oxidation cycling efficiency, Mg rechargeable batteries are still far from a commercial reality. This is partly due to the lack of cathode materials, which can be operated at high positive voltages and support a usefully high energy density. The conventional transition-metal oxide cathodes for Li-based batteries are not effective for Mg^{2+} ion insertion due to the slow diffusion of these divalent cations; their high charge density leads to strong electrostatic interactions with the host lattice [343].

Cathode materials for magnesium-based batteries are summarized in Table 13. We include the highest reported voltage range, capacity, current density, number of cycles, and retention.

Table 13. Cathode materials for Mg-based batteries.

Cathode Material	Voltage Range (V)	Capacity (mA h g^{-1})	Current (mA g^{-1})	Cycles	Retention (%)
Intercalation					
Mo_6S_8 [342]	1.3	122	15 [a]	3000	93
MoO_3 [344]	1.8	210			
MoOF [345]		70			
MoVO [346]	2.1	235			
Ti_2S_4 [347]	1.2	200	0.2 [a]	40	
TiS_2 [348]	1.5	158	24	400	95
TiS_3 [349,350]	1.2	83.7	10	50	
$TiSe_2$ [351]	1	110	5	50	
V_2O_5 [352]	2.56	150			
VSe_2 [351]	1	110			
Polianionic compound					
AgCl [353]	2	178	930	100	
CuS [354,355]	1.6	200	50	30	
Cu_2Se [356]	1.2	230	5	35	
$FePO_4$ [357]	2	15		20	
MnO_2 [358]	2	150			
Silicate [359]	4				
Organic					
DMBQ [360]	2	100	0.2 [a]	30	
PAQ [361]	1.7	150	130	100	
Sulphur					
S [362]	1.77	600	200	100	
S-ACCS [363]	1.5	950		100	48
S-C [364]	1.1	1081		30	76
S-CMK [365]	1.6	800		100 [366]	50 [366]
S-CNT [367]	1.3	1200		100	83
S-rGO [368]	1.5	1028		50	21
Oxygen					
O [369,370]	2.9	1300			
I [371]	2	200			

[a]: C-rate, CNT: carbon nanotubes, DMBQ: dimethoxy-1,4-benzoquinone, PAQ: olyanthraquinone.

4.7. Cathodes for Aluminum-Based Batteries

The cathodes discussed for Al batteries include carbon-based, layered (TMO, TMS, TMF), spinel, polyanionic compounds, and organics.

Carbon paper, which is composed of graphite, has been explored as a cathode material for Al-based batteries [372], showing a voltage plateau of 1.8 V. Furthermore, a discharged capacity of 69.92 mA h g^{-1} is achieved experimentally at a current density of 100 mA g^{-1} during 100 cycles. To improve the low discharge voltage, and low cycle life, pyrolytic graphite (PG) has been studied [373]. This graphite cathode has shown a cycling life of over 7500 cycles without capacity decay, a discharge voltage plateaus around 2 V, and a specific capacity of 60 mA h g^{-1} to 66 mA h g^{-1} at 4000 mA g^{-1} with high Coulombic efficiency (~98%). The use of Al/PG exhibited a limited rate capability with lower specific capacity when charged, and discharged at a rate 65 mA g^{-1}. Another alternative cathode studied is the 3D graphitic foams [374], which have a porous graphitic structure to facilitate ion diffusion and intercalation–deintercalation kinetics, and to increase the battery power density and rate capability. Wu et al. reported a new method of synthesizing a monolithic 3D graphitic foam (3DGF) containing aligned few-layered Graphene sheets with a low density of defects or oxygen groups. The tested Al/3DGF battery showed a high current density up to 12 A g^{-1} and a plateau voltage of 1.8 V. Moreover, it delivers a discharge capacity of 60 mA h g^{-1} with high Coulombic efficiency (~100%) and high capacity retention (~100% after 4000 cycles).

To improve the cathode performance, another material explored is a defect-free graphene aerogel (GA) [375]. The GA cathode design exhibits crystallized carbons in the atomic structure, eliminating the inactive defects and improving the fast intercalation of large-sized anions. The electrochemical tests showed a capacity of 100 mA h g^{-1} at a current density of 5000 mA g^{-1} with an average voltage of 1.95 V and a capacity retention of 97% after 25,000 cycles. Moreover, it can deliver a capacity of 97 mA h g^{-1} at 50 A g^{-1}. The authors suggest that the material quality and cell performance are highly reproducible, favoring large-scale manufacture. Recently, researchers developed a free-standing graphitic nanoribbon interconnect nanocup stack as a cathode to avoid the risk of side reactions and electrode pulverization [376]. This binder-free material is a flexible electrode that exhibits a capacity of 126 mA h g^{-1} at a current density of 1000 mA g^{-1}, and it is capable of delivering a capacity of 95 mA h g^{-1} at a high current density of 50 A g^{-1}. In addition, this material was tested for long-term cycling stability, 20,000 cycles at a current density of 10 A g^{-1}.

In 2017, the fabrication of mesoporous Li$_3$VO$_4$/C hollow spheres composite as a cathode material was reported for the first time. The structure of Li$_3$VO$_4$ can be estimated as a hollow lantern-like 3D structure, which consists of orderly corner-shared VO$_4$ and LiO$_4$ tetrahedrons, and with many empty sites in the hollow lantern-like 3D structure to accommodate ions inserting reversibly. The Al/Li$_3$VO$_4$/C battery showed an initial discharge capacity of is 137 mA h g^{-1} at a current density of 20 mA g^{-1} and remains at 48 mA h g^{-1} after 100 cycles with high Coulombic efficiency (~100%) but a low voltage discharge plateau (0.5 V) [377].

Mo oxide is another alternative studied as a cathode material to improve the voltage in Al-based batteries. For example, a dense Mo oxide layer was fabricated on Ni foam (MoO$_2$/Ni) [378]. Experimentally, the electrochemical performance demonstrated a discharge potential of 1.9 V, which is higher than most of the studied metal oxide cathodes of Al batteries. In addition, MoO$_2$/Ni showed a specific discharge capacity of 90 mA h g^{-1} at a current density of 1000 mA g^{-1}. A disadvantage of using MoO$_2$/Ni is the rapid capacity decay as a result of the MoO$_2$ being dissolved and transferred to the separator after long cycling.

Metal sulfides were also suggested as cathodes in Al batteries. For instance, a 3D hierarchical copper sulfide (CuS) microsphere composed of nanoflakes [379] showed an average discharge voltage of 1.0 V and a reversible specific capacity of about 90 mA h g^{-1} at 20 mA g^{-1} with high Coulombic efficiency (~100% after 100 cycles). The remarkable electrochemical performance results from both the particular crystalline structure and uniform nanoflakes, facilitating the electron and ion transfer. Layered TiS$_2$ and spinel-based cubic Cu$_{0.31}$Ti$_2$S$_4$ are other alternative metal sulfides tested, but they showed low performance and slow diffusion of Al^{3+} [380]. Alternatively, porous microspherical CuO (PM-CuO) composed of stacked nanorods was synthesized as the cathode for improving the electrochemical performance thanks to its porous features [381]. This material delivers a discharge capacity of 250.12 mA h g^{-1}, and it is capable of maintaining 130.49 mA h g^{-1} after 100 cycles, which is better than the Li$_3$VO$_4$.

Sulfur is another attractive conversion material for cathodes in Al-based batteries. However, elemental sulfur, as a positive electrode material, may have two critical challenges: (1) batteries with a notorious dissolution of multisulfide compounds, which results in the loss of electrochemically active species; (2) low kinetics owing to the insulating nature of elemental S, which inhibit the reversible electrochemical reaction between Al and S. Some examples to address these challenges are S/mesoporous carbon (S/CMK-3), which can deliver a capacity of 1500 mA h g^{-1} at a current density of 251 mA g^{-1} [382], and using a Li-ion mediation strategy [383].

Cathode materials for aluminum-based batteries are summarized in Table 14. We include the highest reported voltage range, capacity, current density, number of cycles, and retention.

Table 14. Cathode materials for Al-based batteries.

Cathode Material	Voltage (V)	Capacity (mA h g^{-1})	Current (mA g^{-1})	Cycles	Retention (%)
Carbon					
3D graphitic foams [374]	1.8	60	12,000	7500 [373]	100
C paper [372]	1.8	85	50		
C NF [376]		126	1000	20,000	100
C nanoscrolls [376]		104	1000	55,000	100
Defect-free G [375]	1.94	100	5000	25,000	97
G film [384]	2.3	240 [385]	6000	250,000	91.7
G microflower [386]	1.85	92	100	5000	100
G nanoribbons [387]	2	148	2000	10,000	98
kish Graphite flakes [388]	1.79	142	50	200	100
mesoporous rGO [389]		120	20	100	85
Zeolite-templated C [390]	1.05	382	50	1000	86
Metal Oxides					
Li$_3$VO$_4$/C [377]	0.5	137	20	100	35
MoO$_2$ [378]	1.95	90	100	100	28
Mo$_{2.5+y}$VO$_{9+x}$ [391]	0.75	340	2	25	70
VO$_2$ [392]	0.5	116	50	100	70
V$_2$O$_5$ [393]	0.6	239	44.2		
V$_2$O$_5$/C [394]	1	200	10	15	66
V$_2$O$_5$ nanowires [49]	0.55	305		20	78
Metal sulfide					
CuS/C MS [379]	1	90	20	100	32
Cu$_{0.31}$Ti$_2$S$_4$ [380]		95	5	50	16
Hexagonal NiS NB [395]	0.9	105	20	100	100
Ni$_3$S$_2$/G [396]	1	350	100	100	17
Porous CuO MS [381]	0.6	250	50	100	45
Mo$_6$S$_8$ [397]		80	6	50	47
SnS$_2$/rGO [398]	0.68	392	100	100	25
Sn$_S$ porous film [399]	1.1	406	20	100	91
TiS$_2$ [380]		50	5	50	72
VS$_4$/rGO [399]		407	100	100	20
Sulfur					
S/ACC [400]	0.75	1320	500	20	
S/CNF [382]	0.76	600	21	20	
S/mesoporous C [383]	0.5	400	251		
S powder [401]	1.2	1300	50	20	

ACC: Activate Carbon cloth, C: Carbon, CNT: Carbon, nanofiber, G: Graphene, MS: microsphere, NB: Nanobelts, NF: Nanofibers.

Figure 10 compares the current research in cathode materials for each battery type, presenting the theoretical capacity, experimental capacity, current density, and the number of cycles for each material. In Figure 11, the evolution and progress of the specific capacity and the cathode materials in the research field are shown. The color of the line represents the type to which the cathode belongs. Continuous lines represent intercalation cathode materials, while dot lines represent conversion cathode materials.

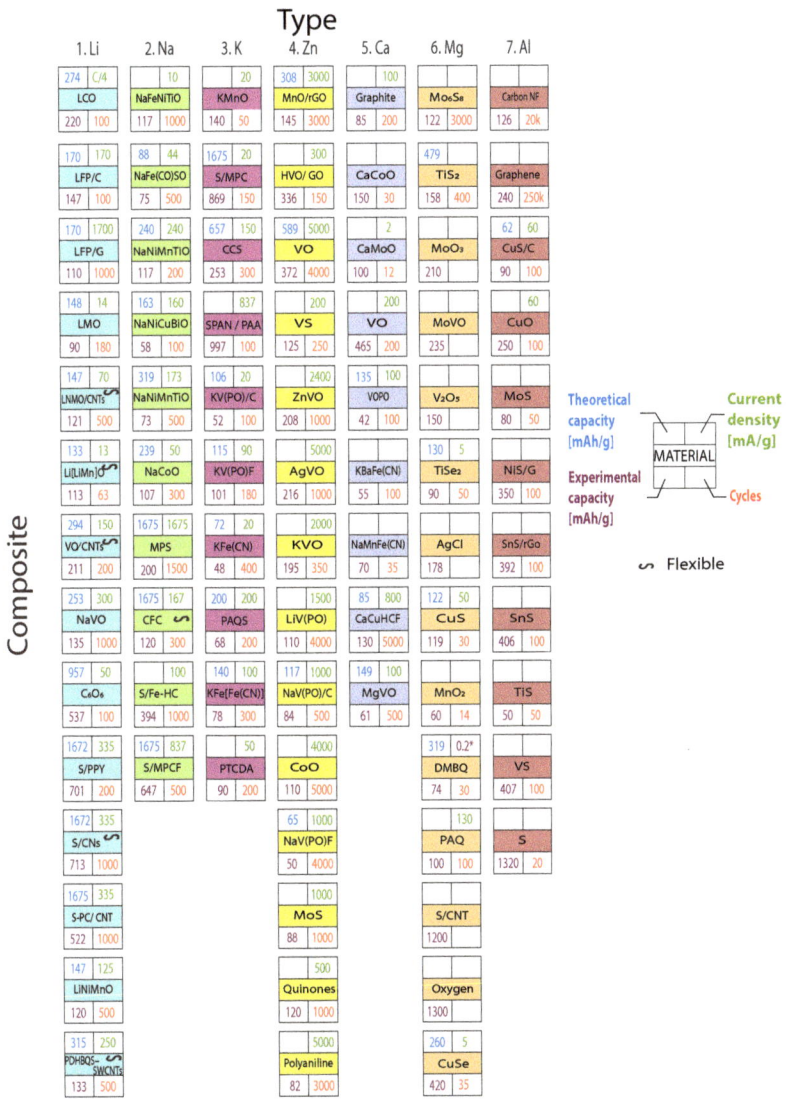

Figure 10. Current research in cathode materials. Columns organize the seven battery types. Rows represent the composite material group. Theoretical capacity, experimental capacity, current density, and the number of cycles are also included. A special symbol highlights flexible anode materials. Graph constructed by the authors from references in Tables 8–14.

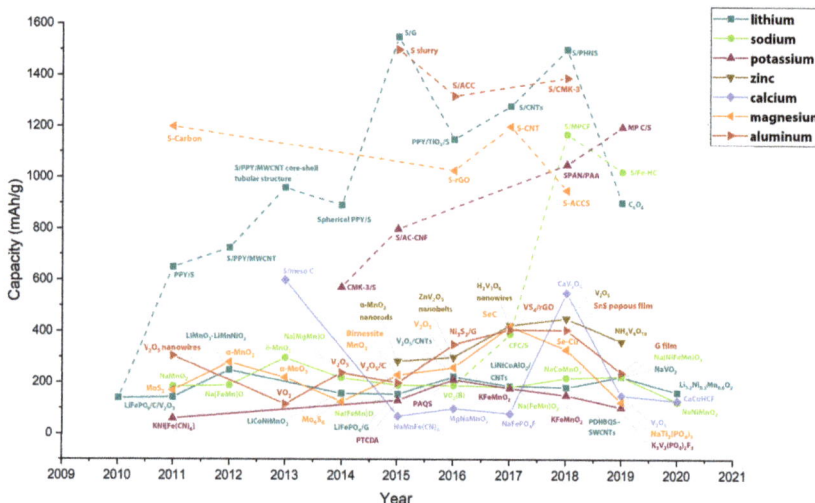

Figure 11. The advance of the cathodes' specific capacity, comparing the seven battery types: Li, Na, K, Zn, Ca, Mg, and Al. Continuous line: traditional intercalation materials. Dot lines conversion materials. Graph constructed by the authors from references in Tables 8–14. NW, nanowires; NR, nanorods; NB, nanobelts.

5. Electrolyte and Separator

The electrolyte has three main functions in a battery. First, it electronically separates the anode and the cathode. Second, the electrolyte must efficiently transport the ionic charge carrier of interest between these electrodes. Third, it should support a continuous reduction/oxidation process, satisfying the needs of both electrodes [15,21]. Therefore, an ideal electrolyte requires the following characteristics to improve battery performance:

1. Chemical inertness toward inactive and active battery components.
2. Thermal stability with low melting and high boiling temperatures.
3. Electrochemical stability window.
4. High ionic conductivity and no electronic conductivity.
5. Environmental friendliness and nontoxicity.
6. Sustainable chemistry.
7. Simple synthesis, preparation, and scaling processes.
8. Tunable interphase property on both electrodes.

Electrolytes can be classified as a liquid [13,191,402,403] or a solid state [404–406]. Liquid electrolytes integrate a liquid solvent (Table 15) and an electrolytic salt (Table 16). The complete dissociation of the electrolytic salt results in excellent compatibility and high ionic conductivity [404]. In addition, it has been demonstrated that salt concentration in electrolytes influences ionic conductivity and stability. Recent research suggests that highly concentrated electrolytes could improve the ionic conductivity due to their low free solvent molecules. This also helps to reduce the formation of dendrites since this improves the interface between the electrolyte and the electrode [407,408]. The challenges of using highly concentrated electrolytes are the high viscosity and high cost.

Liquid electrolytes have some safety issues including short circuits, which could generate an overheating and battery explosion, and an unstable electrolyte/electrode interface results in electrode inactivation. One alternative explored to address these issues is the use of binders and polymers in electrolyte solutions [409] (Table 17). Solid-state electrolytes (SSE) are versatile films with good viscoelasticity. They can solve the liquid electrolytes problems with no exposure to volatile organic solvents, and they have good mechanical stability, flexibility, and prevent dendrite growth [404,406]. Nevertheless, the main drawbacks of SSE are ionic conductivity at room temperature (10^{-8} S cm^{-1} to 10^{-5} S cm^{-1}), which is lower than in liquid electrolytes ($\sim 10^{-3}$ S cm^{-1}), and has a low compatibility with electrodes (Table 18).

Liquid electrolytes can be classified as aqueous [53,191,322] or non-aqueous [256,402,410] electrolytes depending on the type of solvent and salts. Aqueous solvents are based on water, while non-aqueous solvents are based on organic liquids. Water-in-salt electrolytes offer a high ionic conductivity, but they are limited by their narrow electrochemical window [411]. A strategy to increase the electrochemical window consists of increasing the concentration, which can reach a stable voltage window from ~ 1.2 V to ~ 3.0 V with a salt/solvent concentration ratio > 1 by volume or weight.

Organic liquid electrolytes include materials such as standard organic solvents, fluorinated carbonates, sulfones, and nitriles. The solvents must fulfill specific battery requirements, such as high fluidity, a high dielectric constant, large electrochemical stability, a wide liquid range, high thermal stability, low vapor pressure, incombustibility, low viscosity, and high ionic conductivity [17]. Common solvents which have high ionic conductivity, good compatibility with commercial electrodes, and a long cycle include ethylene carbonate (EC), propylene carbonate (PC), diethyl carbonate (DEC), and ethyl methyl carbonate (EMC). These solvents come with some technical challenges, such as low thermal stability, low electrochemical stability, and safety problems. To overcome these challenges, Ionic liquid (IL) electrolytes have been researched. IL incorporates salts that melt at room temperatures or below. These salts have properties such as nonflammability, negligible vapor pressure, remarkable ionic conductivity, high thermal and chemical and electrochemical stabilities, low heat capacity, and the ability to dissolve inorganic, organic, and polymeric material [412].

SSE can be made from polymers [405,406] or ceramics [413–415]. Solid polymer electrolytes (SPEs) have been explored as a replacement for liquid electrolytes to overcome flammable and volatile behavior in liquid electrolytes. Although SPEs are a versatile material to form films with good viscoelasticity (e.g., polyethylene oxide (PEO)) the main challenges of SPEs are low ionic conductivity at room temperature (10^{-8} S cm^{-1} to 10^{-5} S cm^{-1}), low ion transference number (ITN< 0.3), and poor electrochemical stability [416]. On the other hand, solid ceramic electrolytes have a high stability in contact with metal anodes and have wide electrochemical windows. The use of ceramic materials is limited by their inherent fragility, reducing its application as a flexible electrolyte [413]. Solid ceramic electrolytes can be divided into two types: oxide- and sulfide based.

Polymer electrolytes can be classified into gel polymer electrolytes and dry polymer electrolytes. Gel polymer electrolytes are made by dissolving the salts into a polar solvent or an ionic liquid and by adding it to a polymer host, creating a composite. Although this composite does not exhibit porosity, it does have ionic conductivity, electrochemical stability, a melting temperature, wettability, and mechanical properties. The salts in the polymer matrix allow the ionic conduction in polymer electrolytes. Polymer materials include polyvinylidene fluoride (PVDF), polyvinyl chloride (PVC), poly ethylene oxide (PEO), SCT, polyethylene glycol bis-carbamate PEGBCDMA, and polytetrahydrofuran (PTHF). According to the polymer-solvent-ion, gel polymer electrolytes can have two kinds of interactions: (1) strong polymer-solvent interactions, which have high chemical stability and low ion transport; or (2) weak polymer–solvent interactions, which have low chemical stability and fast ion transport. Dry polymer electrolytes are a solvent-free polymer–salt system composed of electrolytic salt dissolved in a polymer matrix. These electrolytes open the door to safety, flexibility, robustness, novel thicknesses and shapes, and new possibilities of use for electrochemical devices.

Table 15. Liquid electrolyte solvents.

Cyclic Carbonates	Organic Fluorinated Carbonates	Sulfones	Ionic Liquid
ACN	MFAa	EMSa	EMI
γ−BL	FPCa	TMS	DMPI
DEC	EMSa	FS	DEDMI
DMC	TFPMSa	BS	TMHE
DME	GLNb	EVS	PYR
DMF	ADNc	ADN	PIP
DMSO	SENa	DMMP	MORP
EC		TMMP	TFSI
EMC		BC	BETI
MF			FSI
NM			TSAC
PC			FSA^-
THF			$TFSA^-$
VC			BF_4^-
diglyme			PF_6^-
tetraglyme			$N(CN)_2^-$
DGM			$[BH_4]^-$
UREA			PEGylated

Table 16. Salts for electrolytes.

Li	Na	K	Zn	Ca	Mg	Al
LiTFSI	NaTFSI	KFTFSI	$Zn(TFSI)_2$	$Ca(TFSI)_2$	$MgTSFI_2$	$AlCl_4$
$LiClO_4$	$NaClO_4$	$KClO_4$	$Zn(ClO_4)_2$	$CA(ClO_4)_2$	$MgCl_2$	
LiBOB	NaBOB	KPF_6	$Zn(CF_3SO_3)_2$	$Ca(PF_6)_2$	$Mg(CB_{11}H_{12})_2$	
$LiPF_6$	$NaPF_6$	KBF_4	$ZnSO_4$	$Ca(BF_4)_2$	$Mg(BBu_2Ph_2)_2$	
$LiBF_4$	$NaBF_4$	KFSI		$Ca(NO_3)_2$	$Mg[B(HFIP)_4]_2$	
$LiAsF_6$	NaFSI	KCF_3SO_3		$Ca(BH_4)_2$	$Mg(BH_4)_2$	
$LiOSO_2CF_3$	NaFTFSI					
	NaOTf					
	NaDFOB					

Table 17. Additive materials for electrolytes. Source from [412].

Additive	Boiling Point (K)	Density (g cm^{-3})
TEPa	488.15	1.072
TMPa	453.15	1.197
TFPa	355.15	1.594
DMMPb	453.15	1.145
DMMEMPc	553.15	
MFEa	333.15	1.529
MFAa	358.15	1.272

The separator is a porous membrane located between negative and positive electrodes, serving as the physical separation between the anode and the cathode [417]. A separator function is a membrane which helps to avoid short circuits and control the movement of ions from/to electrodes. An ideal separator for batteries should have high ionic conductivity and excellent mechanical and thermal stability.

Table 18. Ionic conductivity of common solid-state electrolytes.

Polymer	S cm^{-1}	Ceramic	S cm^{-1}
PEO	10^{-8} to 10^{-6}	LISICON	10^{-5} to 10^{-3}
PMMA	10^{-4} to 10^{-5}	NASICON	10^{-5} to 10^{-3}
PAN		Garnet	10^{-5} to 10^{-3}
PVdF		Perovskite	10^{-5} to 10^{-3}
PVdF-HFP		Sulfide	10^{-7} to 10^{-3}
PVdF-TrFE		LiPON	10^{-6}
PPO			
PVA			
PAM			
PNA			
PAA			
PNIPAM			

Critical separator characteristics that influence properties are: thickness, weight, ionic conductivity, porosity, pore size, and wettability (Figure 12). Thickness, which refers to a film's width, and weight affect energy and power density as well as the swelling process and mechanical properties. Ionic conductivity is a measure of the movement of an ion from one site to another, and it can vary according to porosity and the separator's morphology. Porosity is a measure of the holes (void spaces) in the separator, and it is estimated in three ways: (1) a fraction of voids' volume over the total volume, (2) between 0 and 1, or (3) as a percentage. The porous volume also affects the uptake of the electrolyte and the mechanical strength. Pore size is the dimension of the pore, and it is critical to reducing dendrite growth. Wettability is the separator surface's ability to maintain contact with a liquid, and it is essential for producing good ionic conductivity and rate capability of a battery.

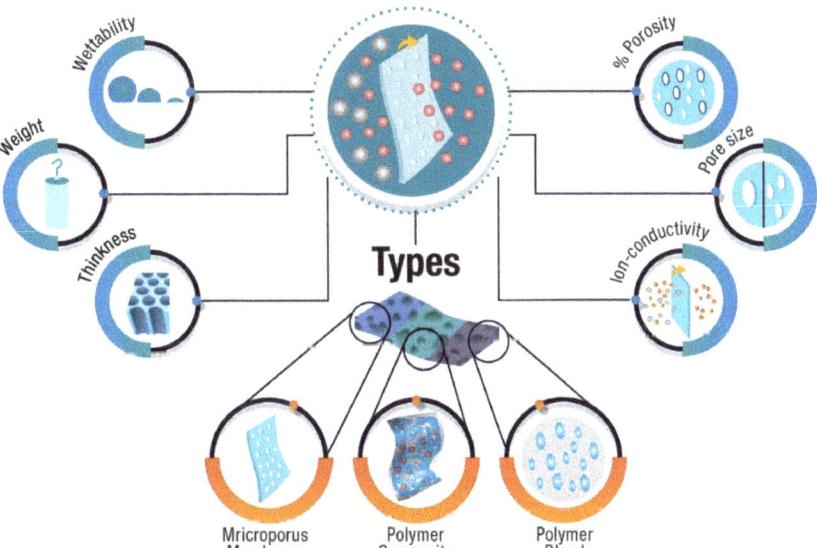

Figure 12. Critical characteristics of separator materials and their general classification. Graph constructed by the authors.

Common materials used for battery separators are polymeric and ceramic materials. Polymer materials are used in commercial applications since the swelling process, essential in obtaining a high ionic conductivity and high ion transport number in the separator membrane, is more efficient in polymers. The use of polymer separators is limited by their poor thermal stability and wetting ability. Ceramic materials have higher thermal stability than polymerics, but their use in commercial applications is limited by poor mechanical stability, scalability, and high production costs [17].

Separators can be divided into three main types: microporous membranes, composite membranes, and polymer blends [17]. Microporous membranes are commonly composed of porous polymer layers. These membranes can be classified into four categories according to their fabrication method, structure (pore size and porosity), and composition. These four categories are: (1) single layer membranes that include membranes with a porosity between 20 % to 80 % and a pore size <2 µm; (2) non-woven membranes with a porosity between 60 % to 75 % and a pore size >6 µm; (3) electrospun membranes that possess a high porosity above 70 % and a pore size >5 µm; and (4) membranes with an external surface modification to develop specific properties by coatings, plasma treatment, polymer grafting, and chemical modification.

In addition, composite membranes are analogous to the microporous membrane but include micro/nanofillers inside the polymer to improve its thermal, mechanical, or electrochemical properties. Common materials used as fillers are inert ceramic oxides (SiO_2, TiO_2, Al_2O_3), ferroelectric materials ($BaTiO_3$), super acid oxides ($CaCO_3$, AlI_3, $AlPO_4$, Fe_2O_3, $Zr-O-SO_4$, BN, SN, NiO, CuO, nano$-ZnO$), clays (MMT), carbonaceous fillers (CNT), molecular sieves and zeolites (ZSM, NaY, SBA-15, MCM-41, MOF-5). Moreover, filler types are divided into two main classes: passive and active classes. The difference is that passive fillers do not directly contribute to the conduction process, while the active fillers participate in the conduction process.

Finally, polymer blend membranes incorporate two or more polymers to improve separator performance through complementary properties. Common polymers for polymer blends are PVDF, P(VDF-HFP), PEO, PAN, PMMA, PE, and PVC.

5.1. Lithium Batteries

Liquid electrolyte solutions for commercial and research Li-based batteries usually contain Li salts, such as lithium hexafluorophosphate ($LiPF_6$), lithium tetrafluoroborate ($LiBF_4$), lithium perchlorate ($LiClO_4$), lithium hexafluroarsenate ($LiAsF_6$), lithium trifluoromethane-sulfonate ($LiCF_3SO_3$) or lithium bis (trifluoromethanesulfonimide $Li(CF_3SO_2)_2N$) (Table 16), and non-aqueous solvents, such as DME, DMC, DEC, PC, EC and γ-butyrolactone (BL) (Table 15).

$LiPF_6$ is used in commercial LIB, and it has been shown to be a good ion conductor (2.4 mS cm^{-1} to 6.2 mS cm^{-1}) with EC:DEC or PC solvents, and has a high level of solubility in organic plasticizers [418,419]. The main challenge of using $LiPF_6$ is poor thermal stability since the decomposition reaction of $LiPF_6 = PF_5+LiF$ is above 333.15 K. To improve $LiPF_6$ thermal and hydrolytic stability, $LiBF_4$ has been studied. The use of $LiBF_4$ is limited in practical applications by a low ionic conductivity (3.7 mS cm^{-1}) due to its poor ion dissociation, a narrow electrochemical stable window (3.5 V), and the lack of a stable solid electrolyte interface (SEI) layer [419].

Electrolytes are also researched to improve the performance of novel materials and chemical composition. For example, N-butyl-N-methylpyrrolidinium bis(fluorosulfonyl) imide ($Pyr_{14}FSI$) has been studied to reduce the capacity and voltage fading in cobalt-free material such as $Li_{1.2}Ni_{0.2}Mn_{0.6}O_2$ [420]. $_{0.8}Pyr_{14}FSI-_{0.2}LiTFSI$ was synthesized to evaluate this impact on the cathode material. As a result, the electrolyte abolished the structural modification of the cathode that is caused by side reactions, improving the cycling stability when compared with conventional carbonate-based electrolytes. The tested battery with this ionic liquid electrolyte showed a capacity of 219 mA h g^{-1} at a current of 25 mA g^{-1} and of 144 mA h g^{-1} at 250 mA g^{-1}, and it maintains a 56 % of retention capacity after 2000 cycles.

Current liquid electrolytes have some technical (thermal stability) and safety challenges (flammability and explosibility) that reduce the performance and cycle life of batteries. Attempts to overcome these challenges, maintaining the ion conductivity, focus on a new generation of IL and SSE. For example, IL crystal-based electrolytic lithium salt (LiBIB) has shown both ionic liquid and liquid crystal behaviors, being an alternative to working as a salt or as a polymer solvent-free electrolyte [402]. Experimental results have shown that LiBIB has a melting point at 316.15 K, a good ionic conductivity (3 mS cm^{-1}), and electrochemical stability at room temperature, which are associated with the formation of fast ion-conductive tunnels.

In addition, to address mechanical requirements for novel applications (flexibility), SSEs have been researched. For example, LiBIB-like ionic liquid crystal lithium salt (LiBMB) has been combined with flexible poly (ethylene oxide) (PEO) to fabricate SSEs [404]. This material shows a maximum ionic conductivity of 0.45 mS cm^{-1}, a Li-ion transference number of 0.54, and a diffusion coefficient 2.33×10^{-7} cm at 303.15 K. SSEs are an alternative to volatile organic solvents because they do not leak, exhibit good mechanical stability and flexibility, and prevent Li dendrite growth. The main obstacle is their lower ion-conductivity than liquids, increasing the internal resistance in the battery.

SSEs researched for Li-based batteries also include ceramic materials that can be classified into oxide- and sulfide-based materials. Oxide-based SSEs have been shown to have good chemical stability against Li metals and good ionic conductivity. For example, Li superionic conductors (LISICON) have high ionic conductivity at an elevated temperature (10^{-1} S cm^{-1} at 773.15 K. The practical application of LISICON is limited at room temperature due to the low ionic conductivity. Other oxide SSEs are Garnet, Perovskite-type, and Na superionic conductors (NASICON). Sulfide-based SSEs have shown superior ionic conductivity and better compatibility with electrode materials due to the lower bonding strength between S and Li$^+$. One problem related to these materials is the generation of gases if they contact oxygen and water molecules (H$_2$S).

Separators for Li-based batteries are common polymers in a single layer membrane: polyethylene (PE), polypropylene (PP), polyethylene oxide (PEO), polyacrylonitrile (PAN), polymethyl methacrylate (PMMA), polyvinylidene fluoride (PVDF), polyvinylidene fluoride-co-hexafluoropropene (PVDF-HFP), and polyvinylidene fluoride-co-trifluoroethylene (PVDF-TrFE). These polymer types do not exhibit high thermal resistances. Alternative polymers have been explored, such as polyimide (PI), polym-phenylene isophthalamide (PMIA), polyether-ether-ketone (PEEK), polybenzimidazole (PBI), polyetherimide (PEI) and polystyrene-b-butadiene-b-styrene (SBS). In addition, eco-friendly polymers, such as cellulose, chitin, silk fibroin, and poly(vinyl alcohol) (PVA), have been proposed. These eco-friendly materials use water as the solvent and regenerate and recyclable eggshell membranes to suppress dendritic lithium growth.

5.2. Sodium Batteries

Electrolytes for Na-based batteries include carbon-based, ionic liquid, and SSE. Similar to electrolytes for Li-based batteries, liquid electrolytes in Na-based batteries join a Na-based salt, such as NaTFSI, NaClO$_4$, NaBOB, NaPF$_6$, NaBF$_4$, NaFSI, NaFTFSI, NaOTf, and NaDFOB Table 16, and a non-aqueous carbonate-ester-based (DMC, DEC, PC, EC and ECM) or ether-based (TEGDME, DEGDME, and DME) solvent Table 15. Carbonate solvents have been used mainly due to their higher electrochemical stability. However, these carbonates decompose at Na metal electrodes, reducing the Coulombic efficiency. Therefore, these solvents are mixed, creating a binary (e.g., EC:DEC, EC:PC, and EC:DMC) solvent, or using an additive (e.g., fluoroethylene carbonate (FEC)) [421]. For example, a multifunctional electrolyte incorporating 2 M NaTFSI in PC:FEC (1:1) has been explored to manufacture room temperature Na-S batteries. FEC and high salt concentration reduce Na polysulfides' solubility and create a robust SEI on the Na anode upon cycling [422]. On the other hand, ether-based solvents have enabled the Na$^+$ co-intercalation in graphite electrodes and the development of a stable SEI [409]. In addition, ether-based electrolytes could improve the reversibility of the Na metal in the stripping/deposition process, helping non dendritic growth.

Ionic liquid electrolytes consist of organic ions that allow unlimited structural variation and tune the properties, and they are focused on imidazolium, and pyrrolidinium [409].

Another alternative explored is aqueous electrolytes since they have a low cost, elevated safety, and are environmentally friendly [409]. The main obstacle to using water as a solvent is its electrochemical decomposition which hinders the selection of the electrode materials for practical applications, and also requires the elimination of residual O_2 in the electrolyte, protecting electrode stability, inhibiting H_3O^+ co-intercalation into the electrode, and keeping efficient internal consumption of O_2 and H_2 produced at the cathode and anode sides.

SSEs researched for Na-based batteries have focused on polymer electrolytes, which can be categorized as dry polymers and gel polymers. The dry polymers which have been studied are poly (ethylene) oxide (PEO), poly(vinyl pyrrolidone) (PVP), polyvinyl chloride (PVC), poly(vinyl alcohol) (PVA), polyacrylonitrile (PAN), and polycarbonates. The main challenge of dry polymers is the low ionic conductivity at room temperature. On the other hand, gel polymers integrate a polymer and a liquid component that serves as a plasticizer, resulting in a material with properties between liquid and dry polymer electrolytes, such as ion conductivity in the range of $10^{-3}\,\text{S cm}^{-1}$. Polymers explored for gel polymer electrolytes include PEO, PAN, perfluorinated sulfonic membranes (NAFION type), poly(methyl methacrylate) (PMMA), and polyvinylidene fluoride (PVDF) [409].

5.3. Potassium Batteries

Electrolytes for K-based batteries face a low ionic conductivity due to the poor interaction between K-ions and solvents that lead to an insufficient solubility of K-salts. The K-salts that have been researched include: potassium bis-(fluorosulfonyl)imide KFTFSI, $KClO_4$, potassium hexafluorophosphate KPF_6, KBF_4, KFSI, and KCF_3SO_3. KPF_6 and KFSI have shown sufficient solubility ($0.5\,\text{mol L}^{-1}$ to $1.0\,\text{mol L}^{-1}$) in solvents such as EC, DEC, PC, and DME. Currently, there continues to be an inefficiency of systematic study on nonaqueous KIB electrolytes [410].

Similar to Li- and Na-based batteries, liquid electrolytes researched in K-based batteries include ester- and ether-based electrolytes. Ester-based electrolytes (e.g., in acetonitrile (AN), EC and DMC) have shown higher ionic conductivity for K^+ than for Li^+ and Na^+ ions, but they have little applicability due to the low solubility of salts such as KPF_6 (~0.8 M) [423]. Efforts to adjust transport properties of KPF_6 are directed at the low percentage of ion-pair formation, its hydrodynamic radius in the solvated state, the shape of the anion, and the KPF_6 low viscosity in EC/DMC and AN.

Another challenge that should be addressed is the dissolution and shuttle reactions of the redox intermediates. One explored strategy to mitigate this effect is to increase the concentration of salt. For example, increasing the concentration of KTFSI to 5M in DEGDME (<1.5M in conventional ether-based electrolytes) suppresses the dissolution and shuttle reactions of K_2S_x intermediates, and enables a full operation of K-S battery chemistry in a voltage range of 1.2 V to 3.0 V [289]. More research needs to be performed to improve the stability and cycle life.

SSE research on K-based batteries includes mostly polymer materials. For example, a poly (methyl methacrylate) (PMMA) polymer-gel infiltrated with KPF_6 in EC/DEC/FEC has been explored [405], showing an ion conductivity similar to that of liquid electrolytes ($4.3 \times 10^{-3}\,\text{S cm}^{-1}$), and preventing the growth of potassium dendrites due to cross-linked PMMA architecture produces adjustable pore sizes to form stable solid-electrolyte interphases. Another polymer studied is the gel-polymer poly (ethylene oxide) (PEO), which is also able to work with different alkali ions (Li^+, Na^+, K^+). As a result of integrating the tailor-made star polymers in a functional PEO matrix, an ion conductivity of $9.84 \times 10^{-4}\,\text{S cm}^{-1}$ at 353.15 K to K^+ was achieved [416]. Finally, a potassium ferrite phase, $K_2Fe_4O_7$ was synthesized under hydrothermal conditions, and it exhibited an ionic conductivity of $5 \times 10^{-2}\,\text{S cm}^{-1}$ and an electrical conductivity of $3.2 \times 10^{-2}\,\text{S cm}^{-1}$.

5.4. Zinc Batteries

Electrolytes for Zn-based batteries are based on Zn salts, such as $Zn(TFSI)_2$, $Zn(ClO_4)_2$, $Zn(CF_3SO_3)_2$, and $ZnSO_4$, dissolved in an aqueous or non-aqueous solvent.

Aqueous electrolytes have been widely studied in rechargeable Zn-based batteries for large-scale energy storage applications since they are very safe, cheap, environmentally friendly, and have a higher ionic conductivity. The use of aqueous solvents results in a short cycle life in the batteries due to side reactions with the Zn electrode (corrosion, hydrogen evolution, passivation, and Zn dendrite growth), limiting their use in practical applications. The main strategy explored to optimize the electrolyte performance has been to include electrolyte additives [424–427]. For example, PEG has been studied as an additive in gel electrolytes since it suppresses the corrosion on Zn surfaces and promotes the (002) crystallographic orientation of Zn, reducing Zn dendrite growth and improving the cycle life [424]. The influence of the different amounts of PEG-200 has also been studied, showing that 20(v)% PEG-200 improves battery cycle life from 10 cycles to up to 100 cycles [425], and it reduces the corrosion current density by ~37 % and increases the capacity retention by ~12 % when compared with a reference battery [424]. Another alternative that has been explored is the use of graphene oxide (GO) as an additive in $ZnSO_4$ electrolyte to achieve a uniformly distributed electric field with reduced nucleation overpotential [426]. The results showed that the uniform electric field generated by GO's increased the battery's cycle life (650 h) compared to a reference battery without GO (96 h). An additional additive studied is diethyl ether (Et_2O) and ethylene glycol (EG), which is able to work at low temperatures (263.15 K to 273.15 K) [427]. Et_2O possess highly polarized molecules that are absorbed by Zn foils' protuberance, assuring a homogeneous Zn deposition and suppressing Zn dendrite growth. EG works as an anti-freezing agent by interacting with H_2O molecules via hydrogen bonds, obstructing the ice crystals formation and decreasing the freezing point. The challenge of using additives has been to find the optimal amount of additives to reduce dendrite growth and anode self-corrosion but still maintain electrolyte performance.

Non-aqueous solvents have been explored as an alternative to address the challenges of using aqueous solvents. For example, the $Zn(ClO_4)_2$ acetonitrile (AN) electrolyte was tested in a battery and showed a high reversibility of the metal deposition with a discharge capacity of 55.6 mA h g^{-1} at 0.2 C and high Coulombic efficiency (99.9 %) [428]. AN has also been tested with $Zn(TFSI)_2$ and has exhibited a good ionic conductivity (28×10^{-3} S cm^{-1}) without Zn dendrite growth that induces cycling stability (~1000 h) and high Coulombic efficiency (>99 %) [429].

The use of liquid electrolytes requires incorporating a separator membrane. Most of the studies on Zn-based batteries employ filter paper or glass fiber membrane as separators. An alternative membrane based on thermal-gated poly(N-isopropylacrylamide) (PNIPAM) has been explored in hydrogel electrolytes [430]. PNIPAM is a porous structure that can be controlled by temperature to restrict the migration of ions between electrodes at high temperatures (333.15 K) due to the formation of intramolecular hydrogen bonds shrinking the porous network. The practical application of PNIPAM is still limited by low-temperature windows, requiring additives to tune the phase transition temperature.

Another separator membrane that has been explored consists of mixed polyacrylonitrile (PAN) and lithium polysulfide (Li_2S_3) to create a cross-linked membrane that suppresses dendrite growth. PAN allows designing a mechanically robust skeleton, while Li_2S_3 enables sulfonyl functional groups to be hygroscopic and cationic selective transport characteristics, ensuring a uniform ionic flux distribution. This separator was tested in Zn/Zn symmetric cells, exhibiting 350 cycles with efficient dendrite growth suppression [431].

SSEs for Zn-based batteries are a promising alternative to address the demand for portable and wearable electronics [172]. Research on SSEs has been focused on integrating a concentrated salt with a gel polymer [298,432]. For example, a concentrated 3 M $Zn(CF_3SO_3)_2$ polyacrylamide gel electrolyte has been explored to achieve a durable and practical Zn battery system [298]. The designed concentrated gel electrolyte has been shown to have a high-voltage window, a wide operating temperature, and has been Zn dendrite-free. This gel electrolyte's thin film has 1 mm of thickness and can be manually stretched to as long as a 600% strain. Another concentrated gel polymer studied as an electrolyte is 21 M LiTFSI + 3 M ZnOTf$_2$ embedded in a PVA matrix [432]. As a result, the electrolyte exhibited an ion conductivity of $\sim$2.1 $\times$ 10^{-3} S cm^{-1} at room temperature (293.15 K), increased anode stability and window stability, and achieved thermal stability operating from 243.15 K to 333.15 K without structure modification.

5.5. Calcium Batteries

Liquid electrolytes studied for Ca batteries include common solvents such as γ-butyrolactone (BL), AN, tetrahydrofuran (THF), and PC, and salts such as LiAsF$_6$, LiClO$_4$, Ca(ClO$_4$)$_2$, Ca(BF$_4$)$_2$, TBA(BF$_4$), and TBA(ClO$_4$) [433]. However, the use of these materials has shown poor reversibility of Ca. The main challenge to develop rechargeable batteries is the lack of electrolytes for reversible Ca deposition since the reversible Ca electrochemistry depends on the electrolyte [433,434]. While Li anodes form an SEI in contact with organic electrolytes that only allow the Li$^+$ conducting, the SEI formed on Ca anodes in most organic electrolytes block the Ca^{2+} conducting, thus limiting the Coulombic efficiency of Ca batteries.

The first reversible Ca deposition was demonstrated in 2016, integrating a Ca(BF$_4$)$_2$ salt and ethylene carbonate EC: PC as the solvent at elevated temperatures (348.15 K to 373.15 K) [50]. In 2019, it was reported that Ca plating/stripping in carbonate solvents could be reversible at room temperature using EC:PC solvent and Ca(BF$_4$)$_2$ salt [435]. EC can form stable and ion-permeable SEI, and the addition of PC creates a stable liquid solvent with a wide electrochemical stability window. The use of Ca(BF$_4$)$_2$ salts supports the formation of stabilizing the ion-permeable SEI.

Additional research studied the Ca deposition/stripping behaviors of Ca[B(hfip)$_4$]$_2$ salts by using different solvents such as THF, DME, and DGM, showing that DGM improved the reversibility of Ca deposition and stripping of the Ca[B(hfip)$_4$]$_2$ when compared with THF and DME [434]. DGM as an electrolyte avoids the Ca dendrite growth and the form of a passivation layer.

In addition, computational analysis helped to engineer electrolytes for Ca batteries. Solubility and solvation of Ca^{2+} ions have been tested in pure carbonate solvents (EC, VC, PC, BC, DMC, EMC, and DEC), providing that the EMC and the binary mixture EC/DEC are the best electrolytes for potential Ca batteries [436].

5.6. Magnesium Batteries

Liquid electrolytes explored for Mg-based batteries integrates Mg salts such as MgTSFI$_2$, MgCl$_2$, Mg(CB$_{11}$H$_{12}$)$_2$, Mg(BBu$_2$Ph$_2$)$_2$, Mg[B(HFIP)$_4$]$_2$, and Mg(BH$_4$)$_2$ Table 16, while solvents are based mainly in non-aqueous liquids. Mg can react with esters; as a consequence, the Mg analogues of some successful Li salts (e.g., Mg((PF$_6$)$_2$, and Mg(ClO$_4$)$_2$) can not be applied since they may be reduced [437].

Finding appropriate electrolytes for Mg batteries that permit efficient reversible Mg reduction and oxidation is still a challenge. Similar to Ca batteries, in Mg batteries the solvents' decomposition and the existence of impurities causes a passivating SEI on the Mg anode that blocks Mg$_{2+}$ ions [343]. Therefore, the electrolyte solution must be stable with Mg and not be reduced by it, such as ether-based electrolytes (glymes and THF) [195].

Others Mg electrolytes researched are organoborate-based materials. Boron-based electrolytes have been shown to have a functional reaction with highly reversible magnesium deposition/dissolution. Moreover, these electrolytes could be integrated with sulfur cathodes [195]. Other electrolytes include: Borohydride-Based Mg(BH$_4$)$_2$ in DME or THF, Mg-hexamethyldisilazane (Mg−HMDS) in THF, Mg aluminate chloride complex solutions (MgCl$_2$ and AlCl$_3$) in THF, and Mg(TFSI)$_2$ in DME.

An alternative studied to conventional electrolytes is a dual layer of liquid and polymer electrolytes [438]. Polymer electrolytes integrate PVDF, TEGDME, and Mg(O$_3$SCF$_3$)$_2$, showing an Mg-ion conductivity up to 4.62×10^{-4} S cm^{-1} at 328.15 K. The use of solid Polymer electrolytes may reduce unwanted chemical reactions with the Mg anode. The PVDF film possesses a microsphere morphology, and the microsphere radius is reduced by adding TEGDME. In addition, the use of salt changes the PVDF morphological structure and increases the number of pores, promoting the diffusion of Mg.

5.7. Aluminum Batteries

The challenge in Al-based batteries is developing electrolytes free from Cl to avoid corrosion of battery components and to be capable of fast cationic Al^{+3} transport [47]. An alternative explored is ionic liquid electrolytes, which have shown to have wide electrochemical potential windows but high viscosities that limit the fast migration of ions in the electrolyte and inhibit performance. The most common ionic liquid electrolytes include the mixtures of 1- ethyl-3-methylimidazolium chloride (EMIC) or 1-butyl-3- methylimidazolium chloride (BMIC) with AlCl$_3$. Another alternative explored is the use of urea with AlCl$_3$ as electrolyte [439]. This test showed a capacity of 73 mA h g^{-1} at a current density of 100 mA g^{-1} with an efficiency of $\sim$100% after 180 cycles. However, the battery stability required $\sim$5 to 10 cycles to achieve a stable capacity, changing the Coulombic efficiency from $\sim$90% in the first cycle to $\sim$100% in the 10th cycle.

Another alternative explored for rechargeable aqueous Al batteries was to use Al(OTF)$_3$–H$_2$O as an electrolyte [440]. This battery has promising applications due to the high safety of aqueous electrolytes, which facilitates cell assembly and reduces material costs. It was demonstrated that this battery allows reversible ex/insertion of Al^{3+} in an aqueous electrolyte and achieves the trivalent reaction at a high redox potential. The Al^{3+} contribution in capacity was identified by using the aqueous HOTF electrolyte, while the ionic liquid AlCl$_3$/[BMIM]Cl was used as a comparison to expose the effect of the aqueous electrolyte on electrochemical performance. In this case, the aqueous electrolyte was essential for enhancing kinetics and keeping the cycle life with an Al$_x$MnO$_2 \cdot$nH$_2$O cathode and an Al anode when compared with other electrolytes.

6. Applications of Batteries

In this section, we focus on two major battery applications: smartphones and electric vehicles. These applications have specific battery requirements that have been addressed by research and commercial sectors. In the research field, developments have focused on solving technical challenges such as dendrite growth and side reactions to improve performance metrics of each single battery part, including capacity, voltage, Coulombic efficiency, retention capacity, and the number of cycles. In the commercial field, batteries require a high energy density and capacity, long cycle life, fast charging time, high safety, and environmental sustainability. In the next paragraphs, we discuss these two applications which do not have the most efficient batteries but are driven by an existing market that naturally evolves.

Smartphones are among the most widely used consumer electronics products, reaching 323 million units shipped in the second quarter of 2021 [441] and a sales forecast of 1535 billion units [442]. Today, smartphones are not only used as a communication device, but also include other features, such as the ability to connect to the web and other multimedia features. Standard smartphones contain a battery that offers a capacity ranging from 2000 mA h to 5000 mA h, allowing $\sim$8 h of continuous operation, which means that users are generally required to charge the battery twice per day. This time-consuming activity of constantly charging your phone is something that users have had to accept. To satisfy the need of the charging and discharging cycles required by these new functionalities, batteries with larger capacities are needed [1].

Electric vehicles (EVs) have recently achieved widespread notice due to progress in electrochemical energy storage technology ($\sim$250 W h kg^{-1}) and cost reduction ($\sim$\$ 100/kWh) [3]. The first trimester of 2021 saw an increase of $\sim$140% in the global electric vehicle sales from 2020 [443]. The increasing EV sales are also being driven by supportive regulatory policies and additional incentives. Therefore, it is necessary to improve the quality of EV batteries if we wish to increase EV adoption rates. The most important features of EV include long-range transport (>186 km), low-cost transport ($ 45/kWh), and high-utilization transport (fast charging <1 h) [444]. Batteries must be able to satisfy all these areas. To have long-range transport, it is necessary to manufacture batteries with energy densities of ~500 W h kg^{-1}. To reduce the cost, abundant materials need to be used. Finally, to achieve high utilization, batteries must be capable of charging quickly in 1 h or under. Tesla, for example, has been able to address these demands of EVs, reporting a range of 650 km on a single charge and 320 km after a 15 min fast charge [445].

For these advanced applications that require high energy efficiency batteries (see Figure 13), a joint effort between the commercial and research sectors is needed, not only for developing more efficient materials and manufacturing techniques, but also for establishing standards to ensure battery quality and proper performance in commercial products.

Figure 13. Current and future battery applications in daily life, including health monitoring, communication, transportation, entertainment, working, lighting, cleaning, and so on. Graph constructed by the authors.

Currently, international efforts around the world connect academia and industry to face the challenge of a transition from research to commercial applications, including the development of cost-effective large-scale manufacturing. Two macro projects sponsored by the EU are mentioned here: "CObalt-free Batteries for FutuRe Automotive Applications" (COBRA) [446] and "ERA-NET for research and innovation in materials and battery technologies, supporting the European Green Deal" (M-ERA.NET3) [447]. COBRA explores using cobalt-free materials for manufacturing a battery with superior energy density, low cost, increased cycles, and reduced critical materials. The project integrates the research (3 universities) and the industrial (4 SMEs and 5 enterprises) fields to ensure an easier adaptation to production lines and the ability to scale up to meet the demands of higher market adoption. M-ERA.NET3 coordinates the efforts of several participating EU Member States in materials research and innovation for future batteries by addressing emerging technologies and related applications areas, such as surfaces, coatings, composites, additive manufacturing, or integrated materials modeling.

We also would like to call attention to the standards that commercial products should fit, and batteries are not an exception. Standards focused on device operability and compatibility have been the target of the International Electrotechnical Commission (IEC), the American National Standards Institute (ANSI), the International Standards Organization (ISO), and the Institute of Electrical and Electronics Engineers (IEEE) [448,449].

These are standards for conventional batteries, and more effort is needed to develop novel battery technologies.

7. Conclusions and Outlook

In this review, we discussed the spectrum of possibilities for anodes, cathodes, electrolytes, and separators, classified into seven types according to the working ions: Li^+, Na^+, K^+, Zn^{2+}, Ca^{2+}, Mg^{2+}, and Al^{3+}. These families allow a material selection understanding of how the working ion affects the material performance: voltage, current density, capacity, and capacity retention. We also presented the advantages and disadvantages of different materials, and which strategies have been studied to enhance their performance.

We examined metal anodes with high capacity. The use of metal anodes is still limited in its practical application due to dendrite growth and because side reactions limit a stable and long cycle life. The two main strategies explored to address some of the issues related to the use of metal anodes are (1) designing structured anodes, and (2) developing novel materials (Figure 14). The first strategy is focused on 1D, 2D, and 3D structures by adding binders or controlling the manufacturing techniques. The second is based on alternative materials that can successfully operate with the corresponding ion carrier. These materials can be classified according to their operation mechanism as alloys, intercalation, or conversion [73].

Cathodes discussed in this review have been studied in the frame of their operational mechanisms: intercalation and conversion. Intercalation cathodes offer good stability but low capacity, while conversion achieves higher capacity, but the side reactions limit their cycle life. The traditional intercalation materials are classified into three major classes according to their crystal structure: layered, spinel, and olivine structure. The next generation of cathodes is exploring conversion mechanisms, which are present in metal–sulfur and metal–oxygen technologies and offer high theoretical energy (Figure 15).

Electrolytes and separators play an important role in the batteries' performance, as they ensure a stable and long cycle life. In this review, we studied electrolytes in two classes: liquid and SSE. Traditional liquid electrolytes consist of a salt dissolved in an organic solvent. Different levels of performance are achieved depending not only on the kinds of salt and solvent used, but also on their concentration. Today, there is a large interest in switching from organic to ionic and aqueous solvents. Ionic solvents can improve battery performance, and aqueous solvents can reduce battery cost and they are more environmentally friendly. The wide range of possibilities makes it difficult to have a clear view of the electrolyte research pathway going forward. Liquid electrolytes have a high ion conductivity, resulting in faster charging but causing safety concerns. In contrast, SSEs are a key component to developing FB. SSEs are safe and possess better mechanical and thermal stability than liquids. Improving the ion conductivity is a crucial factor to manufacture safe and practical SSE for novel battery requirements.

Battery demand for the seven ion families batteries discussed will come from all the emerging technologies envisioned by the authors in Figure 13: Internet of Things (Li^+, Zn^{2+}, Al^{3+}), reconfigurable wearables (Li^+, Zn^{2+}, Al^{3+}), sports tattoos (Zn^{2+}, Al^{3+}), large-scale energy storage (Na^+, K^+, Mg^{2+}, Ca^{2+}), unmanned aerial vehicle (Li^+, K^+, Ca^{2+}, Mg^{2+}), metaverse (Li^+, Na^+, Zn^{2+},), intelligent virtual assistant (Ca^{2+}, Mg^{2+}, Al^{3+}) and pet robotics (K^+, Ca^{2+}, Mg^{2+}, Al^{3+}). Since the applications will grow with the human endlessly needs and creativity, professionals and researchers who are designing and manufacturing novel materials will have to face the need not only to set up the materials' technical criteria for ion batteries, but interconnected criteria that emerge on the sustainability of supply and demand challenges.

This review is expected to be useful for professionals and researchers who are designing and manufacturing novel materials and need to set up technical criteria for batteries. Figures 14 and 15 could give an overview of the materials for anodes and cathodes. Although most of them are in the innovation and high expectations stage, exciting new battery developments are just around the corner and will undoubtedly have an impact on the way we live.

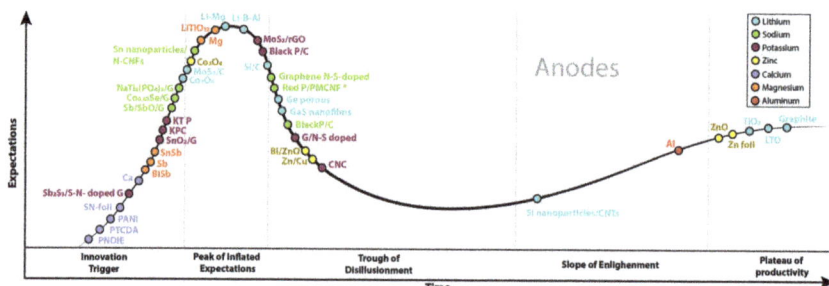

Figure 14. Graph constructed by the authors from Gartner hype cycle for promising anodes materials for batteries. Materials are grouped according to the seven battery types: Li, Na, K, Zn, Ca, Mg, and Al. Each material is organized into a stage according to the estimation of development among innovation and productivity stage.

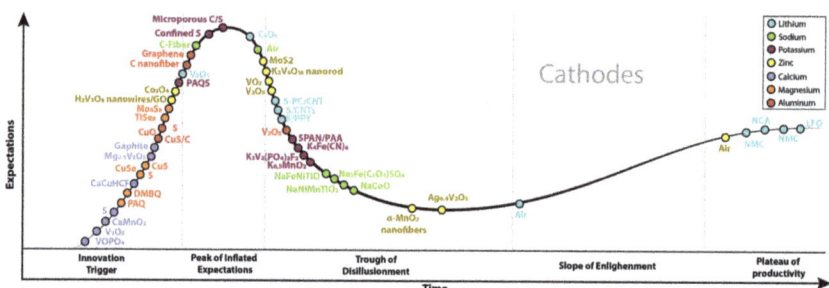

Figure 15. Graph constructed by the authors from Gartner hype cycle for promising cathode materials for batteries. Materials are grouped according to the seven battery types: Li, Na, K, Zn, Ca, Mg, and Al. Each material is organized into a stage according to the estimation of development among innovation and productivity stages.

In our society with 222 years of battery dependency, we cannot imagine the evolution of technology without a power cell. With the proliferation of home appliances, IoT devices acquiring massive amounts of data, artificial intelligence algorithms running on embedded systems, and many applications with many devices, the energy demand appears to proliferate with human technological creativity. However, it seems that nature sets an upper limit; resources are not unlimited. Therefore, we cannot only focus on optimizing battery metrics but also on the sustainability of materials and the availability of exploration sites. Batteries must explore three fronts: material flow analysis (MFA), life cycle assessment (LCA), and battery quantity reduction (RA). In addition, we must start training our disruptive thinking to go beyond the dependence on batteries and integrate a new self-powered option that will optimize the dimensions and resources of the systems. This is the tug-of-war in which we find ourselves as a society. On the one hand, the need to remotely and permanently power billions of devices with high current densities and long life cycles. On the other hand, the enormous pressure in the exploitation of increasingly scarce minerals leads us to the need to be very disruptive in selecting materials for the following battery technologies.

Author Contributions: Conceptualization, W.P., J.A.P.-T. and A.A. (equal contribution); methodology W.P., J.A.P.-T. and A.A. (equal contribution); investigation, W.P., J.A.P.-T. and A.A. (equal contribution); writing—original draft preparation, W.P., J.A.P.-T. and A.A. (equal contribution); writing—review and editing, W.P., J.A.P.-T. and A.A. (equal contribution); visualization, W.P., J.A.P.-T. and A.A. (equal contribution); supervision, J.A.P.-T. and A.A. (equal contribution); funding acquisition, J.A.P.-T. and A.A. (equal contribution). All authors have read and agreed to the published version of the manuscript.

Funding: This research received no external funding.

Institutional Review Board Statement: Not applicable.

Data Availability Statement: Not applicable.

Acknowledgments: The authors acknowledge the financial provided by the Vice Presidency for Research & Creation publication fund at Universidad de los Andes, Colombia. Wendy Pantoja was supported by Center for Interdisciplinary Studies in Basic and Applied Complexity, CEIBA, through Bécate Nariño scholarship. The authors thank the Language Department at Universidad de los Andes, Colombia, for the insightful comments in reviewing the manuscript.

Conflicts of Interest: The authors declare no conflict of interest.

References

1. Liang, Y.; Zhao, C.; Yuan, H.; Chen, Y.; Zhang, W.; Huang, J.; Yu, D.; Liu, Y.; Titirici, M.; Chueh, Y.; et al. A review of rechargeable batteries for portable electronic devices. *InfoMat* **2019**, *1*, 6–32. [CrossRef]
2. Nadeem, F.; Hussain, S.M.; Tiwari, P.K.; Goswami, A.K.; Ustun, T.S. Comparative review of energy storage systems, their roles, and impacts on future power systems. *IEEE Access* **2019**, *7*, 4555–4585. [CrossRef]
3. Ding, Y.; Cano, Z.P.; Yu, A.; Lu, J.; Chen, Z. Automotive Li-Ion Batteries: Current Status and Future Perspectives. *Electrochem. Energy Rev.* **2019**, *2*, 1–28. [CrossRef]
4. Zubi, G.; Dufo-López, R.; Carvalho, M.; Pasaoglu, G. The lithium-ion battery: State of the art and future perspectives. *Renew. Sustain. Energy Rev.* **2018**, *89*, 292–308. [CrossRef]
5. Fan, X.; Liu, X.; Hu, W.; Zhong, C.; Lu, J. Advances in the development of power supplies for the Internet of Everything. *InfoMat* **2019**, *1*, 130–139. [CrossRef]
6. Raj, A.; Steingart, D. Review—Power Sources for the Internet of Things. *J. Electrochem. Soc.* **2018**, *165*, B3130–B3136. [CrossRef]
7. Salama, M.; Rosy.; Attias, R.; Yemini, R.; Gofer, Y.; Aurbach, D.; Noked, M. Metal-Sulfur Batteries: Overview and Research Methods. *ACS Energy Lett.* **2019**, *4*, 436–446. [CrossRef]
8. Mao, M.; Gao, T.; Hou, S.; Wang, C. A critical review of cathodes for rechargeable Mg batteries. *Chem. Soc. Rev.* **2018**, *47*, 8804–8841. [CrossRef]
9. Manthiram, A. A reflection on lithium-ion battery cathode chemistry. *Nat. Commun.* **2020**, *11*, 1550. [CrossRef]
10. Mukherjee, S.; Singh, G. Two-Dimensional Anode Materials for Non-lithium Metal-Ion Batteries. *ACS Appl. Energy Mater.* **2019**, *2*, 932–955. [CrossRef]
11. Liu, H.; Cheng, X.B.; Huang, J.Q.; Kaskel, S.; Chou, S.; Park, H.S.; Zhang, Q. Alloy Anodes for Rechargeable Alkali-Metal Batteries: Progress and Challenge. *ACS Mater. Lett.* **2019**, *1*, 217–229. [CrossRef]
12. Puthusseri, D.; Wahid, M.; Ogale, S. Conversion-type Anode Materials for Alkali-Ion Batteries: State of the Art and Possible Research Directions. *ACS Omega* **2018**, *3*, 4591–4601. [CrossRef] [PubMed]
13. Matsumoto, K.; Hwang, J.; Kaushik, S.; Chen, C.Y.; Hagiwara, R. Advances in sodium secondary batteries utilizing ionic liquid electrolytes. *Energy Environ. Sci.* **2019**, *12*, 3247–3287. [CrossRef]
14. Yu, Z.; Wang, H.; Kong, X.; Huang, W.; Tsao, Y.; Mackanic, D.G.; Wang, K.; Wang, X.; Huang, W.; Choudhury, S.; et al. Molecular design for electrolyte solvents enabling energy-dense and long-cycling lithium metal batteries. *Nat. Energy* **2020**, *5*, 526–533. [CrossRef]
15. Zhao, H.; Xu, J.; Yin, D.; Du, Y. Electrolytes for Batteries with Earth-Abundant Metal Anodes. *Chem. A Eur. J.* **2018**, *24*, 18220–18234. [CrossRef]
16. Weber, C.J.; Geiger, S.; Falusi, S.; Roth, M. *Material Review of Li Ion Battery Separators*; American Institute of Physics: College Park, MD, USA, 2014; Volume 1597, pp. 66–81. [CrossRef]
17. Costa, C.M.; Lee, Y.H.; Kim, J.H.; Lee, S.Y.; Lanceros-Méndez, S. Recent advances on separator membranes for lithium-ion battery applications: From porous membranes to solid electrolytes. *Energy Storage Mater.* **2019**, *22*, 346–375. [CrossRef]
18. Winter, M.; Barnett, B.; Xu, K. Before Li Ion Batteries. *Chem. Rev.* **2018**, *118*, 11433–11456. 8b00422. [CrossRef]
19. "Engineering and Technology History Wiki". Milestones: Volta's Electrical Battery Invention, 1799, 1999. Available online: https://ethw.org/ (accessed on 30 January 2022).
20. Cadex Electronics Inc. *When Was the Battery Invented?*; Cadex Electronics Inc.: Richmond, BC, Canada, 2019. Available online: https://batteryuniversity.com/learn/ (accessed on 30 January 2022).

21. Viswanathan, B. *Energy Sources*; Elsevier: Amsterdam, The Netherlands, 2017. pp. 263–313. [CrossRef]
22. Cheng, X.B.; Zhang, R.; Zhao, C.Z.; Zhang, Q. Toward Safe Lithium Metal Anode in Rechargeable Batteries: A Review. *Chem. Rev.* **2017**, *117*, 10403–10473. [CrossRef]
23. Turcheniuk, K.; Bondarev, D.; Singhal, V.; Yushin, G. Ten years left to redesign lithium-ion batteries. **2018**, *14*, 467–470. [CrossRef]
24. Bernhart, W. Recycling of Lithium-Ion Batteries in the Context of Technology and Price Developments. *ATZelectronics Worldw.* **2019**, *14*, 38–43. [CrossRef]
25. Zhang, C.; Zhao, H.; Lei, Y. Recent Research Progress of Anode Materials for Potassium-ion Batteries. *Energy Environ. Mater.* **2020**, *3*, 105–120. [CrossRef]
26. Yao, Z.; Hegde, V.I.; Aspuru-Guzik, A.; Wolverton, C. Discovery of Calcium-Metal Alloy Anodes for Reversible Ca-Ion Batteries. *Adv. Energy Mater.* **2019**, *9*, 1802994. [CrossRef]
27. Zhao, S.; Qin, B.; Chan, K.; Li, C.V.; Li, F. Recent Development of Aprotic Na-O_2 Batteries. *Batter. Supercaps* **2019**, *2*, 725–742. [CrossRef]
28. Lewis, D. The COVID pandemic has harmed researcher productivity—And mental health. *Nature* **2021**. [CrossRef] [PubMed]
29. Holland, A.; He, X. *Advanced Li-Ion and beyond Lithium Batteries 2022–2032: Technologies, Players, Trends, Markets*; Technical Report; IDTechEx: Cambridge, UK, 2021.
30. Zhu, J.; Zhu, P.; Yan, C.; Dong, X.; Zhang, X. Recent progress in polymer materials for advanced lithium-sulfur batteries. *Prog. Polym. Sci.* **2019**, *90*, 118–163. [CrossRef]
31. Xu, X.L.; Wang, S.J.; Wang, H.; Xu, B.; Hu, C.; Jin, Y.; Liu, J.B.; Yan, H. The suppression of lithium dendrite growth in lithium sulfur batteries: A review. *J. Energy Storage* **2017**, *13*, 387–400. [CrossRef]
32. Zhang, X.; Chen, A.; Jiao, M.; Xie, Z.; Zhou, Z. Understanding Rechargeable Li-O_2 Batteries via First-Principles Computations. *Batter. Supercaps* **2019**, *2*, 498–508. [CrossRef]
33. Huang, J.; Peng, Z. Understanding the Reaction Interface in Lithium-Oxygen Batteries. *Energy Environ.* **2019**, *2*, 37–48. [CrossRef]
34. Kang, J.H.; Lee, J.; Jung, J.W.; Park, J.; Jang, T.; Kim, H.S.; Nam, J.S.; Lim, H.; Yoon, K.R.; Ryu, W.H.; et al. Lithium—Air Batteries: Air-Breathing Challenges and Perspective. *ACS Nano* **2020**, *14*, 14549–14578. [CrossRef]
35. Delmas, C. Sodium and Sodium-Ion Batteries: 50 Years of Research. *Adv. Energy Mater.* **2018**, *8*, 1703137. aenm.201703137. [CrossRef]
36. Peters, J.; Peña Cruz, A.; Weil, M. Exploring the Economic Potential of Sodium-Ion Batteries. *Batteries* **2019**, *5*, 10. [CrossRef]
37. Kumar, D.; Kuhar, S.B.; Kanchan, D. Room temperature sodium-sulfur batteries as emerging energy source. *J. Energy Storage* **2018**, *18*, 133–148. [CrossRef]
38. Konarov, A.; Voronina, N.; Jo, J.H.; Bakenov, Z.; Sun, Y.K.; Myung, S.T. Present and Future Perspective on Electrode Materials for Rechargeable Zinc-Ion Batteries. *ACS Energy Lett.* **2018**, *3*, 2620–2640. [CrossRef]
39. Ding, J.; Zhang, H.; Fan, W.; Zhong, C.; Hu, W.; Mitlin, D. Review of Emerging Potassium–Sulfur Batteries. *Adv. Mater.* **2020**, *32*, 1908007. [CrossRef] [PubMed]
40. Wang, Y.; Sun, Y.; Ren, W.; Zhang, D.; Yang, Y.; Yang, J.; Wang, J.; Zeng, X.; NuLi, Y. Challenges and prospects of Mg-air batteries: A review. *Energy Mater.* **2022**, *2*, 200024. [CrossRef]
41. Lee, B.; Paek, E.; Mitlin, D.; Lee, S.W. Sodium Metal Anodes: Emerging Solutions to Dendrite Growth. *Chem. Rev.* **2019**, *119*, 5416–5460. [CrossRef]
42. Xiao, Y.; Abbasi, N.M.; Zhu, Y.; Li, S.; Tan, S.; Ling, W.; Peng, L.; Yang, T.; Wang, L.; Guo, X.; et al. Layered Oxide Cathodes Promoted by Structure Modulation Technology for Sodium-Ion Batteries. *Adv. Funct. Mater.* **2020**, *30*, 2001334. [CrossRef]
43. Nayak, P.K.; Yang, L.; Brehm, W.; Adelhelm, P. From Lithium-Ion to Sodium-Ion Batteries: Advantages, Challenges, and Surprises. *Angew. Chem. Int. Ed.* **2018**, *57*, 102–120. [CrossRef]
44. Electric VehicleTeam. *A Guide to Understanding Battery Specifications*; Electric VehicleTeam: Cambridge, MA, USA, 2008.
45. Farahani, S. *ZigBee Wireless Networks and Transceivers*; Elsevier: Amsterdam, The Netherlands, 2008; Chapter 6; pp. 207–224. [CrossRef]
46. Muench, S.; Wild, A.; Friebe, C.; Häupler, B.; Janoschka, T.; Schubert, U.S. Polymer-Based Organic Batteries. *Chem. Rev.* **2016**, *116*, 9438–9484. [CrossRef]
47. Ponrouch, A.; Bitenc, J.; Dominko, R.; Lindahl, N.; Johansson, P.; Palacin, M.R. Multivalent rechargeable batteries. *Energy Storage Mater.* **2019**, *20*, 253–262. [CrossRef]
48. Liang, Y.; Dong, H.; Aurbach, D.; Yao, Y. Current status and future directions of multivalent metal-ion batteries. *Nat. Energy* **2020**, *5*, 646–656. [CrossRef]
49. Jayaprakash, N.; Das, S.K.; Archer, L.A. The rechargeable aluminum-ion battery. *Chem. Commun.* **2011**, *47*, 12610. [CrossRef] [PubMed]
50. Ponrouch, A.; Frontera, C.; Bardé, F.; Palacín, M.R. Towards a calcium-based rechargeable battery. *Nat. Mater.* **2016**, *15*, 169–172. [CrossRef] [PubMed]
51. Gao, X.; Dong, Y.; Li, S.; Zhou, J.; Wang, L.; Wang, B. MOFs and COFs for Batteries and Supercapacitors. *Electrochem. Energy Rev.* **2020**, *3*, 81–126. [CrossRef]
52. Poizot, P.; Dolhem, F.; Gaubicher, J. Progress in all-organic rechargeable batteries using cationic and anionic configurations: Toward low-cost and greener storage solutions? *Curr. Opin. Electrochem.* **2018**, *9*, 70–80. [CrossRef]

53. Wan, F.; Zhang, L.; Wang, X.; Bi, S.; Niu, Z.; Chen, J. An Aqueous Rechargeable Zinc-Organic Battery with Hybrid Mechanism. *Adv. Funct. Mater.* **2018**, *28*, 1804975. [CrossRef]
54. Lindahl, N.; Bitenc, J.; Dominko, R.; Johansson, P. Aluminum Metal—Organic Batteries with Integrated 3D Thin Film Anodes. *Adv. Funct. Mater.* **2020**, *30*, 3–9. [CrossRef]
55. Zhang, X.; Dong, P.; Song, M. Metal-Organic Frameworks for High-Energy Lithium Batteries with Enhanced Safety: Recent Progress and Future Perspectives. *Batter. Supercaps* **2019**, *2*, 591–626. [CrossRef]
56. Sabihuddin, S.; Kiprakis, A.E.; Mueller, M. A numerical and graphical review of energy storage technologies. *Energies* **2015**, *8*, 172–216. [CrossRef]
57. Lu, Y.; Zhao, C.Z.; Yuan, H.; Hu, J.K.; Huang, J.Q.; Zhang, Q. Dry electrode technology, the rising star in solid-state battery industrialization. *Matter* **2022**, *5*, 876–898. [CrossRef]
58. Liu, W.; Song, M.S.; Kong, B.; Cui, Y. Flexible and Stretchable Energy Storage: Recent Advances and Future Perspectives. *Adv. Mater.* **2017**, *29*, 1603436. [CrossRef]
59. Isidor Buchmann. Types of Battery Cells, 2019. Available online: https://batteryuniversity.com/article/bu-301a-types-of-battery-cells (accessed on 30 January 2022).
60. Copyright Epec, L. Prismatic and Pouch Battery Packs, 2021. Available online: https://www.epectec.com/batteries/prismatic-pouch-packs.html (accessed on 30 January 2022).
61. Xiaoxi, H. *Flexible, Printed and Thin Film Batteries 2020–2030: Technologies, Markets and Players*; Technical Report; IDTechEx: Cambridge, UK, 2020.
62. Song, W.; Yoo, S.; Song, G.; Lee, S.; Kong, M.; Rim, J.; Jeong, U.; Park, S. Recent Progress in Stretchable Batteries for Wearable Electronics. *Batter. Supercaps* **2019**, *2*, 181–199. [CrossRef]
63. Praveen, S.; Santhoshkumar, P.; Joe, Y.C.; Senthil, C.; Lee, C.W. 3D-printed architecture of Li-ion batteries and its applications to smart wearable electronic devices. *Appl. Mater. Today* **2020**, *20*, 100688. [CrossRef]
64. Qian, G.; Liao, X.; Zhu, Y.; Pan, F.; Chen, X.; Yang, Y. Designing Flexible Lithium-Ion Batteries by Structural Engineering. *ACS Energy Lett.* **2019**, *4*, 690–701. [CrossRef]
65. Qu, S.; Liu, B.; Wu, J.; Zhao, Z.; Liu, J.; Ding, J.; Han, X.; Deng, Y.; Zhong, C.; Hu, W. Kirigami-Inspired Flexible and Stretchable Zinc-Air Battery Based on Metal-Coated Sponge Electrodes. *ACS Appl. Mater. Interfaces* **2020**, *12*, 54833–54841. [CrossRef] [PubMed]
66. Liu, T.; Chen, X.; Tervoort, E.; Kraus, T.; Niederberger, M. Design and Fabrication of Transparent and Stretchable Zinc Ion Batteries. *ACS Appl. Energy Mater.* **2021**, *4*, 6166–6179. [CrossRef]
67. Guo, Z.H.; Liu, M.; Cong, Z.; Guo, W.; Zhang, P.; Hu, W.; Pu, X. Stretchable Textile Rechargeable Zn Batteries Enabled by a Wax Dyeing Method. *Adv. Mater. Technol.* **2020**, *5*, 2000544. [CrossRef]
68. Fu, W.; Turcheniuk, K.; Naumov, O.; Mysyk, R.; Wang, F.; Liu, M.; Kim, D.; Ren, X.; Magasinski, A.; Yu, M.; et al. Materials and technologies for multifunctional, flexible or integrated supercapacitors and batteries. *Mater. Today* **2021**, *48*, 1. [CrossRef]
69. Peng, J.; Jeffrey Snyder, G. A figure of merit for flexibility. *Science* **2019**, *366*, 690–691. [CrossRef]
70. Kong, L.; Tang, C.; Peng, H.; Huang, J.; Zhang, Q. Advanced energy materials for flexible batteries in energy storage: A review. *SmartMat* **2020**, *1*. [CrossRef]
71. Song, J.; Yan, W.; Cao, H.; Song, Q.; Ding, H.; Lv, Z.; Zhang, Y.; Sun, Z. Material flow analysis on critical raw materials of lithium-ion batteries in China. *J. Clean. Prod.* **2019**, *215*, 570–581. [CrossRef]
72. Dehghani-Sanij, A.; Tharumalingam, E.; Dusseault, M.; Fraser, R. Study of energy storage systems and environmental challenges of batteries. *Renew. Sustain. Energy Rev.* **2019**, *104*, 192–208. [CrossRef]
73. Lu, J.; Chen, Z.; Pan, F.; Cui, Y.; Amine, K. High-Performance Anode Materials for Rechargeable Lithium-Ion Batteries. *Electrochem. Energy Rev.* **2018**, *1*, 35–53. [CrossRef]
74. Liu, H.; Cheng, X.B.; Jin, Z.; Zhang, R.; Wang, G.; Chen, L.Q.; Liu, Q.B.; Huang, J.Q.; Zhang, Q. Recent advances in understanding dendrite growth on alkali metal anodes. *EnergyChem* **2019**, *1*, 100003. [CrossRef]
75. Chang, H.; Wu, Y.R.; Han, X.; Yi, T.F. Recent developments in advanced anode materials for lithium-ion batteries. *Energy Mater.* **2021**, *1*, 24. [CrossRef]
76. Xiong, X.; Zhou, Q.; Zhu, Y.; Chen, Y.; Fu, L.; Liu, L.; Yu, N.; Wu, Y.; van Ree, T. In Pursuit of a Dendrite-Free Electrolyte/Electrode Interface on Lithium Metal Anodes: A Minireview. *Energy Fuels* **2020**, *34*, 10503–10512. 0c02211. [CrossRef]
77. Xiao, X.; Yao, W.; Tang, J.; Liu, C.; Lian, R.; Urbankowski, P.; Anayee, M.; He, S.; Li, J.; Wang, H.; et al. Interconnected Two-dimensional Arrays of Niobium Nitride Nanocrystals as Stable Lithium Host. *Batter. Supercaps* **2021**, *4*, 106–111. [CrossRef]
78. Goriparti, S.; Miele, E.; De Angelis, F.; Di Fabrizio, E.; Proietti Zaccaria, R.; Capiglia, C. Review on recent progress of nanostructured anode materials for Li-ion batteries. *J. Power Sources* **2014**, *257*, 421–443. [CrossRef]
79. Yang, J.; Zhou, X.Y.; Li, J.; Zou, Y.L.; Tang, J.J. Study of nano-porous hard carbons as anode materials for lithium ion batteries. *Mater. Chem. Phys.* **2012**, *135*, 445–450. [CrossRef]
80. Chen, Z.; Liu, Y.; Zhang, Y.; Shen, F.; Yang, G.; Wang, L.; Zhang, X.; He, Y.; Luo, L.; Deng, S. Ultrafine layered graphite as an anode material for lithium ion batteries. *Mater. Lett.* **2018**, *229*, 134–137. [CrossRef]
81. Ge, M.; Rong, J.; Fang, X.; Zhou, C. Porous Doped Silicon Nanowires for Lithium Ion Battery Anode with Long Cycle Life. *Nano Lett.* **2012**, *12*, 2318–2323. [CrossRef]

82. Zhang, P.; Zhu, Q.; Guan, Z.; Zhao, Q.; Sun, N.; Xu, B. A Flexible Si@C Electrode with Excellent Stability Employing an MXene as a Multifunctional Binder for Lithium-Ion Batteries. *ChemSusChem* **2020**, *13*, 1621–1628. [CrossRef] [PubMed]
83. Mishra, K.; Liu, X.C.; Ke, F.S.; Zhou, X.D. Porous germanium enabled high areal capacity anode for lithium-ion batteries. *Compos. Part B Eng.* **2019**, *163*, 158–164. [CrossRef]
84. Xiao, X.; Li, X.; Zheng, S.; Shao, J.; Xue, H.; Pang, H. Nanostructured Germanium Anode Materials for Advanced Rechargeable Batteries. *Adv. Mater. Interfaces* **2017**, *4*, 1600798. [CrossRef]
85. Medvedev, A.G.; Mikhaylov, A.A.; Grishanov, D.A.; Yu, D.Y.W.; Gun, J.; Sladkevich, S.; Lev, O.; Prikhodchenko, P.V. GeO_2 Thin Film Deposition on Graphene Oxide by the Hydrogen Peroxide Route: Evaluation for Lithium-Ion Battery Anode. *ACS Appl. Mater. Interfaces* **2017**, *9*, 9152–9160. [CrossRef] [PubMed]
86. Zhong, H.; Wu, Y.; Ding, F.; Sang, L.; Mai, Y. An artificial Li-Al interphase layer on Li-B alloy for stable lithium-metal anode. *Electrochim. Acta* **2019**, *304*, 255–262. [CrossRef]
87. Kong, L.L.; Wang, L.; Ni, Z.C.; Liu, S.; Li, G.R.; Gao, X.P. Lithium—Magnesium Alloy as a Stable Anode for Lithium–Sulfur Battery. *Adv. Funct. Mater.* **2019**, *29*, 1–10. [CrossRef]
88. Li, H.; Wang, Z.; Chen, L.; Huang, X. Research on Advanced Materials for Li-ion Batteries. *Adv. Mater.* **2009**, *21*, 4593–4607. [CrossRef]
89. Coelho, J.; Pokle, A.; Park, S.H.; McEvoy, N.; Berner, N.C.; Duesberg, G.S.; Nicolosi, V. Lithium Titanate/Carbon Nanotubes Composites Processed by Ultrasound Irradiation as Anodes for Lithium Ion Batteries. *Sci. Rep.* **2017**, *7*, 7614. [CrossRef]
90. Chen, Z.; Belharouak, I.; Sun, Y.K.; Amine, K. Titanium-Based Anode Materials for Safe Lithium-Ion Batteries. *Adv. Funct. Mater.* **2013**, *23*, 959–969. [CrossRef]
91. Trang, N.T.H.; Ali, Z.; Kang, D.J. Mesoporous TiO_2 spheres interconnected by multiwalled carbon nanotubes as an anode for high-performance lithium ion batteries. *ACS Appl. Mater. Interfaces* **2015**, *7*, 3676–3683. [CrossRef]
92. Mo, R.; Lei, Z.; Sun, K.; Rooney, D. Facile Synthesis of Anatase TiO_2 Quantum-Dot/Graphene-Nanosheet Composites with Enhanced Electrochemical Performance for Lithium-Ion Batteries. *Adv. Mater.* **2014**, *26*, 2084–2088. [CrossRef] [PubMed]
93. Hwang, I.S.; Kim, J.C.; Seo, S.D.; Lee, S.; Lee, J.H.; Kim, D.W. A binder-free Ge-nanoparticle anode assembled on multiwalled carbon nanotube networks for Li-ion batteries. *Chem. Commun.* **2012**, *48*, 7061–7063. [CrossRef] [PubMed]
94. Shi, Z.; Liu, M.; Naik, D.; Gole, J.L. Electrochemical properties of Li-Mg alloy electrodes for lithium batteries. *J. Power Sources* **2001**, *92*, 70–80. [CrossRef]
95. Chang, J.; Huang, X.; Zhou, G.; Cui, S.; Hallac, P.B.; Jiang, J.; Hurley, P.T.; Chen, J. Multilayered Si Nanoparticle/Reduced Graphene Oxide Hybrid as a High-Performance Lithium-Ion Battery Anode. *Adv. Mater.* **2014**, *26*, 758–764. [CrossRef]
96. Huang, Y.H.; Bao, Q.; Chen, B.H.; Duh, J.G. Nano-to-microdesign of marimo-like carbon nanotubes supported frameworks via in-spaced polymerization for high performance silicon lithium ion battery anodes. *Small* **2015**, *11*, 2314–2322. [CrossRef]
97. Qin, J.; He, C.; Zhao, N.; Wang, Z.; Shi, C.; Liu, E.Z.; Li, J. Graphene networks anchored with Sn@Graphene as lithium ion battery anode. *ACS Nano* **2014**, *8*, 1728–1738. [CrossRef] [PubMed]
98. Zhou, X.; Wan, L.J.; Guo, Y.G. Binding SnO_2 Nanocrystals in Nitrogen-Doped Graphene Sheets as Anode Materials for Lithium-Ion Batteries. *Adv. Mater.* **2013**, *25*, 2152–2157. [CrossRef]
99. Meng, X.; He, K.; Su, D.; Zhang, X.; Sun, C.; Ren, Y.; Wang, H.H.; Weng, W.; Trahey, L.; Canlas, C.P.; et al. Gallium Sulfide-Single-Walled Carbon Nanotube Composites: High-Performance Anodes for Lithium-Ion Batteries. *Adv. Funct. Mater.* **2014**, *24*, 5435–5442. [CrossRef]
100. Hwang, J.Y.; Myung, S.T.; Sun, Y.K. Sodium-ion batteries: Present and future. *Chem. Soc. Rev.* **2017**, *46*, 3529–3614. [CrossRef]
101. Dong, R.; Wu, F.; Bai, Y.; Wu, C. Sodium Storage Mechanism and Optimization Strategies for Hard Carbon Anode of Sodium Ion Batteries. *Acta Chim. Sin.* **2021**, *79*, 1461–1476. [CrossRef]
102. Arie, A.A.; Tekin, B.; Demir, E.; Demir-Cakan, R. Hard carbons derived from waste tea bag powder as anodes for sodium ion battery. *Mater. Technol.* **2019**, *34*, 515–524. [CrossRef]
103. Ding, C.; Huang, L.; Lan, J.; Yu, Y.; Zhong, W.H.; Yang, X. Superresilient Hard Carbon Nanofabrics for Sodium-Ion Batteries. *Small* **2020**, *16*, 1–9. [CrossRef] [PubMed]
104. Wang, Y.; Zhu, W.; Guerfi, A.; Kim, C.; Zaghib, K. Roles of ti in electrode materials for sodium-ion batteries. *Front. Energy Res.* **2019**, *7*, 28. [CrossRef]
105. Bayhan, Z.; Huang, G.; Yin, J.; Xu, X.; Lei, Y.; Liu, Z.; Alshareef, H.N. Two-Dimensional TiO_2/TiS_2 Hybrid Nanosheet Anodes for High-Rate Sodium-Ion Batteries. *ACS Appl. Energy Mater.* **2021**, *4*, 8721–8727. [CrossRef]
106. Morito, H.; Yamada, T.; Ikeda, T.; Yamane, H. Na-Si binary phase diagram and solution growth of silicon crystals. *J. Alloy. Compd.* **2009**, *480*, 723–726. [CrossRef]
107. Xu, Y.; Swaans, E.; Basak, S.; Zandbergen, H.W.; Borsa, D.M.; Mulder, F.M. Reversible na-ion uptake in Si nanoparticles. *Adv. Energy Mater.* **2016**, *6*, 1501436. [CrossRef]
108. Tseng, K.W.; Huang, S.B.; Chang, W.C.; Tuan, H.Y. Synthesis of Mesoporous Germanium Phosphide Microspheres for High-Performance Lithium-Ion and Sodium-Ion Battery Anodes. *Chem. Mater.* **2018**, *30*, 4440–4447. chemmater.8b01922. [CrossRef]
109. Sung, G.K.; Nam, K.H.; Choi, J.H.; Park, C.M. Germanium telluride: Layered high-performance anode for sodium-ion batteries. *Electrochim. Acta* **2020**, *331*, 135393. [CrossRef]
110. Ni, J.; Li, L.; Lu, J. Phosphorus: An Anode of Choice for Sodium-Ion Batteries. *ACS Energy Lett.* **2018**, *3*, 1137–1144. [CrossRef]

111. Liu, Y.; Zhang, N.; Liu, X.; Chen, C.; Fan, L.Z.; Jiao, L. Red phosphorus nanoparticles embedded in porous N-doped carbon nanofibers as high-performance anode for sodium-ion batteries. *Energy Storage Mater.* **2017**, *9*, 170–178. [CrossRef]
112. Wang, L.; Jiang, Z.; Li, W.; Gu, X.; Huang, L. Hybrid phosphorene/graphene nanocomposite as an anode material for Na-ion batteries: A first-principles study. *J. Phys. D Appl. Phys.* **2017**, *50*, 165501. [CrossRef]
113. Sun, X.; Li, W.; Zhong, X.; Yu, Y. Superior sodium storage in phosphorus@porous multichannel flexible freestanding carbon nanofibers. *Energy Storage Mater.* **2017**, *9*, 112–118. [CrossRef]
114. Zhu, J.; He, Q.; Liu, Y.; Key, J.; Nie, S.; Wu, M.; Shen, P.K. Three-dimensional, hetero-structured, $Cu_3P@C$ nanosheets with excellent cycling stability as Na-ion battery anode material. *J. Mater. Chem. A* **2019**, *7*, 16999–17007. [CrossRef]
115. Kaushik, S.; Matsumoto, K.; Sato, Y.; Hagiwara, R. Vanadium phosphide–phosphorus composite as a high-capacity negative electrode for sodium secondary batteries using an ionic liquid electrolyte. *Electrochem. Commun.* **2019**, *102*, 46–51. [CrossRef]
116. Liu, D.; Huang, X.; Qu, D.; Zheng, D.; Wang, G.; Harris, J.; Si, J.; Ding, T.; Chen, J.; Qu, D. Confined phosphorus in carbon nanotube-backboned mesoporous carbon as superior anode material for sodium/potassium-ion batteries. *Nano Energy* **2018**, *52*, 1–10. [CrossRef]
117. Ma, M.; Yao, Y.; Wu, Y.; Yu, Y. Progress and Prospects of Transition Metal Sulfides for Sodium Storage. *Adv. Fiber Mater.* **2020**, *2*, 314–337. [CrossRef]
118. Zhang, X.; Weng, W.; Gu, H.; Hong, Z.; Xiao, W.; Wang, F.; Li, W.; Gu, D. Versatile Preparation of Mesoporous Single-Layered Transition-Metal Sulfide/Carbon Composites for Enhanced Sodium Storage. *Adv. Mater.* **2022**, *34*. [CrossRef]
119. Yun, Y.S.; Park, Y.U.; Chang, S.J.; Kim, B.H.; Choi, J.; Wang, J.; Zhang, D.; Braun, P.V.; Jin, H.J.; Kang, K. Crumpled graphene paper for high power sodium battery anode. *Carbon* **2016**, *99*, 658–664. [CrossRef]
120. Wang, G.Z.; Feng, J.M.; Dong, L.; Li, X.F.; Li, D.J. Porous graphene anchored with Sb/SbOxas sodium-ion battery anode with enhanced reversible capacity and cycle performance. *J. Alloy. Compd.* **2017**, *693*, 141–149. [CrossRef]
121. Wang, G.; Zhang, S.; Li, X.; Liu, X.; Wang, H.; Bai, J. Multi-layer graphene assembled fibers with porous structure as anode materials for highly reversible lithium and sodium storage. *Electrochim. Acta* **2018**, *259*, 702–710. [CrossRef]
122. Zhang, G.; Liu, K.; Liu, S.; Song, H.; Zhou, J. Flexible $Co_{0.85}Se$ nanosheets/graphene composite film as binder-free anode with high Li- and Na-Ion storage performance. *J. Alloy. Compd.* **2018**, *731*, 714–722. [CrossRef]
123. Jian, Z.; Zhao, B.; Liu, P.; Li, F.; Zheng, M.; Chen, M.; Shi, Y.; Zhou, H. Fe_2O_3 nanocrystals anchored onto graphene nanosheets as the anode material for low-cost sodium-ion batteries. *Chem. Commun.* **2014**, *50*, 1215–1217. [CrossRef] [PubMed]
124. Song, J.; Yu, Z.; Gordin, M.L.; Hu, S.; Yi, R.; Tang, D.; Walter, T.; Regula, M.; Choi, D.; Li, X.; et al. Chemically bonded phosphorus/graphene hybrid as a high performance anode for sodium-ion batteries. *Nano Lett.* **2014**, *14*, 6329–6335. [CrossRef]
125. Liu, Y.; Kang, H.; Jiao, L.; Chen, C.; Cao, K.; Wang, Y.; Yuan, H. Exfoliated-SnS_2 restacked on graphene as a high-capacity, high-rate, and long-cycle life anode for sodium ion batteries. *Nanoscale* **2015**, *7*, 1325–1332. [CrossRef] [PubMed]
126. Li, S.; Cao, X.; Schmidt, C.N.; Xu, Q.; Uchaker, E.; Pei, Y.; Cao, G. $TiNb_2O_7$/graphene composites as high-rate anode materials for lithium/sodium ion batteries. *J. Mater. Chem. A* **2016**, *4*, 4242–4251. [CrossRef]
127. Wang, H.G.; Wu, Z.; Meng, F.L.; Ma, D.L.; Huang, X.L.; Wang, L.M.; Zhang, X.B. Nitrogen-doped porous carbon nanosheets as low-cost, high-performance anode material for sodium-ion batteries. *ChemSusChem* **2013**, *6*, 56–60. [CrossRef]
128. Hu, Y.; Ma, X.; Guo, P.; Jaeger, F.; Wang, Z. Design of $NaTi_2(PO_4)_3$ nanocrystals embedded in N-doped graphene sheets for sodium-ion battery anode with superior electrochemical performance. *Ceram. Int.* **2017**, *43*, 12338–12342. [CrossRef]
129. Yue, X.; Huang, N.; Jiang, Z.; Tian, X.; Wang, Z.; Hao, X.; Jiang, Z.J. Nitrogen-rich graphene hollow microspheres as anode materials for sodium-ion batteries with super-high cycling and rate performance. *Carbon* **2018**, *130*, 574–583. [CrossRef]
130. Ma, Y.; Guo, Q.; Yang, M.; Wang, Y.; Chen, T.; Chen, Q.; Zhu, X.; Xia, Q.; Li, S.; Xia, H. Highly doped graphene with multi-dopants for high-capacity and ultrastable sodium-ion batteries. *Energy Storage Mater.* **2018**, *13*, 134–141. [CrossRef]
131. Sultana, I.; Rahman, M.M.; Ramireddy, T.; Chen, Y.; Glushenkov, A.M. High capacity potassium-ion battery anodes based on black phosphorus. *J. Mater. Chem. A* **2017**, *5*, 23506–23512. [CrossRef]
132. Wu, X.; Zhao, W.; Wang, H.; Qi, X.; Xing, Z.; Zhuang, Q.; Ju, Z. Enhanced capacity of chemically bonded phosphorus/carbon composite as an anode material for potassium-ion batteries. *J. Power Sources* **2018**, *378*, 460–467. 2017.12.077. [CrossRef]
133. Qian, J.; Wu, X.; Cao, Y.; Ai, X.; Yang, H. High capacity and rate capability of amorphous phosphorus for sodium ion batteries. *Angew. Chem. Int. Ed.* **2013**, *52*, 4633–4636. [CrossRef] [PubMed]
134. Dahbi, M.; Yabuuchi, N.; Fukunishi, M.; Kubota, K.; Chihara, K.; Tokiwa, K.; Yu, X.F.; Ushiyama, H.; Yamashita, K.; Son, J.Y.; et al. Black Phosphorus as a High-Capacity, High-Capability Negative Electrode for Sodium-Ion Batteries: Investigation of the Electrode/Electrolyte Interface. *Chem. Mater.* **2016**, *28*, 1625–1635. [CrossRef]
135. Peng, B.; Xu, Y.; Liu, K.; Wang, X.; Mulder, F.M. High-Performance and Low-Cost Sodium-Ion Anode Based on a Facile Black Phosphorus-Carbon Nanocomposite. *ChemElectroChem* **2017**, *4*, 2140–2144. [CrossRef]
136. Sun, J.; Lee, H.W.; Pasta, M.; Yuan, H.; Zheng, G.; Sun, Y.; Li, Y.; Cui, Y. A phosphorene-graphene hybrid material as a high-capacity anode for sodium-ion batteries. *Nat. Nanotechnol.* **2015**, *10*, 980–985. [CrossRef]
137. Su, D.; Dou, S.; Wang, G. Ultrathin MoS_2 Nanosheets as Anode Materials for Sodium-Ion Batteries with Superior Performance. *Adv. Energy Mater.* **2015**, *5*, 1401205. [CrossRef]
138. Park, S.K.; Lee, J.; Bong, S.; Jang, B.; Seong, K.D.; Piao, Y. Scalable Synthesis of Few-Layer MoS_2 Incorporated into Hierarchical Porous Carbon Nanosheets for High-Performance Li- and Na-Ion Battery Anodes. *ACS Appl. Mater. Interfaces* **2016**, *8*, 19456–19465. [CrossRef]

139. David, L.; Bhandavat, R.; Singh, G. MoS$_2$/graphene composite paper for sodium-ion battery electrodes. *ACS Nano* **2014**, *8*, 1759–1770. [CrossRef]
140. Wang, H.; Lan, X.; Jiang, D.; Zhang, Y.; Zhong, H.; Zhang, Z.; Jiang, Y. Sodium storage and transport properties in pyrolysis synthesized MoSe$_2$ nanoplates for high performance sodium-ion batteries. *J. Power Sources* **2015**, *283*, 187–194. [CrossRef]
141. Share, K.; Lewis, J.; Oakes, L.; Carter, R.E.; Cohn, A.P.; Pint, C.L. Tungsten diselenide (WSe$_2$) as a high capacity, low overpotential conversion electrode for sodium ion batteries. *RSC Adv.* **2015**, *5*, 101262–101267. [CrossRef]
142. Zhang, Z.; Yang, X.; Fu, Y. Nanostructured WSe$_2$/C composites as anode materials for sodium-ion batteries. *RSC Adv.* **2016**, *6*, 12726–12729. [CrossRef]
143. Zhang, W.; Liu, Y.; Guo, Z. Approaching high-performance potassium-ion batteries via advanced design strategies and engineering. *Sci. Adv.* **2019**, *5*, eaav7412. [CrossRef] [PubMed]
144. Jian, Z.; Luo, W.; Ji, X. Carbon Electrodes for K-Ion Batteries. *J. Am. Chem. Soc.* **2015**, *137*, 11566–11569. jacs.5b06809. [CrossRef]
145. An, Y.; Fei, H.; Zeng, G.; Ci, L.; Xi, B.; Xiong, S.; Feng, J. Commercial expanded graphite as a low–cost, long-cycling life anode for potassium—Ion batteries with conventional carbonate electrolyte. *J. Power Sources* **2018**, *378*, 66–72. [CrossRef]
146. Jian, Z.; Xing, Z.; Bommier, C.; Li, Z.; Ji, X. Hard Carbon Microspheres: Potassium-Ion Anode Versus Sodium-Ion Anode. *Adv. Energy Mater.* **2016**, *6*, 1501874. [CrossRef]
147. Kishore, B.; G, V.; Munichandraiah, N. K$_2$Ti$_4$O$_9$: A Promising Anode Material for Potassium Ion Batteries. *J. Electrochem. Soc.* **2016**, *163*, A2551–A2554. [CrossRef]
148. Han, J.; Xu, M.; Niu, Y.; Li, G.N.; Wang, M.; Zhang, Y.; Jia, M.; Li, C.M. Exploration of K$_2$Ti$_8$O$_{17}$ as an anode material for potassium-ion batteries. *Chem. Commun.* **2016**, *52*, 11274–11276. [CrossRef]
149. Sha, M.; Liu, L.; Zhao, H.; Lei, Y. Anode materials for potassium-ion batteries: Current status and prospects. *Carbon Energy* **2020**, *2*, 350–369. [CrossRef]
150. Sultana, I.; Rahman, M.M.; Mateti, S.; Ahmadabadi, V.G.; Glushenkov, A.M.; Chen, Y. K-ion and Na-ion storage performances of Co$_3$O$_4$–Fe$_2$O$_3$ nanoparticle-decorated super P carbon black prepared by a ball milling process. *Nanoscale* **2017**, *9*, 3646–3654. [CrossRef]
151. Gao, H.; Zhou, T.; Zheng, Y.; Zhang, Q.; Liu, Y.; Chen, J.; Liu, H.; Guo, Z. CoS Quantum Dot Nanoclusters for High-Energy Potassium-Ion Batteries. *Adv. Funct. Mater.* **2017**, *27*, 1702634. [CrossRef]
152. Dong, Y.; Wu, Z.S.; Zheng, S.; Wang, X.; Qin, J.; Wang, S.; Shi, X.; Bao, X. Ti$_3$C$_2$ MXene-Derived Sodium/Potassium Titanate Nanoribbons for High-Performance Sodium/Potassium Ion Batteries with Enhanced Capacities. *ACS Nano* **2017**, *11*, 4792–4800. [CrossRef] [PubMed]
153. Han, J.; Niu, Y.; Bao, S.J.; Yu, Y.N.; Lu, S.Y.; Xu, M. Nanocubic KTi$_2$(PO$_4$)$_3$ electrodes for potassium-ion batteries. *Chem. Commun.* **2016**, *52*, 11661–11664. [CrossRef] [PubMed]
154. Ren, X.; Zhao, Q.; McCulloch, W.D.; Wu, Y. MoS$_2$ as a long-life host material for potassium ion intercalation. *Nano Res.* **2017**, *10*, 1313–1321. [CrossRef]
155. Xie, K.; Yuan, K.; Li, X.; Lu, W.; Shen, C.; Liang, C.; Vajtai, R.; Ajayan, P.; Wei, B. Superior Potassium Ion Storage via Vertical MoS$_2$ "Nano-Rose" with Expanded Interlayers on Graphene. *Small* **2017**, *13*, 1701471. [CrossRef] [PubMed]
156. Lu, Y.; Chen, J. Robust self-supported anode by integrating Sb$_2$S$_3$ nanoparticles with S,N-codoped graphene to enhance K-storage performance. *Sci. China Chem.* **2017**, *60*, 1533–1539. [CrossRef]
157. Lei, K.; Wang, C.; Liu, L.; Luo, Y.; Mu, C.; Li, F.; Chen, J. A Porous Network of Bismuth Used as the Anode Material for High-Energy-Density Potassium-Ion Batteries. *Angew. Chem.* **2018**, *130*, 4777–4781. [CrossRef]
158. Huang, J.; Chen, Z.; Ding, S.; Chen, C.; Zhang, M. Enhanced conductivity and properties of SnO$_2$-graphene-carbon nanofibers for potassium-ion batteries by graphene modification. *Mater. Lett.* **2018**, *219*, 19–22. [CrossRef]
159. Zhang, W.; Pang, W.K.; Sencadas, V.; Guo, Z. Understanding High-Energy-Density Sn$_4$P$_3$ Anodes for Potassium-Ion Batteries. *Joule* **2018**, *2*, 1534–1547. [CrossRef]
160. Verma, R.; Didwal, P.N.; Ki, H.S.; Cao, G.; Park, C.J. SnP$_3$/Carbon Nanocomposite as an Anode Material for Potassium-Ion Batteries. *ACS Appl. Mater. Interfaces* **2019**, *11*, 26976–26984. [CrossRef]
161. Lakshmi, V.; Chen, Y.; Mikhaylov, A.A.; Medvedev, A.G.; Sultana, I.; Rahman, M.M.; Lev, O.; Prikhodchenko, P.V.; Glushenkov, A.M. Nanocrystalline SnS$_2$ coated onto reduced graphene oxide: Demonstrating the feasibility of a non-graphitic anode with sulfide chemistry for potassium-ion batteries. *Chem. Commun.* **2017**, *53*, 8272–8275. [CrossRef]
162. Lian, P.; Dong, Y.; Wu, Z.S.; Zheng, S.; Wang, S.; Sun, C.; Qin, J.; Shi, X.; Bao, X. Alkalized Ti$_3$C$_2$ MXene nanoribbons with expanded interlayer spacing for high-capacity sodium and potassium ion batteries. *Nano Energy* **2017**, *40*, 1–8. [CrossRef]
163. Naguib, M.; Adams, R.A.; Zhao, Y.; Zemlyanov, D.; Varma, A.; Nanda, J.; Pol, V.G. Electrochemical performance of MXenes as K-ion battery anodes. *Chem. Commun.* **2017**, *53*, 6883–6886. [CrossRef] [PubMed]
164. Li, P.; Zheng, X.; Yu, H.; Zhao, G.; Shu, J.; Xu, X.; Sun, W.; Dou, S.X. Electrochemical potassium/lithium-ion intercalation into TiSe$_2$: Kinetics and mechanism. *Energy Storage Mater.* **2019**, *16*, 512–518. [CrossRef]
165. Yang, C.; Feng, J.; Lv, F.; Zhou, J.; Lin, C.; Wang, K.; Zhang, Y.; Yang, Y.; Wang, W.; Li, J.; et al. Metallic Graphene-Like VSe$_2$ Ultrathin Nanosheets: Superior Potassium-Ion Storage and Their Working Mechanism. *Adv. Mater.* **2018**, *30*, 1800036. [CrossRef]
166. Kim, H.; Yoon, G.; Lim, K.; Kang, K. A comparative study of graphite electrodes using the co-intercalation phenomenon for rechargeable Li, Na and K batteries. *Chem. Commun.* **2016**, *52*, 12618–12621. [CrossRef]

167. Zhao, J.; Zou, X.; Zhu, Y.; Xu, Y.; Wang, C. Electrochemical Intercalation of Potassium into Graphite. *Adv. Funct. Mater.* **2016**, *26*, 8103–8110. [CrossRef]
168. Komaba, S.; Hasegawa, T.; Dahbi, M.; Kubota, K. Potassium intercalation into graphite to realize high-voltage/high-power potassium-ion batteries and potassium-ion capacitors. *Electrochem. Commun.* **2015**, *60*, 172–175. j.elecom.2015.09.002. [CrossRef]
169. Xu, Z.; Lv, X.; Chen, J.; Jiang, L.; Lai, Y.; Li, J. Dispersion-corrected DFT investigation on defect chemistry and potassium migration in potassium-graphite intercalation compounds for potassium ion batteries anode materials. *Carbon* **2016**, *107*, 885–894. [CrossRef]
170. Deng, Q.; Pei, J.; Fan, C.; Ma, J.; Cao, B.; Li, C.; Jin, Y.; Wang, L.; Li, J. Potassium salts of para-aromatic dicarboxylates as the highly efficient organic anodes for low-cost K-ion batteries. *Nano Energy* **2017**, *33*, 350–355. [CrossRef]
171. Lei, K.; Li, F.; Mu, C.; Wang, J.; Zhao, Q.; Chen, C.; Chen, J. High K-storage performance based on the synergy of dipotassium terephthalate and ether-based electrolytes. *Energy Environ. Sci.* **2017**, *10*, 552–557. [CrossRef]
172. Xu, W.; Wang, Y. Recent Progress on Zinc-Ion Rechargeable Batteries. *Nano-Micro Lett.* **2019**, *11*, 90. [CrossRef] [PubMed]
173. Stock, D.; Dongmo, S.; Janek, J.; Schröder, D. Benchmarking Anode Concepts: The Future of Electrically Rechargeable Zinc–Air Batteries. *ACS Energy Lett.* **2019**, *4*, 1287–1300. [CrossRef]
174. Yan, Y.; Zhang, Y.; Wu, Y.; Wang, Z.; Mathur, A.; Yang, H.; Chen, P.; Nair, S.; Liu, N. A Lasagna-Inspired Nanoscale ZnO Anode Design for High-Energy Rechargeable Aqueous Batteries. *ACS Appl. Energy Mater.* **2018**, *1*, 6345–6351. [CrossRef]
175. Wu, Y.; Zhang, Y.; Ma, Y.; Howe, J.D.; Yang, H.; Chen, P.; Aluri, S.; Liu, N. Ion-Sieving Carbon Nanoshells for Deeply Rechargeable Zn-Based Aqueous Batteries. *Adv. Energy Mater.* **2018**, *8*, 1802470. [CrossRef]
176. Stock, D.; Dongmo, S.; Miyazaki, K.; Abe, T.; Janek, J.; Schröder, D. Towards zinc-oxygen batteries with enhanced cycling stability: The benefit of anion-exchange ionomer for zinc sponge anodes. *J. Power Sources* **2018**, *395*, 195–204. [CrossRef]
177. Chamoun, M.; Hertzberg, B.J.; Gupta, T.; Davies, D.; Bhadra, S.; Van Tassell, B.; Erdonmez, C.; Steingart, D.A. Hyper-dendritic nanoporous zinc foam anodes. *NPG Asia Mater.* **2015**, *7*, e178. [CrossRef]
178. Yan, Z.; Wang, E.; Jiang, L.; Sun, G. Superior cycling stability and high rate capability of three-dimensional Zn/Cu foam electrodes for zinc-based alkaline batteries. *RSC Adv.* **2015**, *5*, 83781–83787. [CrossRef]
179. Hwang, H.J.; Chi, W.S.; Kwon, O.; Lee, J.G.; Kim, J.H.; Shul, Y.G. Selective Ion Transporting Polymerized Ionic Liquid Membrane Separator for Enhancing Cycle Stability and Durability in Secondary Zinc-Air Battery Systems. *ACS Appl. Mater. Interfaces* **2016**, *8*, 26298–26308. [CrossRef]
180. Huang, J.; Yang, Z.; Feng, Z.; Xie, X.; Wen, X. A novel ZnO@Ag@Polypyrrole hybrid composite evaluated as anode material for zinc-based secondary cell. *Sci. Rep.* **2016**, *6*, 24471. [CrossRef]
181. Lee, J.; Hwang, B.; Park, M.S.; Kim, K. Improved reversibility of Zn anodes for rechargeable Zn-air batteries by using alkoxide and acetate ions. *Electrochim. Acta* **2016**, *199*, 164–171. [CrossRef]
182. Kakeya, T.; Nakata, A.; Arai, H.; Ogumi, Z. Enhanced zinc electrode rechargeability in alkaline electrolytes containing hydrophilic organic materials with positive electrode compatibility. *J. Power Sources* **2018**, *407*, 180–184. j.jpowsour.2018.08.026. [CrossRef]
183. Yan, X.; Chen, Z.; Wang, Y.; Li, H.; Zhang, J. In-situ growth of ZnO nanoplates on graphene for the application of high rate flexible quasi-solid-state Ni-Zn secondary battery. *J. Power Sources* **2018**, *407*, 137–146. [CrossRef]
184. Stock, D.; Dongmo, S.; Damtew, D.; Stumpp, M.; Konovalova, A.; Henkensmeier, D.; Schlettwein, D.; Schröder, D. Design Strategy for Zinc Anodes with Enhanced Utilization and Retention: Electrodeposited Zinc Oxide on Carbon Mesh Protected by Ionomeric Layers. *ACS Appl. Energy Mater.* **2018**, *1*, 5579–5588. [CrossRef]
185. Parker, J.F.; Chervin, C.N.; Pala, I.R.; Machler, M.; Burz, M.F.; Long, J.W.; Rolison, D.R. Rechargeable nickel–3D zinc batteries: An energy-dense, safer alternative to lithium-ion. *Science* **2017**, *356*, 415–418. [CrossRef]
186. Arroyo-De Dompablo, M.E.; Ponrouch, A.; Johansson, P.; Palacín, M.R. Achievements, Challenges, and Prospects of Calcium Batteries. *Chem. Rev.* **2020**, *120*, 6331–6357. [CrossRef]
187. Wu, S.; Zhang, F.; Tang, Y. A Novel Calcium-Ion Battery Based on Dual-Carbon Configuration with High Working Voltage and Long Cycling Life. *Adv. Sci.* **2018**, *5*, 1701082. [CrossRef]
188. Ponrouch, A.; Tchitchekova, D.; Frontera, C.; Bardé, F.; Dompablo, M.A.d.; Palacín, M. Assessing Si-based anodes for Ca-ion batteries: Electrochemical decalciation of CaSi$_2$. *Electrochem. Commun.* **2016**, *66*, 75–78. [CrossRef]
189. Wang, M.; Jiang, C.; Zhang, S.; Song, X.; Tang, Y.; Cheng, H.M. Reversible calcium alloying enables a practical room-temperature rechargeable calcium-ion battery with a high discharge voltage. *Nat. Chem.* **2018**, *10*, 667–672. [CrossRef]
190. Adil, M.; Sarkar, A.; Roy, A.; Panda, M.R.; Nagendra, A.; Mitra, S. Practical Aqueous Calcium-Ion Battery Full-Cells for Future Stationary Storage. *ACS Appl. Mater. Interfaces* **2020**, *12*, 11489–11503. [CrossRef]
191. Gheytani, S.; Liang, Y.; Wu, F.; Jing, Y.; Dong, H.; Rao, K.K.; Chi, X.; Fang, F.; Yao, Y. An Aqueous Ca-Ion Battery. *Adv. Sci.* **2017**, *4*, 1700465. [CrossRef]
192. Wu, N.; Yao, W.; Song, X.; Zhang, G.; Chen, B.; Yang, J.; Tang, Y. A Calcium-Ion Hybrid Energy Storage Device with High Capacity and Long Cycling Life under Room Temperature. *Adv. Energy Mater.* **2019**, *9*, 1803865. [CrossRef]
193. Tsai, P.c.; Chung, S.C.; Lin, S.k.; Yamada, A. Ab initio study of sodium intercalation into disordered carbon. *J. Mater. Chem. A* **2015**, *3*, 9763–9768. [CrossRef]

194. Rodríguez-Pérez, I.A.; Yuan, Y.; Bommier, C.; Wang, X.; Ma, L.; Leonard, D.P.; Lerner, M.M.; Carter, R.G.; Wu, T.; Greaney, P.A.; et al. Mg-Ion Battery Electrode: An Organic Solid's Herringbone Structure Squeezed upon Mg-Ion Insertion. *J. Am. Chem. Soc.* **2017**, *139*, 13031–13037. [CrossRef] [PubMed]
195. Attias, R.; Salama, M.; Hirsch, B.; Goffer, Y.; Aurbach, D. Anode-Electrolyte Interfaces in Secondary Magnesium Batteries. *Joule* **2019**, *3*, 27–52. [CrossRef]
196. Wu, N.; Lyu, Y.C.; Xiao, R.J.; Yu, X.; Yin, Y.X.; Yang, X.Q.; Li, H.; Gu, L.; Guo, Y.G. A highly reversible, low-strain Mg-ion insertion anode material for rechargeable Mg-ion batteries. *NPG Asia Mater.* **2014**, *6*, e120–e120. [CrossRef]
197. Meng, Z.; Foix, D.; Brun, N.; Dedryvère, R.; Stievano, L.; Morcrette, M.; Berthelot, R. Alloys to Replace Mg Anodes in Efficient and Practical Mg-Ion/Sulfur Batteries. *ACS Energy Lett.* **2019**, *4*, 2040–2044. [CrossRef]
198. Arthur, T.S.; Singh, N.; Matsui, M. Electrodeposited Bi, Sb and Bi1-xSbx alloys as anodes for Mg-ion batteries. *Electrochem. Commun.* **2012**, *16*, 103–106. [CrossRef]
199. Shao, Y.; Gu, M.; Li, X.; Nie, Z.; Zuo, P.; Li, G.; Liu, T.; Xiao, J.; Cheng, Y.; Wang, C.; et al. Highly reversible Mg insertion in nanostructured Bi for Mg ion batteries. *Nano Lett.* **2014**, *14*, 255–260. [CrossRef]
200. Cheng, Y.; Shao, Y.; Parent, L.R.; Sushko, M.L.; Li, G.; Sushko, P.V.; Browning, N.D.; Wang, C.; Liu, J. Interface Promoted Reversible Mg Insertion in Nanostructured Tin-Antimony Alloys. *Adv. Mater.* **2015**, *27*, 6598–6605. [CrossRef]
201. Murgia, F.; Weldekidan, E.T.; Stievano, L.; Monconduit, L.; Berthelot, R. First investigation of indium-based electrode in Mg battery. *Electrochem. Commun.* **2015**, *60*, 56–59. [CrossRef]
202. Faegh, E.; Ng, B.; Hayman, D.; Mustain, W.E. Practical assessment of the performance of aluminium battery technologies. *Nat. Energy* **2021**, *6*, 21–29. [CrossRef]
203. Zhang, K.; Kirlikovali, K.O.; Suh, J.M.; Choi, J.W.; Jang, H.W.; Varma, R.S.; Farha, O.K.; Shokouhimehr, M. Recent Advances in Rechargeable Aluminum-Ion Batteries and Considerations for Their Future Progress. *ACS Appl. Energy Mater.* **2020**, *3*, 6019–6035. [CrossRef]
204. Liu, C.; Neale, Z.G.; Cao, G. Understanding electrochemical potentials of cathode materials in rechargeable batteries. *Mater. Today* **2016**, *19*, 109–123. [CrossRef]
205. Liu, M.; Rong, Z.; Malik, R.; Canepa, P.; Jain, A.; Ceder, G.; Persson, K.A. Spinel compounds as multivalent battery cathodes: A systematic evaluation based on ab initio calculations. *Energy Environ. Sci.* **2015**, *8*, 964–974. [CrossRef]
206. Wang, B.; Han, Y.; Wang, X.; Bahlawane, N.; Pan, H.; Yan, M.; Jiang, Y. Prussian Blue Analogs for Rechargeable Batteries. *iScience* **2018**, *3*, 110–133. [CrossRef]
207. Wang, H.F.; Xu, Q. Materials Design for Rechargeable Metal-Air Batteries. *Matter* **2019**, *1*, 565–595. 2019.05.008. [CrossRef]
208. Xing, W. High Energy Density Li-Ion Batteries with ALD Multi-Functional Modified LiCoO$_2$ Cathode. *ECS Trans.* **2017**, *80*, 55–63. [CrossRef]
209. Lu, L.; Jiang, G.; Gu, C.; Ni, J. Revisiting polyanionic LiFePO$_4$ battery material for electric vehicles. *Funct. Mater. Lett.* **2021**, *14*, 2130006. [CrossRef]
210. Wang, B.; Liu, T.; Liu, A.; Liu, G.; Wang, L.; Gao, T.; Wang, D.; Zhao, X.S. A Hierarchical Porous C@LiFePO$_4$/Carbon Nanotubes Microsphere Composite for High-Rate Lithium-Ion Batteries: Combined Experimental and Theoretical Study. *Adv. Energy Mater.* **2016**, *6*, 1600426. [CrossRef]
211. Zhang, L.; Fu, J.; Zhang, C. Mechanical Composite of LiNi$_{0.8}$Co$_{0.15}$Al$_{0.05}$O$_2$/Carbon Nanotubes with Enhanced Electrochemical Performance for Lithium-Ion Batteries. *Nanoscale Res. Lett.* **2017**, *12*, 376. [CrossRef]
212. Muralidharan, N.; Essehli, R.; Hermann, R.P.; Amin, R.; Jafta, C.; Zhang, J.; Liu, J.; Du, Z.; Meyer, H.M.; Self, E.; et al. Lithium Iron Aluminum Nickelate, LiNi$_x$Fe$_y$Al$_z$O$_2$ —New Sustainable Cathodes for Next-Generation Cobalt-Free Li-Ion Batteries. *Adv. Mater.* **2020**, *32*, 2002960. [CrossRef]
213. Zhao, T.; Chang, L.; Ji, R. Controllable preparation of Fe-containing Li-rich cathode material Li[Li$_{1/5}$Fe$_{1/10}$Ni$_{3/20}$Mn$_{11/20}$]O$_2$ with stable high-rate properties for Li-ion batteries. *Funct. Mater. Lett.* **2021**, *14*, 2150004. [CrossRef]
214. Ould Ely, T.; Kamzabek, D.; Chakraborty, D.; Doherty, M.F. Lithium-Sulfur Batteries: State of the Art and Future Directions. *ACS Appl. Energy Mater.* **2018**, *1*, 1783–1814. [CrossRef]
215. Fang, R.; Chen, K.; Yin, L.; Sun, Z.; Li, F.; Cheng, H.M. The Regulating Role of Carbon Nanotubes and Graphene in Lithium-Ion and Lithium-Sulfur Batteries. *Adv. Mater.* **2019**, *31*, 1800863. [CrossRef] [PubMed]
216. Yuan, H.; Li, H.; Zhang, T.; Li, G.; He, T.; Du, F.; Feng, S. A K$_2$Fe$_4$O$_7$ superionic conductor for all-solid-state potassium metal batteries. *J. Mater. Chem. A* **2018**, *6*, 8413–8418. [CrossRef]
217. Sun, Q.; Lau, K.C.; Geng, D.; Meng, X. Atomic and Molecular Layer Deposition for Superior Lithium-Sulfur Batteries: Strategies, Performance, and Mechanisms. *Batter. Supercaps* **2018**, *1*, 41–68. [CrossRef]
218. Deng, C.; Wang, Z.; Wang, S.; Yu, J. Inhibition of polysulfide diffusion in lithium-sulfur batteries: Mechanism and improvement strategies. *J. Mater. Chem. A* **2019**, *7*, 12381–12413. [CrossRef]
219. Yang, W.; Yang, W.; Song, A.; Sun, G.; Shao, G. 3D interconnected porous carbon nanosheets/carbon nanotubes as a polysulfide reservoir for high performance lithium–sulfur batteries. *Nanoscale* **2018**, *10*, 816–824. [CrossRef]
220. Gueon, D.; Hwang, J.T.; Yang, S.B.; Cho, E.; Sohn, K.; Yang, D.K.; Moon, J.H. Spherical Macroporous Carbon Nanotube Particles with Ultrahigh Sulfur Loading for Lithium—Sulfur Battery Cathodes. *ACS Nano* **2018**, *12*, 226–233. [CrossRef]

221. Ummethala, R.; Fritzsche, M.; Jaumann, T.; Balach, J.; Oswald, S.; Nowak, R.; Sobczak, N.; Kaban, I.; Rümmeli, M.H.; Giebeler, L. Lightweight, free-standing 3D interconnected carbon nanotube foam as a flexible sulfur host for high performance lithium-sulfur battery cathodes. *Energy Storage Mater.* **2018**, *10*, 206–215. [CrossRef]
222. Amin, K.; Meng, Q.; Ahmad, A.; Cheng, M.; Zhang, M.; Mao, L.; Lu, K.; Wei, Z. A Carbonyl Compound-Based Flexible Cathode with Superior Rate Performance and Cyclic Stability for Flexible Lithium-Ion Batteries. *Adv. Mater.* **2018**, *30*, 1703868. [CrossRef] [PubMed]
223. Wu, T.; Qi, J.; Xu, M.; Zhou, D.; Xiao, Z. Selective S/Li_2S Conversion via in-Built Crystal Facet Self-Mediation: Toward High Volumetric Energy Density Lithium–Sulfur Batteries. *ACS Nano* **2020**, *14*, 15011–15022. [CrossRef] [PubMed]
224. Zhao, W.; Mu, X.; He, P.; Zhou, H. Advances and Challenges for Aprotic Lithium-Oxygen Batteries using Redox Mediators. *Batter. Supercaps* **2019**, *2*, 803–819. [CrossRef]
225. Jin, Y.; Yang, C.P.; Rui, X.H.; Cheng, T.; Chen, C.H. V_2O_3 modified $LiFePO_4$/C composite with improved electrochemical performance. *J. Power Sources* **2011**, *196*, 5623–5630. [CrossRef]
226. Wang, B.; Al Abdulla, W.; Wang, D.; Zhao, X.S. A three-dimensional porous $LiFePO_4$ cathode material modified with a nitrogen-doped graphene aerogel for high-power lithium ion batteries. *Energy Environ. Sci.* **2015**, *8*, 869–875. [CrossRef]
227. Fang, X.; Shen, C.; Ge, M.; Rong, J.; Liu, Y.; Zhang, A.; Wei, F.; Zhou, C. High-power lithium ion batteries based on flexible and light-weight cathode of $LiNi_{0.5}Mn_{1.5}O_4$/carbon nanotube film. *Nano Energy* **2015**, *12*, 43–51. [CrossRef]
228. Babu, G.; Kalaiselvi, N.; Bhuvaneswari, D. Synthesis and surface modification of $LiCo_{1/3}Ni_{1/3}Mn_{1/3}O_2$ for lithium battery applications. *J. Electron. Mater.* **2014**, *43*, 1062–1070. [CrossRef]
229. Kong, D.; Li, X.; Zhang, Y.; Hai, X.; Wang, B.; Qiu, X.; Song, Q.; Yang, Q.H.; Zhi, L. Encapsulating V_2O_4 into carbon nanotubes enables the synthesis of flexible high-performance lithium ion batteries. *Energy Environ. Sci.* **2016**, *9*, 906–911. [CrossRef]
230. Kim, J.H.; Fu, K.; Choi, J.; Sun, S.; Kim, J.; Hu, L.; Paik, U. Hydroxylated carbon nanotube enhanced sulfur cathodes for improved electrochemical performance of lithium-sulfur batteries. *Chem. Commun.* **2015**, *51*, 13682–13685. [CrossRef] [PubMed]
231. Carter, R.; Davis, B.; Oakes, L.; Maschmann, M.R.; Pint, C.L. A high areal capacity lithium–sulfur battery cathode prepared by site-selective vapor infiltration of hierarchical carbon nanotube arrays. *Nanoscale* **2017**, *9*, 15018–15026. [CrossRef]
232. Pan, H.; Cheng, Z.; Xiao, Z.; Li, X.; Wang, R. The Fusion of Imidazolium-Based Ionic Polymer and Carbon Nanotubes: One Type of New Heteroatom-Doped Carbon Precursors for High-Performance Lithium-Sulfur Batteries. *Adv. Funct. Mater.* **2017**, *27*, 1703936. [CrossRef]
233. Huang, Y.; Zheng, Y.; Li, X.; Adams, F.; Luo, W.; Huang, Y.; Hu, L. Electrode Materials of Sodium-Ion Batteries toward Practical Application. *ACS Energy Lett.* **2018**, *3*, 1604–1612. [CrossRef]
234. Martinez De Ilarduya, J.; Otaegui, L.; López del Amo, J.M.; Armand, M.; Singh, G. NaN_3 addition, a strategy to overcome the problem of sodium deficiency in P_2-$Na_{0.67}[Fe_{0.5}Mn_{0.5}]O_2$ cathode for sodium-ion battery. *J. Power Sources* **2017**, *337*, 197–203. [CrossRef]
235. Wang, P.F.; You, Y.; Yin, Y.X.; Guo, Y.G. An O3-type $NaNi_{0.5}Mn_{0.5}O_2$ cathode for sodium-ion batteries with improved rate performance and cycling stability. *J. Mater. Chem. A* **2016**, *4*, 17660–17664. [CrossRef]
236. Wang, P.F.; Yao, H.R.; Liu, X.Y.; Yin, Y.X.; Zhang, J.N.; Wen, Y.; Yu, X.; Gu, L.; Guo, Y.G. Na+/vacancy disordering promises high-rate Na-ion batteries. *Sci. Adv.* **2018**, *4*, eaar6018. [CrossRef] [PubMed]
237. Manzi, J.; Paolone, A.; Palumbo, O.; Corona, D.; Massaro, A.; Cavaliere, R.; Muñoz-García, A.B.; Trequattrini, F.; Pavone, M.; Brutti, S. Monoclinic and Orthorhombic $NaMnO_2$ for Secondary Batteries: A Comparative Study. *Energies* **2021**, *14*, 1230. [CrossRef]
238. Song, T.; Yao, W.; Kiadkhunthod, P.; Zheng, Y.; Wu, N.; Zhou, X.; Tunmee, S.; Sattayaporn, S.; Tang, Y. A Low-Cost and Environmentally Friendly Mixed Polyanionic Cathode for Sodium-Ion Storage. *Angew. Chem.* **2020**, *132*, 750–755. [CrossRef]
239. Xie, B.; Sun, B.; Gao, T.; Ma, Y.; Yin, G.; Zuo, P. Recent progress of Prussian blue analogues as cathode materials for nonaqueous sodium-ion batteries. *Coord. Chem. Rev.* **2022**, *460*, 214478. [CrossRef]
240. Hu, P.; Peng, W.; Wang, B.; Xiao, D.; Ahuja, U.; Réthoré, J.; Aifantis, K.E. Concentration-Gradient Prussian Blue Cathodes for Na-Ion Batteries. *ACS Energy Lett.* **2020**, *5*, 100–108. [CrossRef]
241. Zhu, J.; Li, Y.; Huang, Y.; Ou, C.; Yuan, X.; Yan, L.; Li, W.; Zhang, H.; Shen, P.K. General Strategy To Synthesize Highly Dense Metal Oxide Quantum Dots-Anchored Nitrogen-Rich Graphene Compact Monoliths To Enable Fast and High-Stability Volumetric Lithium/Sodium Storage. *ACS Appl. Energy Mater.* **2019**, *2*, 3500–3512. [CrossRef]
242. Mao, Q.; Gao, R.; Li, Q.; Ning, D.; Zhou, D.; Schuck, G.; Schumacher, G.; Hao, Y.; Liu, X. O3-type $NaNi_{0.5}Mn_{0.5}O_2$ hollow microbars with exposed {0 1 0} facets as high performance cathode materials for sodium-ion batteries. *Chem. Eng. J.* **2020**, *382*, 122978. [CrossRef]
243. Zhao, C.; Ding, F.; Lu, Y.; Chen, L.; Hu, Y. High-Entropy Layered Oxide Cathodes for Sodium-Ion Batteries. *Angew. Chem.* **2020**, *132*, 270–275. [CrossRef]
244. Wang, P.F.; Yao, H.R.; Liu, X.Y.; Zhang, J.N.; Gu, L.; Yu, X.Q.; Yin, Y.X.; Guo, Y.G. Ti-Substituted $NaNi_{0.5}Mn_{0.5-x}Ti_xO_2$ Cathodes with Reversible O3-P3 Phase Transition for High-Performance Sodium-Ion Batteries. *Adv. Mater.* **2017**, *29*, 1700210. [CrossRef]
245. Wang, P.F.; Guo, Y.J.; Duan, H.; Zuo, T.T.; Hu, E.; Attenkofer, K.; Li, H.; Zhao, X.S.; Yin, Y.X.; Yu, X.; et al. Honeycomb-Ordered $Na_3Ni_{1.5}M_{0.5}BiO_6$ (M = Ni, Cu, Mg, Zn) as High-Voltage Layered Cathodes for Sodium-Ion Batteries. *ACS Energy Lett.* **2017**, *2*, 2715–2722. [CrossRef]

246. Talaie, E.; Duffort, V.; Smith, H.L.; Fultz, B.; Nazar, L.F. Structure of the high voltage phase of layered P2-$Na_{2/3-z}[Mn_{1/2}Fe_{1/2}]O_2$ and the positive effect of Ni substitution on its stability. *Energy Environ. Sci.* **2015**, *8*, 2512–2523. [CrossRef]
247. Fang, Y.; Yu, X.Y.; Lou, X.W.D. A Practical High-Energy Cathode for Sodium-Ion Batteries Based on Uniform P2-$Na_{0.7}CoO_2$ Microspheres. *Angew. Chem.* **2017**, *129*, 5895–5899. [CrossRef]
248. Mu, L.; Xu, S.; Li, Y.; Hu, Y.S.; Li, H.; Chen, L.; Huang, X. Prototype Sodium-Ion Batteries Using an Air-Stable and Co/Ni-Free O3-Layered Metal Oxide Cathode. *Adv. Mater.* **2015**, *27*, 6928–6933. [CrossRef] [PubMed]
249. Wang, H.; Liao, X.Z.; Yang, Y.; Yan, X.; He, Y.S.; Ma, Z.F. Large-Scale Synthesis of $NaNi_{1/3}Fe_{1/3}Mn_{1/3}O_2$ as High Performance Cathode Materials for Sodium Ion Batteries. *J. Electrochem. Soc.* **2016**, *163*, A565–A570. [CrossRef]
250. Wang, P.F.; You, Y.; Yin, Y.X.; Wang, Y.S.; Wan, L.J.; Gu, L.; Guo, Y.G. Suppressing the P2-O2 Phase Transition of $Na_{0.67}Mn_{0.67}Ni_{0.33}O_2$ by Magnesium Substitution for Improved Sodium-Ion Batteries. *Angew. Chem.* **2016**, *128*, 7571–7575. [CrossRef]
251. Wang, L.; Song, J.; Qiao, R.; Wray, L.A.; Hossain, M.A.; Chuang, Y.D.; Yang, W.; Lu, Y.; Evans, D.; Lee, J.J.; et al. Rhombohedral Prussian White as Cathode for Rechargeable Sodium-Ion Batteries. *J. Am. Chem. Soc.* **2015**, *137*, 2548–2554. [CrossRef]
252. Bianchini, M.; Xiao, P.; Wang, Y.; Ceder, G. Additional Sodium Insertion into Polyanionic Cathodes for Higher-Energy Na-Ion Batteries. *Adv. Energy Mater.* **2017**, *7*, 1700514. [CrossRef]
253. Carter, R.; Oakes, L.; Douglas, A.; Muralidharan, N.; Cohn, A.P.; Pint, C.L. A Sugar-Derived Room-Temperature Sodium Sulfur Battery with Long Term Cycling Stability. *Nano Lett.* **2017**, *17*, 1863–1869. [CrossRef] [PubMed]
254. Zhang, B.W.; Sheng, T.; Wang, Y.X.; Chou, S.; Davey, K.; Dou, S.X.; Qiao, S.Z. Long-Life Room-Temperature Sodium–Sulfur Batteries by Virtue of Transition-Metal-Nanocluster—Sulfur Interactions. *Angew. Chem. Int. Ed.* **2019**, *58*, 1484–1488. [CrossRef] [PubMed]
255. Kim, H.; Kim, J.C.; Bo, S.; Shi, T.; Kwon, D.; Ceder, G. K-Ion Batteries Based on a P_2-Type $K_{0.6}CoO_2$ Cathode. *Adv. Energy Mater.* **2017**, *7*, 1700098. [CrossRef]
256. Vaalma, C.; Giffin, G.A.; Buchholz, D.; Passerini, S. Non-Aqueous K-Ion Battery Based on Layered $K_{0.3}MnO_2$ and Hard Carbon/Carbon Black. *J. Electrochem. Soc.* **2016**, *163*, A1295–A1299. [CrossRef]
257. Kim, H.; Seo, D.; Kim, J.C.; Bo, S.; Liu, L.; Shi, T.; Ceder, G. Investigation of Potassium Storage in Layered P3-Type $K_{0.5}MnO_2$ Cathode. *Adv. Mater.* **2017**, *29*, 1702480. [CrossRef]
258. Xiao, Z.; Meng, J.; Xia, F.; Wu, J.; Liu, F.; Zhang, X.; Xu, L.; Lin, X.; Mai, L. K^+ modulated K^+/vacancy disordered layered oxide for high-rate and high-capacity potassium-ion batteries. *Energy Environ. Sci.* **2020**, *13*, 3129–3137. [CrossRef]
259. Chihara, K.; Katogi, A.; Kubota, K.; Komaba, S. $KVPO_4F$ and $KVOPO_4$ toward 4 volt-class potassium-ion batteries. *Chem. Commun.* **2017**, *53*, 5208–5211. [CrossRef]
260. Park, W.B.; Han, S.C.; Park, C.; Hong, S.U.; Han, U.; Singh, S.P.; Jung, Y.H.; Ahn, D.; Sohn, K.S.; Pyo, M. KVP_2O_7 as a Robust High-Energy Cathode for Potassium-Ion Batteries: Pinpointed by a Full Screening of the Inorganic Registry under Specific Search Conditions. *Adv. Energy Mater.* **2018**, *8*, 1703099. [CrossRef]
261. Lin, X.; Huang, J.; Tan, H.; Huang, J.; Zhang, B. $K_3V_2(PO_4)_2F_3$ as a robust cathode for potassium-ion batteries. *Energy Storage Mater.* **2019**, *16*, 97–101. [CrossRef]
262. Bie, X.; Kubota, K.; Hosaka, T.; Chihara, K.; Komaba, S. A novel K-ion battery: Hexacyanoferrate(ii)/graphite cell. *J. Mater. Chem. A* **2017**, *5*, 4325–4330. [CrossRef]
263. Xue, L.; Li, Y.; Gao, H.; Zhou, W.; Lü, X.; Kaveevivitchai, W.; Manthiram, A.; Goodenough, J.B. Low-Cost High-Energy Potassium Cathode. *J. Am. Chem. Soc.* **2017**, *139*, 2164–2167. [CrossRef]
264. Zhao, J.; Yang, J.; Sun, P.; Xu, Y. Sodium sulfonate groups substituted anthraquinone as an organic cathode for potassium batteries. *Electrochem. Commun.* **2018**, *86*, 34–37. [CrossRef]
265. Zhao, Q.; Wang, J.; Lu, Y.; Li, Y.; Liang, G.; Chen, J. Oxocarbon Salts for Fast Rechargeable Batteries. *Angew. Chem. Int. Ed.* **2016**, *55*, 12528–12532. [CrossRef] [PubMed]
266. Zhang, Q.; Wang, Z.; Zhang, S.; Zhou, T.; Mao, J.; Guo, Z. Cathode Materials for Potassium-Ion Batteries: Current Status and Perspective. *Electrochem. Energy Rev.* **2018**, *1*, 625–658. [CrossRef]
267. Jian, Z.; Liang, Y.; Rodríguez-Pérez, I.A.; Yao, Y.; Ji, X. Poly(anthraquinonyl sulfide) cathode for potassium-ion batteries. *Electrochem. Commun.* **2016**, *71*, 5–8. [CrossRef]
268. Liu, Y.; Wang, W.; Wang, J.; Zhang, Y.; Zhu, Y.; Chen, Y.; Fu, L.; Wu, Y. Sulfur nanocomposite as a positive electrode material for rechargeable potassium—Sulfur batteries. *Chem. Commun.* **2018**, *54*, 2288–2291. [CrossRef]
269. Ma, R.; Fan, L.; Wang, J.; Lu, B. Confined and covalent sulfur for stable room temperature potassium-sulfur battery. *Electrochim. Acta* **2019**, *293*, 191–198. [CrossRef]
270. Xiong, P.; Han, X.; Zhao, X.; Bai, P.; Liu, Y.; Sun, J.; Xu, Y. Room-Temperature Potassium–Sulfur Batteries Enabled by Microporous Carbon Stabilized Small-Molecule Sulfur Cathodes. *ACS Nano* **2019**, *13*, acsnano.8b09503. [CrossRef]
271. Sada, K.; Senthilkumar, B.; Barpanda, P. Electrochemical potassium-ion intercalation in $Na:XCoO_2$: A novel cathode material for potassium-ion batteries. *Chem. Commun.* **2017**, *53*, 8588–8591. [CrossRef]
272. Naveen, N.; Park, W.B.; Han, S.C.; Singh, S.P.; Jung, Y.H.; Ahn, D.; Sohn, K.S.; Pyo, M. Reversible K+-Insertion/Deinsertion and Concomitant Na+-Redistribution in P'3-$Na_{0.52}CrO_2$ for High-Performance Potassium-Ion Battery Cathodes. *Chem. Mater.* **2018**, *30*, 2049–2057. [CrossRef]

273. Deng, T.; Fan, X.; Luo, C.; Chen, J.; Chen, L.; Hou, S.; Eidson, N.; Zhou, X.; Wang, C. Self-Templated Formation of P2-type $K_{0.6}CoO_2$ Microspheres for High Reversible Potassium-Ion Batteries. *Nano Lett.* **2018**, *18*, 1522–1529. [CrossRef] [PubMed]
274. Wang, X.; Xu, X.; Niu, C.; Meng, J.; Huang, M.; Liu, X.; Liu, Z.; Mai, L. Earth Abundant Fe/Mn-Based Layered Oxide Interconnected Nanowires for Advanced K-Ion Full Batteries. *Nano Lett.* **2017**, *17*, 544–550. [CrossRef] [PubMed]
275. Deng, T.; Fan, X.; Chen, J.; Chen, L.; Luo, C.; Zhou, X.; Yang, J.; Zheng, S.; Wang, C. Layered P2-Type $K_{0.65}Fe_{0.5}Mn_{0.5}O_2$ Microspheres as Superior Cathode for High-Energy Potassium-Ion Batteries. *Adv. Funct. Mater.* **2018**, *28*, 1800219. [CrossRef]
276. Liu, C.; Luo, S.; Huang, H.; Wang, Z.; Hao, A.; Zhai, Y.; Wang, Z. $K_{0.67}Ni_{0.17}Co_{0.17}Mn_{0.66}O_2$: A cathode material for potassium-ion battery. *Electrochem. Commun.* **2017**, *82*, 150–154. [CrossRef]
277. Deng, L.; Niu, X.; Ma, G.; Yang, Z.; Zeng, L.; Zhu, Y.; Guo, L. Layered Potassium Vanadate $K_{0.5}V_2O_5$ as a Cathode Material for Nonaqueous Potassium Ion Batteries. *Adv. Funct. Mater.* **2018**, *28*, 1800670. [CrossRef]
278. Wessells, C.D.; Huggins, R.A.; Cui, Y. Copper hexacyanoferrate battery electrodes with long cycle life and high power. *Nat. Commun.* **2011**, *2*, 550. [CrossRef]
279. Pei, Y.; Mu, C.; Li, H.; Li, F.; Chen, J. Low-Cost K4Fe(CN)6 as a High-Voltage Cathode for Potassium-Ion Batteries. *ChemSusChem* **2018**, *11*, 1285–1289. [CrossRef]
280. Eftekhari, A. Potassium secondary cell based on Prussian blue cathode. *J. Power Sources* **2004**, *126*, 221–228. [CrossRef]
281. Chong, S.; Chen, Y.; Zheng, Y.; Tan, Q.; Shu, C.; Liu, Y.; Guo, Z. Potassium ferrous ferricyanide nanoparticles as a high capacity and ultralong life cathode material for nonaqueous potassium-ion batteries. *J. Mater. Chem. A* **2017**, *5*, 22465–22471. [CrossRef]
282. Jiang, X.; Zhang, T.; Yang, L.; Li, G.; Lee, J.Y. A Fe/Mn-Based Prussian Blue Analogue as a K-Rich Cathode Material for Potassium-Ion Batteries. *ChemElectroChem* **2017**, *4*, 2237–2242. [CrossRef]
283. Wessells, C.D.; Peddada, S.V.; Huggins, R.A.; Cui, Y. Nickel hexacyanoferrate nanoparticle electrodes for aqueous sodium and potassium ion batteries. *Nano Lett.* **2011**, *11*, 5421–5425. [CrossRef] [PubMed]
284. Nossol, E.; Souza, V.H.; Zarbin, A.J. Carbon nanotube/Prussian blue thin films as cathodes for flexible, transparent and ITO-free potassium secondary battery. *J. Colloid Interface Sci.* **2016**, *478*, 107–116. [CrossRef]
285. Zhu, Y.h.; Yin, Y.b.; Yang, X.; Sun, T.; Wang, S.; Jiang, Y.s.; Yan, J.m.; Zhang, X.b. Transformation of Rusty Stainless-Steel Meshes into Stable, Low-Cost, and Binder-Free Cathodes for High-Performance Potassium-Ion Batteries. *Angew. Chem.* **2017**, *129*, 7989–7993. [CrossRef]
286. Han, J.; Li, G.N.; Liu, F.; Wang, M.; Zhang, Y.; Hu, L.; Dai, C.; Xu, M. Investigation of $K_3V_2(PO_4)_3$/C nanocomposites as high-potential cathode materials for potassium-ion batteries. *Chem. Commun.* **2017**, *53*, 1805–1808. [CrossRef]
287. Chen, Y.; Luo, W.; Carter, M.; Zhou, L.; Dai, J.; Fu, K.; Lacey, S.; Li, T.; Wan, J.; Han, X.; et al. Organic electrode for non-aqueous potassium-ion batteries. *Nano Energy* **2015**, *18*, 205–211. [CrossRef]
288. Lu, X.; Bowden, M.E.; Sprenkle, V.L.; Liu, J. A Low Cost, High Energy Density, and Long Cycle Life Potassium-Sulfur Battery for Grid-Scale Energy Storage. *Adv. Mater.* **2015**, *27*, 5915–5922. [CrossRef]
289. Wang, L.; Bao, J.; Liu, Q.; Sun, C.F. Concentrated electrolytes unlock the full energy potential of potassium-sulfur battery chemistry. *Energy Storage Mater.* **2019**, *18*, 470–475. [CrossRef]
290. Yu, X.; Manthiram, A. Performance Enhancement and Mechanistic Studies of Room-Temperature Sodium-Sulfur Batteries with a Carbon-Coated Functional Nafion Separator and a Na_2S/Activated Carbon Nanofiber Cathode. *Chem. Mater.* **2016**, *28*, 896–905. [CrossRef]
291. Hwang, J.Y.; Kim, H.M.; Sun, Y.K. High performance potassium-sulfur batteries based on a sulfurized polyacrylonitrile cathode and polyacrylic acid binder. *J. Mater. Chem. A* **2018**, *6*, 14587–14593. [CrossRef]
292. Alfaruqi, M.H.; Mathew, V.; Gim, J.; Kim, S.; Song, J.; Baboo, J.P.; Choi, S.H.; Kim, J. Electrochemically Induced Structural Transformation in a γ-MnO_2 Cathode of a High Capacity Zinc-Ion Battery System. *Chem. Mater.* **2015**, *27*, 3609–3620. [CrossRef]
293. Yan, M.; He, P.; Chen, Y.; Wang, S.; Wei, Q.; Zhao, K.; Xu, X.; An, Q.; Shuang, Y.; Shao, Y.; et al. Water-Lubricated Intercalation in $V_2O_5·nH_2O$ for High-Capacity and High-Rate Aqueous Rechargeable Zinc Batteries. *Adv. Mater.* **2018**, *30*, 1703725. [CrossRef] [PubMed]
294. Wang, F.; Hu, E.; Sun, W.; Gao, T.; Ji, X.; Fan, X.; Han, F.; Yang, X.Q.; Xu, K.; Wang, C. A rechargeable aqueous Zn^{2+}-battery with high power density and a long cycle-life. *Energy Environ. Sci.* **2018**, *11*, 3168–3175. [CrossRef]
295. Wei, T.; Li, Q.; Yang, G.; Wang, C. An electrochemically induced bilayered structure facilitates long-life zinc storage of vanadium dioxide. *J. Mater. Chem. A* **2018**, *6*, 8006–8012. [CrossRef]
296. Ding, J.; Du, Z.; Gu, L.; Li, B.; Wang, L.; Wang, S.; Gong, Y.; Yang, S. Ultrafast Zn^{2+} Intercalation and Deintercalation in Vanadium Dioxide. *Adv. Mater.* **2018**, *30*, 1800762. [CrossRef]
297. Pang, Q.; Sun, C.; Yu, Y.; Zhao, K.; Zhang, Z.; Voyles, P.M.; Chen, G.; Wei, Y.; Wang, X. $H_2V_3O_8$ Nanowire/Graphene Electrodes for Aqueous Rechargeable Zinc Ion Batteries with High Rate Capability and Large Capacity. *Adv. Energy Mater.* **2018**, *8*, 1800144. [CrossRef]
298. Deng, W.; Zhou, Z.; Li, Y.; Zhang, M.; Yuan, X.; Hu, J.; Li, Z.; Li, C.; Li, R. High-Capacity Layered Magnesium Vanadate with Concentrated Gel Electrolyte toward High-Performance and Wide-Temperature Zinc-Ion Battery. *ACS Nano* **2020**, *14*, 15776–15785. [CrossRef]
299. Trócoli, R.; La Mantia, F. An aqueous zinc-ion battery based on copper hexacyanoferrate. *ChemSusChem* **2015**, *8*, 481–485. [CrossRef]

300. Zhang, L.; Chen, L.; Zhou, X.; Liu, Z. Towards High-Voltage Aqueous Metal-Ion Batteries Beyond 1.5 V: The Zinc/Zinc Hexacyanoferrate System. *Adv. Energy Mater.* **2015**, *5*, 1400930. [CrossRef]
301. Zhao, H.B.; Hu, C.J.; Cheng, H.W.; Fang, J.H.; Xie, Y.P.; Fang, W.Y.; Doan, T.N.L.; Hoang, T.K.A.; Xu, J.Q.; Chen, P. Novel Rechargeable $M_3V_2(PO_4)_3$//Zinc (M = Li, Na) Hybrid Aqueous Batteries with Excellent Cycling Performance. *Sci. Rep.* **2016**, *6*, 25809. [CrossRef]
302. Li, G.; Yang, Z.; Jiang, Y.; Jin, C.; Huang, W.; Ding, X.; Huang, Y. Towards polyvalent ion batteries: A zinc-ion battery based on NASICON structured $Na_3V_2(PO_4)_3$. *Nano Energy* **2016**, *25*, 211–217. [CrossRef]
303. Li, W.; Wang, K.; Cheng, S.; Jiang, K. A long-life aqueous Zn-ion battery based on $Na_3V_2(PO_4)_2F_3$ cathode. *Energy Storage Mater.* **2018**, *15*, 14–21. [CrossRef]
304. Xu, W.; Zhao, K.; Huo, W.; Wang, Y.; Yao, G.; Gu, X.; Cheng, H.; Mai, L.; Hu, C.; Wang, X. Diethyl ether as self-healing electrolyte additive enabled long-life rechargeable aqueous zinc ion batteries. *Nano Energy* **2019**, *62*, 275–281. [CrossRef]
305. Pan, H.; Shao, Y.; Yan, P.; Cheng, Y.; Han, K.S.; Nie, Z.; Wang, C.; Yang, J.; Li, X.; Bhattacharya, P.; et al. Reversible aqueous zinc/manganese oxide energy storage from conversion reactions. *Nat. Energy* **2016**, *1*, 16039. nenergy.2016.39. [CrossRef]
306. Wu, B.; Zhang, G.; Yan, M.; Xiong, T.; He, P.; He, L.; Xu, X.; Mai, L. Graphene Scroll-Coated α-MnO_2 Nanowires as High-Performance Cathode Materials for Aqueous Zn-Ion Battery. *Small* **2018**, *14*, 1703850. [CrossRef]
307. Sun, W.; Wang, F.; Hou, S.; Yang, C.; Fan, X.; Ma, Z.; Gao, T.; Han, F.; Hu, R.; Zhu, M.; et al. Zn/MnO_2 Battery Chemistry with H+ and Zn^{2+} Coinsertion. *J. Am. Chem. Soc.* **2017**, *139*, 9775–9778. [CrossRef]
308. Huang, J.; Wang, Z.; Hou, M.; Dong, X.; Liu, Y.; Wang, Y.; Xia, Y. Polyaniline-intercalated manganese dioxide nanolayers as a high-performance cathode material for an aqueous zinc-ion battery. *Nat. Commun.* **2018**, *9*, 2906. [CrossRef]
309. He, P.; Quan, Y.; Xu, X.; Yan, M.; Yang, W.; An, Q.; He, L.; Mai, L. High-Performance Aqueous Zinc-Ion Battery Based on Layered $H_2V_3O_8$ Nanowire Cathode. *Small* **2017**, *13*, 1702551. [CrossRef]
310. Yang, Y.; Tang, Y.; Fang, G.; Shan, L.; Guo, J.; Zhang, W.; Wang, C.; Wang, L.; Zhou, J.; Liang, S. Li + intercalated $V_2O_5 \cdot n H_2O$ with enlarged layer spacing and fast ion diffusion as an aqueous zinc-ion battery cathode. *Energy Environ. Sci.* **2018**, *11*, 3157–3162. [CrossRef]
311. Alfaruqi, M.H.; Mathew, V.; Song, J.; Kim, S.; Islam, S.; Pham, D.T.; Jo, J.; Kim, S.; Baboo, J.P.; Xiu, Z.; et al. Electrochemical Zinc Intercalation in Lithium Vanadium Oxide: A High-Capacity Zinc-Ion Battery Cathode. *Chem. Mater.* **2017**, *29*, 1684–1694. [CrossRef]
312. Zhang, N.; Dong, Y.; Jia, M.; Bian, X.; Wang, Y.; Qiu, M.; Xu, J.; Liu, Y.; Jiao, L.; Cheng, F. Rechargeable Aqueous Zn–V_2O_5 Battery with High Energy Density and Long Cycle Life. *ACS Energy Lett.* **2018**, *3*, 1366–1372. [CrossRef]
313. Shin, J.; Choi, D.S.; Lee, H.J.; Jung, Y.; Choi, J.W. Hydrated Intercalation for High-Performance Aqueous Zinc Ion Batteries. *Adv. Energy Mater.* **2019**, *9*, 1900083. [CrossRef]
314. Shen, C.; Li, X.; Li, N.; Xie, K.; Wang, J.G.; Liu, X.; Wei, B. Graphene-Boosted, High-Performance Aqueous Zn-Ion Battery. *ACS Appl. Mater. Interfaces* **2018**, *10*, 25446–25453. [CrossRef] [PubMed]
315. Dai, X.; Wan, F.; Zhang, L.; Cao, H.; Niu, Z. Freestanding graphene/VO_2 composite films for highly stable aqueous zn-ion batteries with superior rate performance. *Energy Storage Mater.* **2019**, *17*, 143–150. [CrossRef]
316. Wang, Z.; Hu, J.; Han, L.; Wang, Z.; Wang, H.; Zhao, Q.; Liu, J.; Pan, F. A MOF-based single-ion Zn^{2+} solid electrolyte leading to dendrite-free rechargeable Zn batteries. *Nano Energy* **2019**, *56*, 92–99. [CrossRef]
317. Kundu, D.; Adams, B.D.; Duffort, V.; Vajargah, S.H.; Nazar, L.F. A high-capacity and long-life aqueous rechargeable zinc battery using a metal oxide intercalation cathode. *Nat. Energy* **2016**, *1*, 16119. [CrossRef]
318. Shan, L.; Yang, Y.; Zhang, W.; Chen, H.; Fang, G.; Zhou, J.; Liang, S. Observation of combination displacement/intercalation reaction in aqueous zinc-ion battery. *Energy Storage Mater.* **2019**, *18*, 10–14. [CrossRef]
319. Xia, C.; Guo, J.; Li, P.; Zhang, X.; Alshareef, H.N. Highly Stable Aqueous Zinc-Ion Storage Using a Layered Calcium Vanadium Oxide Bronze Cathode. *Angew. Chem.* **2018**, *130*, 4007–4012. [CrossRef]
320. Sambandam, B.; Soundharrajan, V.; Kim, S.; Alfaruqi, M.H.; Jo, J.; Kim, S.; Mathew, V.; Sun, Y.K.; Kim, J. $K_2V_6O_{16}\cdot 2.7H_2O$ nanorod cathode: An advanced intercalation system for high energy aqueous rechargeable Zn-ion batteries. *J. Mater. Chem. A* **2018**, *6*, 15530–15539. [CrossRef]
321. He, P.; Zhang, G.; Liao, X.; Yan, M.; Xu, X.; An, Q.; Liu, J.; Mai, L. Sodium Ion Stabilized Vanadium Oxide Nanowire Cathode for High-Performance Zinc-Ion Batteries. *Adv. Energy Mater.* **2018**, *8*, 1702463. [CrossRef]
322. Wan, F.; Zhang, L.; Dai, X.; Wang, X.; Niu, Z.; Chen, J. Aqueous rechargeable zinc/sodium vanadate batteries with enhanced performance from simultaneous insertion of dual carriers. *Nat. Commun.* **2018**, *9*, 1656. [CrossRef]
323. Hu, P.; Zhu, T.; Wang, X.; Wei, X.; Yan, M.; Li, J.; Luo, W.; Yang, W.; Zhang, W.; Zhou, L.; et al. Highly Durable $Na_2V_6O_{16}\cdot 1.63H_2O$ Nanowire Cathode for Aqueous Zinc-Ion Battery. *Nano Lett.* **2018**, *18*, 1758–1763. [CrossRef] [PubMed]
324. Cai, Y.; Liu, F.; Luo, Z.; Fang, G.; Zhou, J.; Pan, A.; Liang, S. Pilotaxitic $Na_{1.1}V_3O_{7.9}$ nanoribbons/graphene as high-performance sodium ion battery and aqueous zinc ion battery cathode. *Energy Storage Mater.* **2018**, *13*, 168–174. [CrossRef]
325. Tang, B.; Zhou, J.; Fang, G.; Liu, F.; Zhu, C.; Wang, C.; Pan, A.; Liang, S. Engineering the interplanar spacing of ammonium vanadates as a high-performance aqueous zinc-ion battery cathode. *J. Mater. Chem. A* **2019**, *7*, 940–945. [CrossRef]
326. Xu, W.; Zhao, K.; Wang, Y. Electrochemical activated MoO_2/Mo_2N heterostructured nanobelts as superior zinc rechargeable battery cathode. *Energy Storage Mater.* **2018**, *15*, 374–379. [CrossRef]

327. Li, H.; Yang, Q.; Mo, F.; Liang, G.; Liu, Z.; Tang, Z.; Ma, L.; Liu, J.; Shi, Z.; Zhi, C. MoS2 nanosheets with expanded interlayer spacing for rechargeable aqueous Zn-ion batteries. *Energy Storage Mater.* **2019**, *19*, 94–101. [CrossRef]
328. Kasiri, G.; Trócoli, R.; Bani Hashemi, A.; La Mantia, F. An electrochemical investigation of the aging of copper hexacyanoferrate during the operation in zinc-ion batteries. *Electrochim. Acta* **2016**, *222*, 74–83. [CrossRef]
329. Zhang, L.; Chen, L.; Zhou, X.; Liu, Z. Morphology-Dependent Electrochemical Performance of Zinc Hexacyanoferrate Cathode for Zinc-Ion Battery. *Sci. Rep.* **2015**, *5*, 18263. [CrossRef]
330. Li, G.; Yang, Z.; Jiang, Y.; Zhang, W.; Huang, Y. Hybrid aqueous battery based on $Na_3V_2(PO_4)_3$/C cathode and zinc anode for potential large-scale energy storage. *J. Power Sources* **2016**, *308*, 52–57. [CrossRef]
331. Zhao, Q.; Huang, W.; Luo, Z.; Liu, L.; Lu, Y.; Li, Y.; Li, L.; Hu, J.; Ma, H.; Chen, J. High-capacity aqueous zinc batteries using sustainable quinone electrodes. *Sci. Adv.* **2018**, *4*, eaao1761. [CrossRef]
332. Jain, A.; Payal, R. Rechargeable Batteries: History, Progress, and Applications. In *Wiley*; Boddula, R.; Pothu, R.; Asiri, A.M., Eds.; John Wiley and Sons.: Hoboken, NJ, USA, 2020; Chapter 10, pp. 195–216.
333. Cabello, M.; Nacimiento, F.; González, J.R.; Ortiz, G.; Alcántara, R.; Lavela, P.; Pérez-Vicente, C.; Tirado, J.L. Advancing towards a veritable calcium-ion battery: $CaCo_2O_4$ positive electrode material. *Electrochem. Commun.* **2016**, *67*, 59–64. [CrossRef]
334. Tchitchekova, D.S.; Frontera, C.; Ponrouch, A.; Krich, C.; Bardé, F.; Palacín, M.R. Electrochemical calcium extraction from 1D-$Ca_3Co_2O_6$. *Dalton Trans.* **2018**, *47*, 11298–11302. [CrossRef] [PubMed]
335. Tchitchekova, D.S.; Ponrouch, A.; Verrelli, R.; Broux, T.; Frontera, C.; Sorrentino, A.; Bardé, F.; Biskup, N.; Arroyo-de Dompablo, M.E.; Palacín, M.R. Electrochemical Intercalation of Calcium and Magnesium in TiS_2: Fundamental Studies Related to Multivalent Battery Applications. *Chem. Mater.* **2018**, *30*, 847–856. [CrossRef]
336. Lipson, A.L.; Pan, B.; Lapidus, S.H.; Liao, C.; Vaughey, J.T.; Ingram, B.J. Rechargeable Ca-Ion Batteries: A New Energy Storage System. *Chem. Mater.* **2015**, *27*, 8442–8447. [CrossRef]
337. Cabello, M.; Nacimiento, F.; Alcántara, R.; Lavela, P.; Pérez Vicente, C.; Tirado, J.L. Applicability of Molybdite as an Electrode Material in Calcium Batteries: A Structural Study of Layer-type Ca_xMoO_3. *Chem. Mater.* **2018**, *30*, 5853–5861. [CrossRef]
338. Hayashi, M.; Arai, H.; Ohtsuka, H.; Sakurai, Y. Electrochemical characteristics of calcium in organic electrolyte solutions and vanadium oxides as calcium hosts. *J. Power Sources* **2003**, *119–121*, 617–620. [CrossRef]
339. Xu, X.; Duan, M.; Yue, Y.; Li, Q.; Zhang, X.; Wu, L.; Wu, P.; Song, B.; Mai, L. Bilayered $Mg_{0.25}V_2O_5 \cdot H_2O$ as a Stable Cathode for Rechargeable Ca-Ion Batteries. *ACS Energy Lett.* **2019**, *4*, 1328–1335. [CrossRef]
340. Wang, J.; Tan, S.; Xiong, F.; Yu, R.; Wu, P.; Cui, L.; An, Q. $VOPO_4 \cdot 2H_2O$ as a new cathode material for rechargeable Ca-ion batteries. *Chem. Commun.* **2020**, *56*, 3805–3808. [CrossRef] [PubMed]
341. See, K.A.; Gerbec, J.A.; Jun, Y.S.; Wudl, F.; Stucky, G.D.; Seshadri, R. A High Capacity Calcium Primary Cell Based on the Ca-S System. *Adv. Energy Mater.* **2013**, *3*, 1056–1061. [CrossRef]
342. Cheng, Y.; Shao, Y.; Zhang, J.G.; Sprenkle, V.L.; Liu, J.; Li, G. High performance batteries based on hybrid magnesium and lithium chemistry. *Chem. Commun.* **2014**, *50*, 9644–9646. [CrossRef]
343. Ma, Z.; MacFarlane, D.R.; Kar, M. Mg Cathode Materials and Electrolytes for Rechargeable Mg Batteries: A Review. *Batter. Supercaps* **2019**, *2*, 115–127. [CrossRef]
344. Wan, L.F.; Incorvati, J.T.; Poeppelmeier, K.R.; Prendergast, D. Building a fast lane for Mg diffusion in α-MoO_3 by fluorine doping. *Chem. Mater.* **2016**, *28*, 6900–6908. [CrossRef]
345. Incorvati, J.T.; Wan, L.F.; Key, B.; Zhou, D.; Liao, C.; Fuoco, L.; Holland, M.; Wang, H.; Prendergast, D.; Poeppelmeier, K.R.; et al. Reversible Magnesium Intercalation into a Layered Oxyfluoride Cathode. *Chem. Mater.* **2016**, *28*, 17–20. [CrossRef]
346. Kaveevivitchai, W.; Jacobson, A.J. High capacity microporous molybdenum-vanadium oxide electrodes for rechargeable lithium batteries. *Chem. Mater.* **2013**, *25*, 2708–2715. [CrossRef]
347. Sun, X.; Bonnick, P.; Duffort, V.; Liu, M.; Rong, Z.; Persson, K.A.; Ceder, G.; Nazar, L.F. A high capacity thiospinel cathode for Mg batteries. *Energy Environ. Sci.* **2016**, *9*, 2273–2277. [CrossRef]
348. Gao, T.; Han, F.; Zhu, Y.; Suo, L.; Luo, C.; Xu, K.; Wang, C. Hybrid Mg^{2+}/Li^+ Battery with Long Cycle Life and High Rate Capability. *Adv. Energy Mater.* **2015**, *5*, 1401507. [CrossRef]
349. Taniguchi, K.; Gu, Y.; Katsura, Y.; Yoshino, T.; Takagi, H. Rechargeable Mg battery cathode TiS_3 with d–p orbital hybridized electronic structures. *Appl. Phys. Express* **2016**, *9*, 011801. [CrossRef]
350. Arsentev, M.; Missyul, A.; Petrov, A.V.; Hammouri, M. TiS_3 Magnesium Battery Material: Atomic-Scale Study of Maximum Capacity and Structural Behavior. *J. Phys. Chem. C* **2017**, *121*, 15509–15515. [CrossRef]
351. Gu, Y.; Katsura, Y.; Yoshino, T.; Takagi, H.; Taniguchi, K. Rechargeable magnesium-ion battery based on a $TiSe_2$-cathode with d-p orbital hybridized electronic structure. *Sci. Rep.* **2015**, *5*, 12486. [CrossRef]
352. Gershinsky, G.; Yoo, H.D.; Gofer, Y.; Aurbach, D. Electrochemical and spectroscopic analysis of Mg^{2+} intercalation into thin film electrodes of layered oxides: V_2O_5 and MoO_3. *Langmuir* **2013**, *29*, 10964–10972. [CrossRef]
353. Zhang, R.; Ling, C.; Mizuno, F. A Conceptual Magnesium Battery with Ultrahigh Rate Capability. *Chem. Commun.* **2015**, *51*, 1487–1490. [CrossRef] [PubMed]
354. Xiong, F.; Fan, Y.; Tan, S.; Zhou, L.; Xu, Y.; Pei, C.; An, Q.; Mai, L. Magnesium storage performance and mechanism of CuS cathode. *Nano Energy* **2018**, *47*, 210–216. [CrossRef]
355. Duffort, V.; Sun, X.; Nazar, L.F. Screening for positive electrodes for magnesium batteries: A protocol for studies at elevated temperatures. *Chem. Commun.* **2016**, *52*, 12458–12461. [CrossRef] [PubMed]

356. Tashiro, Y.; Taniguchi, K.; Miyasaka, H. Copper Selenide as a New Cathode Material based on Displacement Reaction for Rechargeable Magnesium Batteries. *Electrochim. Acta* **2016**, *210*, 655–661. [CrossRef]
357. Zhang, R.; Ling, C. Unveil the Chemistry of Olivine FePO$_4$ as Magnesium Battery Cathode. *ACS Appl. Mater. Interfaces* **2016**, *8*, 18018–18026. [CrossRef]
358. Zhang, R.; Arthur, T.S.; Ling, C.; Mizuno, F. Manganese dioxides as rechargeable magnesium battery cathode; Synthetic approach to understand magnesiation process. *J. Power Sources* **2015**, *282*, 630–638. [CrossRef]
359. Chen, X.; Bleken, F.L.; Løvvik, O.M.; Vullum-Bruer, F. Comparing electrochemical performance of transition metal silicate cathodes and chevrel phase Mo$_6$S$_8$ in the analogous rechargeable Mg-ion battery system. *J. Power Sources* **2016**, *321*, 76–86. [CrossRef]
360. Pan, B.; Zhou, D.; Huang, J.; Zhang, L.; Burrell, A.K.; Vaughey, J.T.; Zhang, Z.; Liao, C. 2,5-Dimethoxy-1,4-Benzoquinone (DMBQ) as Organic Cathode for Rechargeable Magnesium-Ion Batteries. *J. Electrochem. Soc.* **2016**, *163*, A580–A583. [CrossRef]
361. Pan, B.; Huang, J.; Feng, Z.; Zeng, L.; He, M.; Zhang, L.; Vaughey, J.T.; Bedzyk, M.J.; Fenter, P.; Zhang, Z.; et al. Polyanthraquinone-Based Organic Cathode for High-Performance Rechargeable Magnesium-Ion Batteries. *Adv. Energy Mater.* **2016**, *6*, 1600140. [CrossRef]
362. Gao, T.; Hou, S.; Wang, F.; Ma, Z.; Li, X.; Xu, K.; Wang, C. Reversible S^0/MgS$_x$ Redox Chemistry in a MgTFSI$_2$/MgCl$_2$/DME Electrolyte for Rechargeable Mg/S Batteries. *Angew. Chem. Int. Ed.* **2017**, *56*, 13526–13530. [CrossRef]
363. Zhao-Karger, Z.; Liu, R.; Dai, W.; Li, Z.; Diemant, T.; Vinayan, B.P.; Bonatto Minella, C.; Yu, X.; Manthiram, A.; Behm, R.J.; et al. Toward Highly Reversible Magnesium–Sulfur Batteries with Efficient and Practical Mg[B(hfip)$_4$]$_2$ Electrolyte. *ACS Energy Lett.* **2018**, *3*, 2005–2013. [CrossRef]
364. Zhang, Z.; Cui, Z.; Qiao, L.; Guan, J.; Xu, H.; Wang, X.; Hu, P.; Du, H.; Li, S.; Zhou, X.; et al. Novel Design Concepts of Efficient Mg-Ion Electrolytes toward High-Performance Magnesium-Selenium and Magnesium-Sulfur Batteries. *Adv. Energy Mater.* **2017**, *7*, 1602055. [CrossRef]
365. Zhao-Karger, Z.; Zhao, X.; Wang, D.; Diemant, T.; Behm, R.J.; Fichtner, M. Performance Improvement of Magnesium Sulfur Batteries with Modified Non-Nucleophilic Electrolytes. *Adv. Energy Mater.* **2015**, *5*, 1401155. [CrossRef]
366. Zhao-Karger, Z.; Gil Bardaji, M.E.; Fuhr, O.; Fichtner, M. A new class of non-corrosive, highly efficient electrolytes for rechargeable magnesium batteries. *J. Mater. Chem. A* **2017**, *5*, 10815–10820. [CrossRef]
367. Du, A.; Zhang, Z.; Qu, H.; Cui, Z.; Qiao, L.; Wang, L.; Chai, J.; Lu, T.; Dong, S.; Dong, T.; et al. An efficient organic magnesium borate-based electrolyte with non-nucleophilic characteristics for magnesium–sulfur battery. *Energy Environ. Sci.* **2017**, *10*, 2616–2625. [CrossRef]
368. Vinayan, B.P.; Zhao-Karger, Z.; Diemant, T.; Chakravadhanula, V.S.K.; Schwarzburger, N.I.; Cambaz, M.A.; Behm, R.J.; Kübel, C.; Fichtner, M. Performance study of magnesium-sulfur battery using a graphene based sulfur composite cathode electrode and a non-nucleophilic Mg electrolyte. *Nanoscale* **2016**, *8*, 3296–3306. [CrossRef]
369. Vardar, G.; Nelson, E.G.; Smith, J.G.; Naruse, J.; Hiramatsu, H.; Bartlett, B.M.; Sleightholme, A.E.; Siegel, D.J.; Monroe, C.W. Identifying the Discharge Product and Reaction Pathway for a Secondary Mg/O$_2$ Battery. *Chem. Mater.* **2015**, *27*, 7564–7568. [CrossRef]
370. Smith, J.G.; Naruse, J.; Hiramatsu, H.; Siegel, D.J. Theoretical Limiting Potentials in Mg/O$_2$ Batteries. *Chem. Mater.* **2016**, *28*, 1390–1401. [CrossRef]
371. Tian, H.; Gao, T.; Li, X.; Wang, X.; Luo, C.; Fan, X.; Yang, C.; Suo, L.; Ma, Z.; Han, W.; et al. High power rechargeable magnesium/iodine battery chemistry. *Nat. Commun.* **2017**, *8*, 14083. [CrossRef]
372. Sun, H.; Wang, W.; Yu, Z.; Yuan, Y.; Wang, S.; Jiao, S. A new aluminium-ion battery with high voltage, high safety and low cost. *Chem. Commun.* **2015**, *51*, 11892–11895. [CrossRef]
373. Lin, M.C.; Gong, M.; Lu, B.; Wu, Y.; Wang, D.Y.; Guan, M.; Angell, M.; Chen, C.; Yang, J.; Hwang, B.J.; et al. An ultrafast rechargeable aluminium-ion battery. *Nature* **2015**, *520*, 324–328. [CrossRef] [PubMed]
374. Wu, Y.; Gong, M.; Lin, M.C.; Yuan, C.; Angell, M.; Huang, L.; Wang, D.Y.; Zhang, X.; Yang, J.; Hwang, B.J.; et al. 3D Graphitic Foams Derived from Chloroaluminate Anion Intercalation for Ultrafast Aluminum-Ion Battery. *Adv. Mater.* **2016**, *28*, 9218–9222. [CrossRef] [PubMed]
375. Chen, H.; Guo, F.; Liu, Y.; Huang, T.; Zheng, B.; Ananth, N.; Xu, Z.; Gao, W.; Gao, C. A Defect-Free Principle for Advanced Graphene Cathode of Aluminum-Ion Battery. *Adv. Mater.* **2017**, *29*, 1605958. [CrossRef]
376. Hu, Y.; Debnath, S.; Hu, H.; Luo, B.; Zhu, X.; Wang, S.; Hankel, M.; Searles, D.J.; Wang, L. Unlocking the potential of commercial carbon nanofibers as free-standing positive electrodes for flexible aluminum ion batteries. *J. Mater. Chem. A* **2019**, *7*, 15123–15130. [CrossRef]
377. Jiang, J.; Li, H.; Huang, J.; Li, K.; Zeng, J.; Yang, Y.; Li, J.; Wang, Y.; Wang, J.; Zhao, J. Investigation of the Reversible Intercalation/Deintercalation of Al into the Novel Li$_3$VO$_4$@C Microsphere Composite Cathode Material for Aluminum-Ion Batteries. *ACS Appl. Mater. Interfaces* **2017**, *9*, 28486–28494. [CrossRef]
378. Wei, J.; Chen, W.; Chen, D.; Yang, K. Molybdenum Oxide as Cathode for High Voltage Rechargeable Aluminum Ion Battery. *J. Electrochem. Soc.* **2017**, *164*, A2304–A2309. [CrossRef]
379. Wang, S.; Jiao, S.; Wang, J.; Chen, H.S.; Tian, D.; Lei, H.; Fang, D.N. High-Performance Aluminum-Ion Battery with CuS@C Microsphere Composite Cathode. *ACS Nano* **2017**, *11*, 469–477. [CrossRef] [PubMed]

380. Geng, L.; Scheifers, J.P.; Fu, C.; Zhang, J.; Fokwa, B.P.T.; Guo, J. Titanium Sulfides as Intercalation-Type Cathode Materials for Rechargeable Aluminum Batteries. *ACS Appl. Mater. Interfaces* **2017**, *9*, 21251–21257. [CrossRef]
381. Zhang, X.; Zhang, G.; Wang, S.; Li, S.; Jiao, S. Porous CuO microsphere architectures as high-performance cathode materials for aluminum-ion batteries. *J. Mater. Chem. A* **2018**, *6*, 3084–3090. [CrossRef]
382. Yu, X.; Boyer, M.J.; Hwang, G.S.; Manthiram, A. Room-Temperature Aluminum-Sulfur Batteries with a Lithium-Ion-Mediated Ionic Liquid Electrolyte. *Chem* **2018**, *4*, 586–598. [CrossRef]
383. Yang, H.; Yin, L.; Liang, J.; Sun, Z.; Wang, Y.; Li, H.; He, K.; Ma, L.; Peng, Z.; Qiu, S.; et al. An Aluminum-Sulfur Battery with a Fast Kinetic Response. *Angew. Chem. Int. Ed.* **2018**, *57*, 1898–1902. [CrossRef] [PubMed]
384. Chen, H.; Xu, H.; Wang, S.; Huang, T.; Xi, J.; Cai, S.; Guo, F.; Xu, Z.; Gao, W.; Gao, C. Ultrafast all-climate aluminum-graphene battery with quarter-million cycle life. *Sci. Adv.* **2017**, *3*, eaao7233. [CrossRef]
385. Huang, H.; Zhou, F.; Shi, X.; Qin, J.; Zhang, Z.; Bao, X.; Wu, Z.S. Graphene aerogel derived compact films for ultrafast and high-capacity aluminum ion batteries. *Energy Storage Mater.* **2019**, *23*, 664–669. [CrossRef]
386. Zhao, X.; Yao, W.; Gao, W.; Chen, H.; Gao, C. Wet-Spun Superelastic Graphene Aerogel Millispheres with Group Effect. *Adv. Mater.* **2017**, *29*, 1701482. [CrossRef] [PubMed]
387. Yu, X.; Wang, B.; Gong, D.; Xu, Z.; Lu, B. Graphene Nanoribbons on Highly Porous 3D Graphene for High-Capacity and Ultrastable Al-Ion Batteries. *Adv. Mater.* **2017**, *29*, 1604118. [CrossRef]
388. Wang, S.; Kravchyk, K.V.; Krumeich, F.; Kovalenko, M.V. Kish Graphite Flakes as a Cathode Material for an Aluminum Chloride-Graphite Battery. *ACS Appl. Mater. Interfaces* **2017**, *9*, 28478–28485. [CrossRef] [PubMed]
389. Smajic, J.; Alazmi, A.; Batra, N.; Palanisamy, T.; Anjum, D.H.; Costa, P.M.F.J. Mesoporous Reduced Graphene Oxide as a High Capacity Cathode for Aluminum Batteries. *Small* **2018**, *14*, 1803584. [CrossRef]
390. Stadie, N.P.; Wang, S.; Kravchyk, K.V.; Kovalenko, M.V. Zeolite-Templated Carbon as an Ordered Microporous Electrode for Aluminum Batteries. *ACS Nano* **2017**, *11*, 1911–1919. [CrossRef]
391. Kaveevivitchai, W.; Huq, A.; Wang, S.; Park, M.J.; Manthiram, A. Rechargeable Aluminum-Ion Batteries Based on an Open-Tunnel Framework. *Small* **2017**, *13*, 1701296. [CrossRef]
392. Wang, W.; Jiang, B.; Xiong, W.; Sun, H.; Lin, Z.; Hu, L.; Tu, J.; Hou, J.; Zhu, H.; Jiao, S. A new cathode material for super-valent battery based on aluminium ion intercalation and deintercalation. *Sci. Rep.* **2013**, *3*, 3383. [CrossRef]
393. Wang, H.; Bai, Y.; Chen, S.; Luo, X.; Wu, C.; Wu, F.; Lu, J.; Amine, K. Binder-free V_2O_5 cathode for greener rechargeable aluminum battery. *ACS Appl. Mater. Interfaces* **2015**, *7*, 80–84. [CrossRef] [PubMed]
394. Chiku, M.; Takeda, H.; Matsumura, S.; Higuchi, E.; Inoue, H. Amorphous Vanadium Oxide/Carbon Composite Positive Electrode for Rechargeable Aluminum Battery. *ACS Appl. Mater. Interfaces* **2015**, *7*, 24385–24389. [CrossRef] [PubMed]
395. Yu, Z.; Kang, Z.; Hu, Z.; Lu, J.; Zhou, Z.; Jiao, S. Hexagonal NiS nanobelts as advanced cathode materials for rechargeable Al-ion batteries. *Chem. Commun.* **2016**, *52*, 10427–10430. [CrossRef]
396. Wang, S.; Yu, Z.; Tu, J.; Wang, J.; Tian, D.; Liu, Y.; Jiao, S. A Novel Aluminum-Ion Battery: $Al/AlCl_3$-[EMIm]Cl/Ni_3S_2@Graphene. *Adv. Energy Mater.* **2016**, *6*, 1600137. [CrossRef]
397. Geng, L.; Lv, G.; Xing, X.; Guo, J. Reversible Electrochemical Intercalation of Aluminum in Mo_6S_8. *Chem. Mater.* **2015**, *27*, 4926–4929. [CrossRef]
398. Hu, Y.; Luo, B.; Ye, D.; Zhu, X.; Lyu, M.; Wang, L. An Innovative Freeze-Dried Reduced Graphene Oxide Supported SnS_2 Cathode Active Material for Aluminum-Ion Batteries. *Adv. Mater.* **2017**, *29*, 1606132. [CrossRef]
399. Zhang, X.; Wang, S.; Tu, J.; Zhang, G.; Li, S.; Tian, D.; Jiao, S. Flower-like Vanadium Suflide/Reduced Graphene Oxide Composite: An Energy Storage Material for Aluminum-Ion Batteries. *ChemSusChem* **2018**, *11*, 709–715. [CrossRef]
400. Gao, T.; Li, X.; Wang, X.; Hu, J.; Han, F.; Fan, X.; Suo, L.; Pearse, A.J.; Lee, S.B.; Rubloff, G.W.; et al. A Rechargeable Al/S Battery with an Ionic-Liquid Electrolyte. *Angew. Chem.* **2016**, *128*, 10052–10055. [CrossRef]
401. Cohn, G.; Ma, L.; Archer, L.A. A novel non-aqueous aluminum sulfur battery. *J. Power Sources* **2015**, *283*, 416–422. [CrossRef]
402. Yuan, F.; Chi, S.; Dong, S.; Zou, X.; Lv, S.; Bao, L.; Wang, J. Ionic liquid crystal with fast ion-conductive tunnels for potential application in solvent-free Li-ion batteries. *Electrochim. Acta* **2019**, *294*, 249–259. [CrossRef]
403. Wang, F.; Fan, X.; Gao, T.; Sun, W.; Ma, Z.; Yang, C.; Han, F.; Xu, K.; Wang, C. High-Voltage Aqueous Magnesium Ion Batteries. *ACS Cent. Sci.* **2017**, *3*, 1121–1128. [CrossRef]
404. Yuan, F.; Yang, L.; Zou, X.; Dong, S.; Chi, S.; Xie, J.; Xing, H.; Bian, L.; Bao, L.; Wang, J. Flexible all-solid-state electrolytes with ordered fast Li-ion-conductive nano-pathways for rechargeable lithium batteries. *J. Power Sources* **2019**, *444*, 227305. [CrossRef]
405. Gao, H.; Xue, L.; Xin, S.; Goodenough, J.B. A High-Energy-Density Potassium Battery with a Polymer-Gel Electrolyte and a Polyaniline Cathode. *Angew. Chem. Int. Ed.* **2018**, *57*, 5449–5453. [CrossRef] [PubMed]
406. Zhang, Z.; Zuo, C.; Liu, Z.; Yu, Y.; Zuo, Y.; Song, Y. All-solid-state Al—Air batteries with polymer alkaline gel electrolyte. *J. Power Sources* **2014**, *251*, 470–475. [CrossRef]
407. Jiang, G.; Li, F.; Wang, H.; Wu, M.; Qi, S.; Liu, X.; Yang, S.; Ma, J. Perspective on High-Concentration Electrolytes for Lithium Metal Batteries. *Small Struct.* **2021**, *2*, 2000122. [CrossRef]
408. Cao, X.; Jia, H.; Xu, W.; Zhang, J.G. Review—Localized High-Concentration Electrolytes for Lithium Batteries. *J. Electrochem. Soc.* **2021**, *168*, 010522. [CrossRef]
409. Eshetu, G.G.; Elia, G.A.; Armand, M.; Forsyth, M.; Komaba, S.; Rojo, T.; Passerini, S. Electrolytes and Interphases in Sodium-Based Rechargeable Batteries: Recent Advances and Perspectives. *Adv. Energy Mater.* **2020**, *10*, 2000093. [CrossRef]

410. Wu, X.; Leonard, D.P.; Ji, X. Emerging Non-Aqueous Potassium-Ion Batteries: Challenges and Opportunities. *Chem. Mater.* **2017**, *29*, 5031–5042. [CrossRef]
411. Wang, Y.; Meng, X.; Sun, J.; Liu, Y.; Hou, L. Recent Progress in "Water-in-Salt" Electrolytes Toward Non-lithium Based Rechargeable Batteries. *Front. Chem.* **2020**, *8*, 595. [CrossRef]
412. Montanino, M.; Passerini, S.; Appetecchi, G. *Rechargeable Lithium Batteries*; Elsevier: Amsterdam, The Netherlands, 2015; Chapter Four, pp. 73–116. [CrossRef]
413. Yan, J.; Zhao, Y.; Wang, X.; Xia, S.; Zhang, Y.; Han, Y.; Yu, J.; Ding, B. Polymer Template Synthesis of Soft, Light, and Robust Oxide Ceramic Films. *iScience* **2019**, *15*, 185–195. [CrossRef]
414. Le Ruyet, R.; Berthelot, R.; Salager, E.; Florian, P.; Fleutot, B.; Janot, R. Investigation of $Mg(BH_4)(NH_2)$-Based Composite Materials with Enhanced Mg^{2+} Ionic Conductivity. *J. Phys. Chem. C* **2019**, *123*, 10756–10763. [CrossRef]
415. Kotobuki, M. Recent progress of ceramic electrolytes for post Li and Na batteries. *Funct. Mater. Lett.* **2021**, *14*, 2130003. [CrossRef]
416. Xiao, Z.; Zhou, B.; Wang, J.; Zuo, C.; He, D.; Xie, X.; Xue, Z. PEO-based electrolytes blended with star polymers with precisely imprinted polymeric pseudo-crown ether cavities for alkali metal ion batteries. *J. Membr. Sci.* **2019**, *576*, 182–189. [CrossRef]
417. Arora, P.; Zhang, Z.J. Battery Separators. *Chem. Rev.* **2004**, *104*, 4419–4462. [CrossRef]
418. Septiana, A.R.; Honggowiranto, W.; Sudaryanto; Kartini, E.; Hidayat, R. Comparative study on the ionic conductivities and redox properties of LiPF6 and LiTFSI electrolytes and the characteristics of their rechargeable lithium ion batteries. *IOP Conf. Ser. Mater. Sci. Eng.* **2018**, *432*, 012061. [CrossRef]
419. Hwang, S.; Kim, D.H.; Shin, J.H.; Jang, J.E.; Ahn, K.H.; Lee, C.; Lee, H. Ionic Conduction and Solution Structure in LiPF 6 and LiBF 4 Propylene Carbonate Electrolytes. *J. Phys. Chem. C* **2018**, *122*, 19438–19446. [CrossRef]
420. Wu, F.; Kim, G.; Diemant, T.; Kuenzel, M.; Schür, A.R.; Gao, X.; Qin, B.; Alwast, D.; Jusys, Z.; Behm, R.J.; et al. Reducing Capacity and Voltage Decay of Co-Free $Li_{1.2}Ni_{0.2}Mn_{0.6}O_2$ as Positive Electrode Material for Lithium Batteries Employing an Ionic Liquid-Based Electrolyte. *Adv. Energy Mater.* **2020**, *10*, 2001830. [CrossRef]
421. Bommier, C.; Ji, X. Electrolytes, SEI Formation, and Binders: A Review of Nonelectrode Factors for Sodium-Ion Battery Anodes. *Small* **2018**, *14*, 1703576. [CrossRef]
422. Xu, X.; Zhou, D.; Qin, X.; Lin, K.; Kang, F.; Li, B.; Shanmukaraj, D.; Rojo, T.; Armand, M.; Wang, G. A room-temperature sodium–sulfur battery with high capacity and stable cycling performance. *Nat. Commun.* **2018**, *9*, 3870. [CrossRef]
423. Amara, S.; Toulc'Hoat, J.; Timperman, L.; Biller, A.; Galiano, H.; Marcel, C.; Ledigabel, M.; Anouti, M. Comparative Study of Alkali-Cation-Based (Li^+, Na^+, K^+) Electrolytes in Acetonitrile and Alkylcarbonates. *ChemPhysChem* **2019**, *20*, 581. [CrossRef]
424. Xiong, W.; Hoang, T.K.; Yang, D.; Liu, Y.; Ahmed, M.; Xu, J.; Qiu, X.; Chen, P. Electrolyte engineering for a highly stable, rechargeable hybrid aqueous battery. *J. Energy Storage* **2019**, *26*, 100920. [CrossRef]
425. Lyu, L.; Gao, Y.; Wang, Y.; Xiao, L.; Lu, J.; Zhuang, L. Improving the cycling performance of silver-zinc battery by introducing PEG-200 as electrolyte additive. *Chem. Phys. Lett.* **2019**, *723*, 102–110. [CrossRef]
426. Abdulla, J.; Cao, J.; Zhang, D.; Zhang, X.; Sriprachuabwong, C.; Kheawhom, S.; Wangyao, P.; Qin, J. Elimination of Zinc Dendrites by Graphene Oxide Electrolyte Additive for Zinc-Ion Batteries. *ACS Appl. Energy Mater.* **2021**, *4*, 4602–4609. [CrossRef]
427. Wang, A.; Zhou, W.; Huang, A.; Chen, M.; Tian, Q.; Chen, J. Developing improved electrolytes for aqueous zinc-ion batteries to achieve excellent cyclability and antifreezing ability. *J. Colloid Interface Sci.* **2021**, *586*, 362–370. 10.099. [CrossRef]
428. Chae, M.S.; Heo, J.W.; Kwak, H.H.; Lee, H.; Hong, S.T. Organic electrolyte-based rechargeable zinc-ion batteries using potassium nickel hexacyanoferrate as a cathode material. *J. Power Sources* **2017**, *337*, 204–211. [CrossRef]
429. Zhang, N.; Dong, Y.; Wang, Y.; Wang, Y.; Li, J.; Xu, J.; Liu, Y.; Jiao, L.; Cheng, F. Ultrafast Rechargeable Zinc Battery Based on High-Voltage Graphite Cathode and Stable Nonaqueous Electrolyte. *ACS Appl. Mater. Interfaces* **2019**, *11*, 32978–32986. [CrossRef]
430. Zhu, J.; Yao, M.; Huang, S.; Tian, J.; Niu, Z. Thermal-Gated Polymer Electrolytes for Smart Zinc-Ion Batteries. *Angew. Chem.* **2020**, *132*, 16622–16626. [CrossRef]
431. Lee, B.S.; Cui, S.; Xing, X.; Liu, H.; Yue, X.; Petrova, V.; Lim, H.D.; Chen, R.; Liu, P. Dendrite Suppression Membranes for Rechargeable Zinc Batteries. *ACS Appl. Mater. Interfaces* **2018**, *10*, 38928–38935. [CrossRef]
432. Zhang, H.; Liu, X.; Li, H.; Qin, B.; Passerini, S. High-Voltage Operation of a V_2O_5 Cathode in a Concentrated Gel Polymer Electrolyte for High-Energy Aqueous Zinc Batteries. *ACS Appl. Mater. Interfaces* **2020**, *12*, 15305–15312. [CrossRef]
433. Melemed, A.M.; Khurram, A.; Gallant, B.M. Current Understanding of Nonaqueous Electrolytes for Calcium-Based Batteries. *Batter. Supercaps* **2020**, *3*, 570–580. [CrossRef]
434. Nielson, K.V.; Luo, J.; Liu, T.L. Optimizing Calcium Electrolytes by Solvent Manipulation for Calcium Batteries. *Batter. Supercaps* **2020**, *3*, 766–772. [CrossRef]
435. Biria, S.; Pathreeker, S.; Li, H.; Hosein, I.D. Plating and stripping of calcium in an alkyl carbonate electrolyte at room temperature. *ACS Appl. Energy Mater.* **2019**, *2*, 7738–7743. [CrossRef]
436. Shakourian-Fard, M.; Kamath, G.; Taimoory, S.M.; Trant, J.F. Calcium-Ion Batteries: Identifying Ideal Electrolytes for Next-Generation Energy Storage Using Computational Analysis. *J. Phys. Chem. C* **2019**, *123*, 15885–15896. acs.jpcc.9b01655. [CrossRef]
437. Lu, Y.; Wang, C.; Liu, Q.; Li, X.; Zhao, X.; Guo, Z. Progress and Perspective on Rechargeable Magnesium–Sulfur Batteries. *Small Methods* **2021**, *5*, 2001303. [CrossRef] [PubMed]
438. Sheha, E.; Liu, F.; Wang, T.; Farrag, M.; Liu, J.; Yacout, N.; Kebede, M.A.; Sharma, N.; Fan, L.Z. Dual Polymer/Liquid Electrolyte with $BaTiO_3$ Electrode for Magnesium Batteries. *ACS Appl. Energy Mater.* **2020**, *3*, 5882–5892. 0c00810. [CrossRef]

439. Angell, M.; Pan, C.J.; Rong, Y.; Yuan, C.; Lin, M.C.; Hwang, B.J.; Dai, H. High Coulombic efficiency aluminum-ion battery using an AlCl$_3$-urea ionic liquid analog electrolyte. *Proc. Natl. Acad. Sci. USA* **2017**, *114*, 834–839. [CrossRef]
440. Wu, C.; Gu, S.; Zhang, Q.; Bai, Y.; Li, M.; Yuan, Y.; Wang, H.; Liu, X.; Yuan, Y.; Zhu, N.; et al. Electrochemically activated spinel manganese oxide for rechargeable aqueous aluminum battery. *Nat. Commun.* **2019**, *10*, 73. [CrossRef]
441. Counterpoint. Global Smartphone Market Share: By Quarter. Available online: https://www.counterpointresearch.com/global-smartphone-share/ (accessed on 30 January 2022).
442. Statista. *Number of Smartphones Sold to End Users Worldwide from 2007 to 2021*; Statista: Hamburg, Germany, 2021.
443. IEA. *Global EV Outlook 2021*; IEA: Paris, France, 2021.
444. Cano, Z.P.; Banham, D.; Ye, S.; Hintennach, A.; Lu, J.; Fowler, M.; Chen, Z. Batteries and fuel cells for emerging electric vehicle markets. *Nat. Energy* **2018**, *3*, 279–289. [CrossRef]
445. Tesla. Model S. 2021. Available online: https://www.tesla.com/models (accessed on 30 January 2022).
446. Fundacio Institut De Recerca De L'Energia De Catalunya. *CObalt-Free Batteries for FutuRe Automotive Applications*; Fundacio Institut De Recerca De L'Energia De Catalunya: Barcelona, Spain, 2021.
447. Osterreichische Forschungsforderungsgesellschaft MBH. *ERA-NET for Research and Innovation on Materials and Battery Technologies, Supporting the European Green Deal*; Osterreichische Forschungsforderungsgesellschaft MBH: Vienna, Austria, 2021.
448. Epec, L. Battery Standards. 2021. Available online: https://www.epectec.com/batteries/battery-standards.html (accessed on 30 January 2022).
449. Woodbank Communications Ltd. *International Standards and Testing Applicable to Batteries*; Woodbank Communications Ltd.: Chester, UK, 2005.

Article

Comparative Performances of Natural Dyes Extracted from Mentha Leaves, Helianthus Annuus Leaves, and Fragaria Fruit for Dye-Sensitized Solar Cells

Zainab Haider Abdulrahman [1], Dhafer Manea Hachim [2], Ahmed Salim Naser Al-murshedi [2], Furkan Kamil [2], Ahmed Al-Manea [3] and Talal Yusaf [4,5,*]

1. Department of Power Mechanics, Najaf Technical College, Al-Furat Al-Awsat Technical University, Najaf 31001, Iraq
2. Najaf Technical College, Al-Furat Al-Awsat Technical University, Najaf 31001, Iraq
3. Al-Samawah Technical Institute, Al-Furat Al-Awsat Technical University, Al-Samawah 66001, Iraq
4. School of Engineering and Technology, Central Queensland University, Brisbane 4000, Australia
5. Institute of Sustainable Energy, University Tenaga Nasional, Putrajaya Campus, Kajang 43000, Malaysia
* Correspondence: t.yusaf@cqu.edu.au

Citation: Abdulrahman, Z.H.; Hachim, D.M.; Al-murshedi, A.S.N.; Kamil, F.; Al-Manea, A.; Yusaf, T. Comparative Performances of Natural Dyes Extracted from Mentha Leaves, Helianthus Annuus Leaves, and Fragaria Fruit for Dye-Sensitized Solar Cells. *Designs* **2022**, *6*, 100. https://doi.org/10.3390/designs6060100

Academic Editors: Xia Lu and Xueyi Lu

Received: 14 September 2022
Accepted: 19 October 2022
Published: 25 October 2022

Publisher's Note: MDPI stays neutral with regard to jurisdictional claims in published maps and institutional affiliations.

Copyright: © 2022 by the authors. Licensee MDPI, Basel, Switzerland. This article is an open access article distributed under the terms and conditions of the Creative Commons Attribution (CC BY) license (https://creativecommons.org/licenses/by/4.0/).

Abstract: During the last four centuries, there have been extensive research activities looking for green and clean sources of energy instead of traditional (fossil) energy in order to reduce the accumulation of gases and environmental pollution. Natural dye-sensitized solar cells (DSSCs) are one of the most promising types of photovoltaic cells for generating clean energy at a low cost. In this study, DSSCs were collected and experimentally tested using four different dyes extracted from Mentha leaves, Helianthus annuus leaves, Fragaria, and a mixture of the above extracts in equal proportions as natural stimuli for TiO_2 films. The result show that solar energy was successfully turned into electricity. Additionally, DSSCs based on mixtures of dyes showed better results than those based on single dyes. Efficiency (η) was 0.714%, and the fill factor (FF) was 83.3% for the cell area.

Keywords: natural dyes; dye-sensitized solar cells; DSSCs

1. Introduction

Nowadays, the used energy in the industry sector is transitioning towards a more environmentally-friendly future. Fossil fuels are an environmental menace that will be depleted sooner. To triumph over this situation, alternatives to these fossil fuel reserves must be found [1]. Photovoltaic cells are one of the promising technologies for harvesting solar energy [2,3]. Unnatural solar cells are widely used because they have good efficiency, but they are high in cost [1,4]. Therefore, the increasing need for sustainable energy prompted researchers to focus their efforts on the development of photovoltaic technologies to meet the need at a lower cost [5–7]. Dye-based solar cells are used as a viable alternative to conventional (inorganic) solar cells as they are an environmentally and economically beneficial technology, as well as being easy to manufacture and providing a method for tuning optical properties through molecular design; they were first invented by O'Regan and Gratzel in 1991 [8–10]. DSSCs are photovoltaic cells based on semiconductors and light sensors that convert sunlight (photons) or artificial radiation into electrical energy [11]. DSSCs consists of a photodiode, a counter electrode (conductor glass covered with graphite, pt, or carbon), and electrolytes (including redox pairs as iodide or tri-iodide) [12]. At the moment, the maximum efficiency (η) observed for these cells (DSSCs) is 13% [13–16]. The operating concept of DSSCs is based on the natural photosynthesis of plants, and pigments play a vital role in widening the cell spectrum's sensitivity. After photoexcitation of the sensitizer dye, the cascade transmission of excited electrons results in charge separation and rapid regeneration of oxidizing dyes [17–19].

A phototype is an inorganic substance that is sensitized by a donor substance called a sensitizer (dye) [20]. Titanium dioxide (TiO_2) is a broadband-gap semiconductor widely used in DSSCs due to the physical properties that make it suitable for use in DSSCs. They meet one of the criteria for effective electron injection by having a conduction band edge that is just below the excited state energy level of many dyes. When the DSSC is highlighted, photons are absorbed by the dye, electrons are transferred and excited, and the dye is oxidized. The excited electrons are sent into the range where TiO_2 can conduct electricity, and then they spread out through the porous film of the TiO_2 conductive glass fluorine-doped tin oxide (FTO) [21].

To obtain cells that are less costly and environmentally friendly, studies in the literature have suggested using DSSCs dyed with natural dyes or synthetic dyes [22,23]. On the other hand, synthetic dyes cause serious environmental difficulties because they are toxic, carcinogenic, refractory, and difficult to dissolve using water treatment technology [24]. To fight the health risks and environmental problems caused by synthetic dyes around the world, preventive steps must be taken to cut down on how much of these dyes are used in the environment [25]. Different research institutions and government groups have come up with different ideas for how to solve the problems of making water treatment technologies. These ideas include enhanced oxidation processes, reducing the amount of dye used, and using natural dyes instead [26,27]. Replacement with natural dyes has been suggested as a possible green solution for dealing with wastewater problems and reducing the environmental impact of these industrial operations. Natural dyes made from renewable bioresources are safe for the environment, biodegradable, and cheaper than synthetic dyes. They can be used as a good replacement for synthetic dyes [28,29]. Natural dyes derived from plants have been employed in recent years to improve the performance of DSSCs and photo-stimulation therapies [30,31]. To be classified as a sensitizer, a natural dye must meet the following criteria: possessing a broad and robust capacity to absorb light in the visible and near-infrared ranges; ensuring the stability and effective charge injection of the system by firmly adhering to the semiconductor surface, and containing optimal lowest unoccupied molecular orbital (LUMO) and highest occupied molecular orbital (HOMO) energy levels for effective charge injection into the semiconductor conduction band (CB) [31,32]. Sensitizer development has been a major focus of recent years, since it plays an important role in DSSCs [26,27]. Chlorophylls seem to be in charge of collecting and delivering light energy to photosynthesis reaction centers, according to the research [33].

This paper presents an experimental investigation on the performance of four natural dyes extracted from mint, Helianthus annuus, Fragaria, and a mixture of the three dyes in equal proportions. The dye extracted from Helianthus annuus has been shown to contain Helianthus annuus auxin, which is the main reason for the sensitivity of the Helianthus annuus flower and its rotation with the sun, whereas mint leaves contain chlorophyll, which is one of the most important elements that help plants in absorbing sunlight. Fragaria extract was chosen because it contained anthocyanins, and DSSCs were made from these dyes. We found that these dyes are good for the environment and are a good alternative to the typical dyes. The inhibition of TiO_2 particle crystallization was investigated by energy dispersion X-ray analysis of the efficacy of natural photosensitizers, and the DSSCs were manufactured by adding the extracted dyes as the photosensitizers. Maximum power point (P_{max}), fill factor (FF), energy conversion efficiency (η), and short circuit current (I_{SC}) are the primary characteristics of a solar cell.

2. Materials and Methods

2.1. Materials

The materials used in this research included Mentha leaves, Helianthus annuus, Fragaria, and acetone to extract the natural colors from plants. A liquid platinum catalyst and sealing tape were used in this process. An electrolyte (iodide and triiodide) (I^-/I^{+3}), FTO (fluorine-doped tin oxide) glass as a transparent conducting oxide was used; it has a surface resistance of 10–20 Ω and also needs nitric acid (HNO_3) and TiO_2 powder.

2.2. Preparation of Natural Dye Sensitizers

The dyes extracted from the leaves of mint and the leaves of the annual plant Helianthus were prepared in the same way. The leaves of both plants were first washed with distilled water, and then the leaves were dried at room temperature in the dark. Then the leaves were crushed separately using an electric grinder, and 1 g of mint leaves was dissolved in 60 mL of acetone, and also 1 g of Helianthus annuus was dissolved in 60 mL of acetone in an airtight package and left for 24 h, after which it was filtered. Using filter paper to get rid of large portions, the extracts were kept in a container away from light until use. These steps are illustrated in Figure 1.

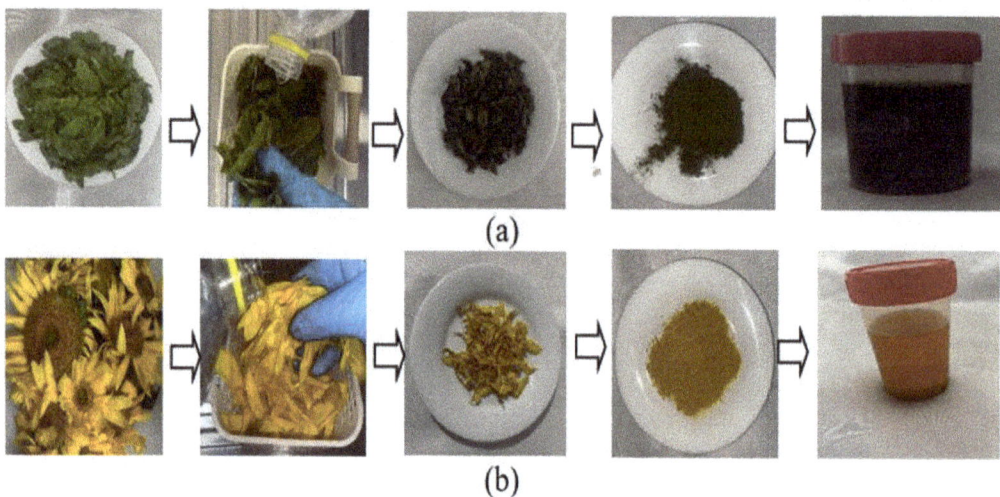

Figure 1. Steps for extracting dye from (**a**) Mentha leaves; (**b**) Helianthus annuus.

As an alternative, the Fragaria fruit extract was made by crushing the fruit in a pestle and extracting the fruit's juices from 1 mg of fruit extract in 6 mL of acetone. After 24 h, they were filtered using filter paper to get rid of large pieces and kept in a tightly closed container in the dark until used. The fourth dye was prepared by mixing equal amounts of all of the aforementioned dyes, and the same method of preserving the other dyes was followed.

2.3. The TiO_2 Paste and Photoelectrode Preparation

The amount of 2 g of TiO_2 powder was dissolved in 6 mL of dilute nitric acid (HNO_3) (PH = 3) to make a TiO_2 paste. The components were mixed with a mortar and pestle for 30 min, followed by magnetic stirring for another 30 min, to create a homogeneous mixture. Ethylene glucose and glucose X-100 were added, which enhance the paste's adhesion to FTO and the doughs' adhesion to each other. Magnetic stirring was then carried out for an hour and a half to fasten the dissolution of the TiO_2 paste. Figure 2 shows the texture of the TiO_2 paste.

Figure 2. Photograph for TiO$_2$ preparation process and its texture.

2.4. Fabrication of DSSCs

The FTO glass was thoroughly cleaned, and then the FTO piece was wrapped with tape on all four sides so that the thickness of the TiO$_2$ layer was controlled and the active area of the substrate was determined, which was equivalent to 1 cm^2. Drops of TiO$_2$ paste were applied and spread on the FTO substrate. Then, the FTO-coated TiO$_2$ was annealed at 500 °C for 80 min. After that, it was left to cool at room temperature. Then, four FTO substrates were soaked in each natural dye for 24 h. The excess dye was then removed by rinsing the substrate with ethanol. Using another FTO glass, the counter electrode was prepared with a Pt distribution and then heated at 450 °C, and these steps are illustrated in Figures 3–5.

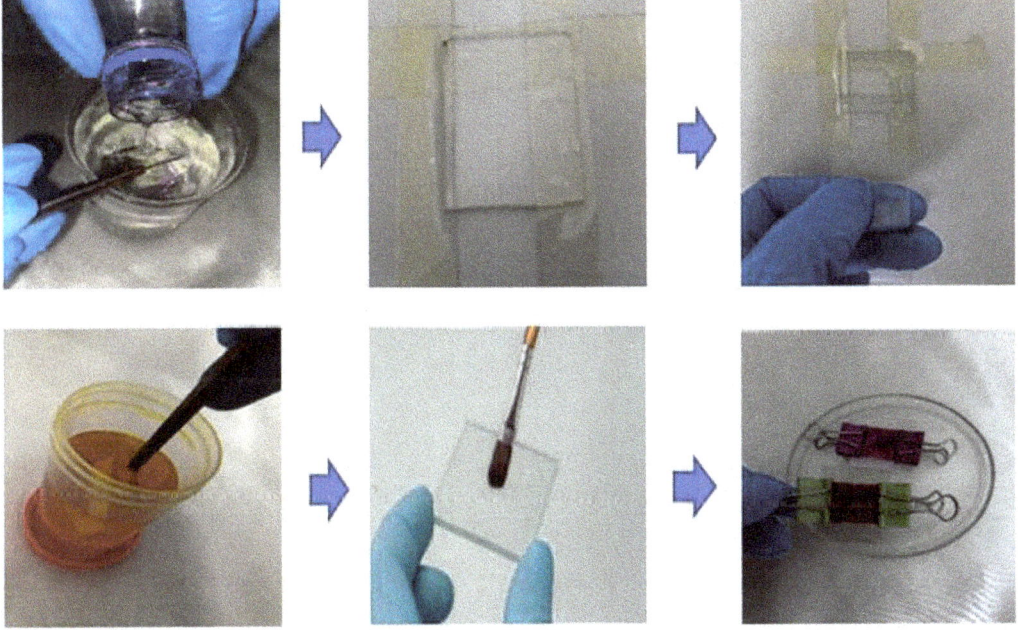

Figure 3. Photographs of DSSCs fabrication stages.

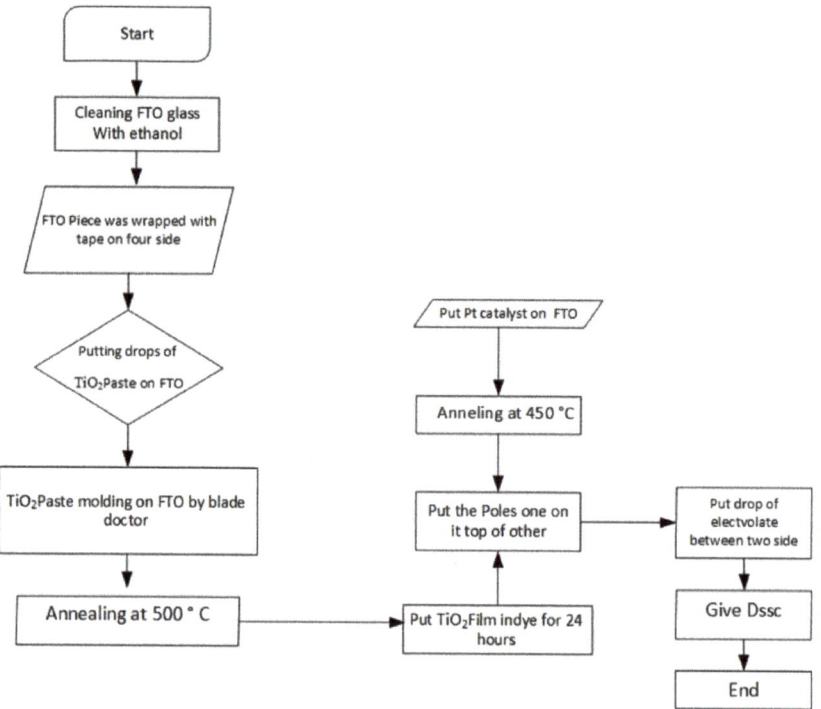

Figure 4. Schematic diagram of DSSCs fabrication process.

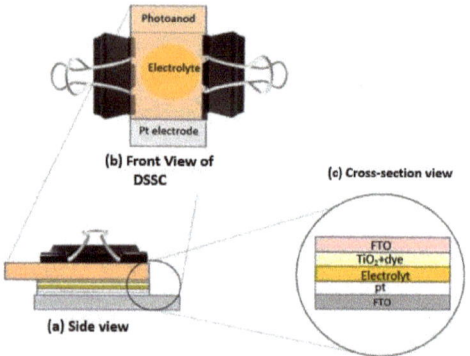

Figure 5. The assembly procedure of the FTO/TiO$_2$-dye/electrolyte/Pt DSSCs from several viewpoints: (**a**) side view; (**b**) front view; (**c**) cross-section view.

2.5. *Measurement of the DSSC's Photoelectric Conversion Rate*

Analysis of the absorption spectra of natural dye solution and the combined solution of TiO$_2$ and natural dye was carried out using a UV-VIS spectrophotometer (Jasco, V650). The DSSCs' photoelectric conversion efficiency was also examined in a lab environment using a source of artificial sunlight (AM1.5). Starting with the current–voltage (I–V) curve, the fill factor (FF) was specified at the time of the foundation as follows:

$$FF = \frac{P_{max}}{I_{sc} \times V_{OC}} \quad (1)$$

The maximum current and voltage values are denoted as I_{max} and V_{max}, respectively, while the short circuit current and open voltage are denoted as I_{sc} and V_{OC}, respectively. Here is how to calculate the total energy conversion efficiency:

$$\eta = \frac{I_{sc} \times V_{OC} \times FF}{P_{in}} \qquad (2)$$

where P_{in} specifies the energy of the incident photon.

The parameters of DSSCs were measured using a device (Keithley 2400), where a digital ammeter and voltage meter are connected at both ends of the DSSC, as shown in Figure 6.

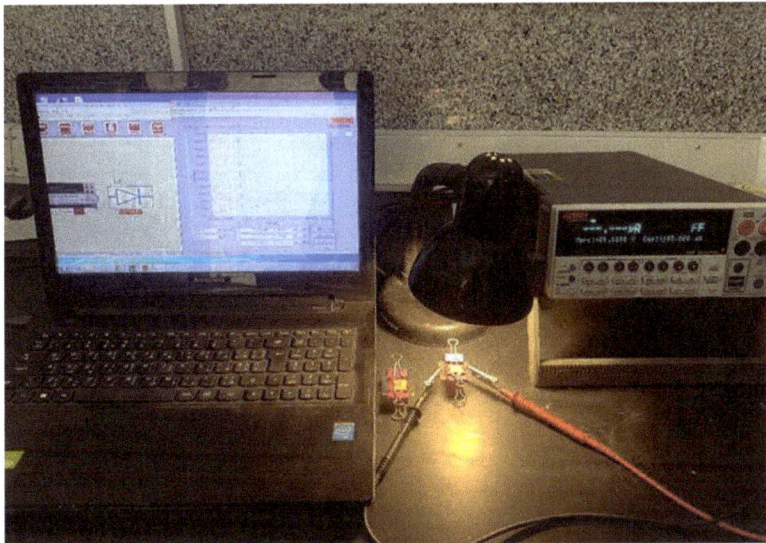

Figure 6. Photograph of solar cell measurement configuration.

3. Results and Discussion

3.1. X-ray Diffraction Analysis

X-ray investigations were carried out on the films prepared on FTO glass, since the nature of TiO_2 plays a very important role in dye adsorption, and controls the efficiency and photocatalytic processes due to the details of different binding patterns. The results of the X-ray show that all of the films are in the anatase-phase through peaks (20.144, 25.005, 37.795, 47.776, 54.212, 53.658, and 62.370) with the respect to TiO_2. This phase is considered an active phase due to its surface chemistry and its high conduction band. This phase also shows better performance for DSSCs than the Rutile phase [34–37]. From the X-rays, a noticeable increase in the height of the peaks was noted, as shown in Figures 7 and 8. These results are consistent with the tables of the American Society of Mechanical Engineers (A.S.T.M). It is also observed from Figures 7 and 8 that there were no phases of impurities or other oxides in the crystal structure. The crystal size, which is the primary determinant in influencing the electron transport properties of materials, can be determined through the following equation [38]:

$$d = \frac{0.9\lambda}{\beta \cos(\theta)} \qquad (3)$$

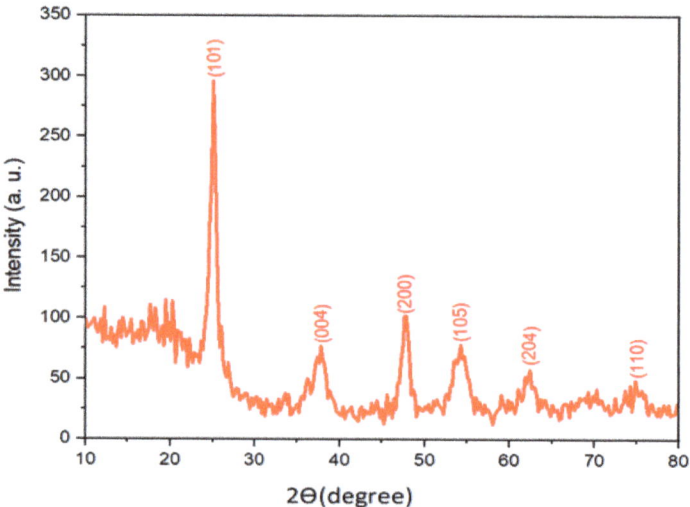

Figure 7. X-ray pattern of TiO$_2$ nanorods before heat treatment.

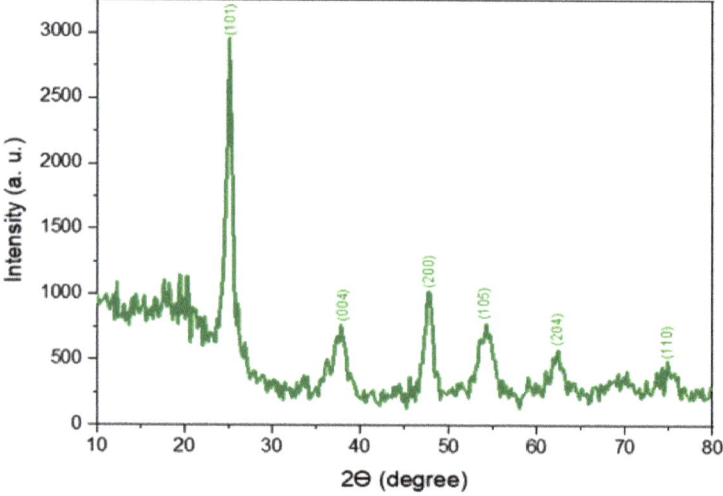

Figure 8. X-ray pattern of TiO$_2$ nanorods after heat treatment.

In this equation, θ is the diffraction angle, β is the full width at half maximum for each peak, and λ is the employed X-ray wavelength.

3.2. Field Emission Scanning Electron Microscopy (FESEM)

To confirm the creation of TiO$_2$ nanostructured films, FESEM measurements are often performed [39]. As TiO$_2$ films were heated to less than 450 °C, they were found to produce an imperceptible current even in the A range, making them ideal for solar cells [40]. According to the FESEM images in Figure 9, the TiO$_2$ grains were formed as spherical particles covering the FTO substrate with an average size of 27 ± 3 nm. Good agglomeration is responsible for this change, and it increases the ability of the FTO glass cover to stick together, which increases the quality of the conversion of light photons into electrons and the improved flow of electricity through the particle grid.

Figure 9. Photographs of FESEM TiO$_2$ layer at 5 µm (**left**) and 10 µm (**right**) at 500 °C.

3.3. Absorbance Behavior of the Prepared TiO$_2$ Film

Figure 10 shows the absorption behavior of the prepared TiO$_2$. It shows that the annealed TiO$_2$ film at 500 °C has an absorption peak of 318 nm, while the maximum absorption wavelength of 304 nm occurs with a power band gap of 4.079 eV. The energy band gap (E_g) of the dyes absorbed by the TiO$_2$ surface was calculated as follows [41]:

$$Eg = \frac{1240}{\lambda} \qquad (4)$$

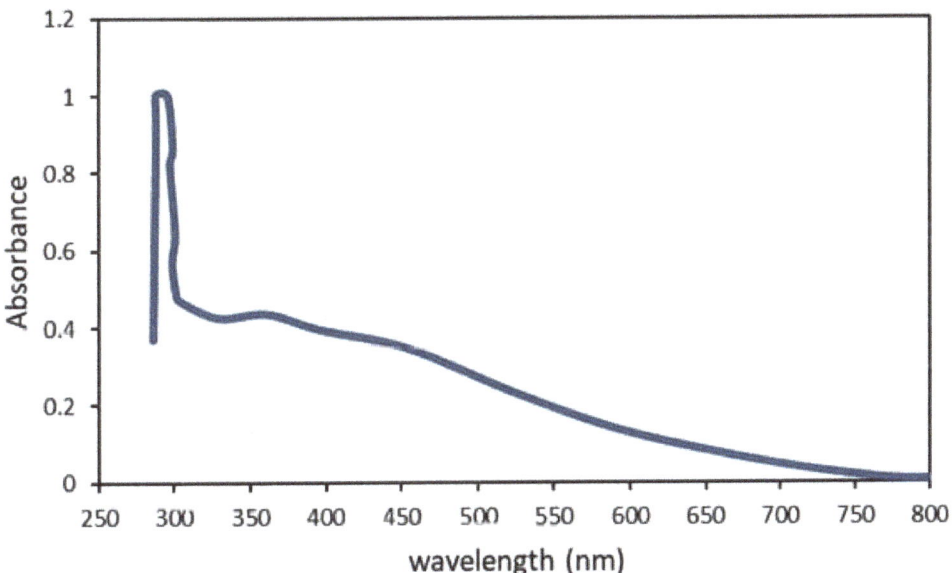

Figure 10. Absorption spectrum of TiO$_2$.

The results of this research are in good agreement with other investigations on the optical energy band gap [42,43].

Figure 11 shows that the TiO$_2$ film that was poured on the FTO glass allows the transmission of visible light and absorbs ultraviolet rays. The average visible light transmission of TiO$_2$ after casting on the FTO substrate reduced to about 74.34%. However, coated FTO allows maximum transmission of visible light to the dye (sensitive material) for optimal absorption and conversion to electricity in DSSCs, while blocking harmful UV rays.

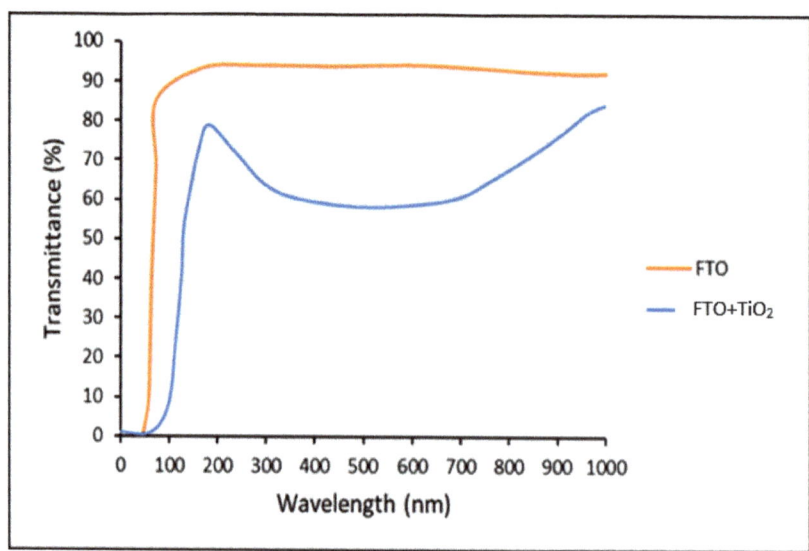

Figure 11. Transmitter spectra of FTO and TiO$_2$ films cast on FTO.

4. UV-Vis Analysis

The optical measurements were made using UV-VIS spectroscopy for the prepared dyes (Mentha leaves, Helianthus annuus leaves, Fragaria, and a dye consisting of a mixture of each of the extracted pigments in equal proportions). The absorption spectra showed that each type of dye had its absorption peak in the visible range.

In the Mentha leaves case, the absorption spectrum of mint leaves showed a peak absorption rate in the visible region at wavelengths of 450–500 and 580–700 nm, showing three peaks at 655, 430, and 485 nm, as shown in Figure 12.

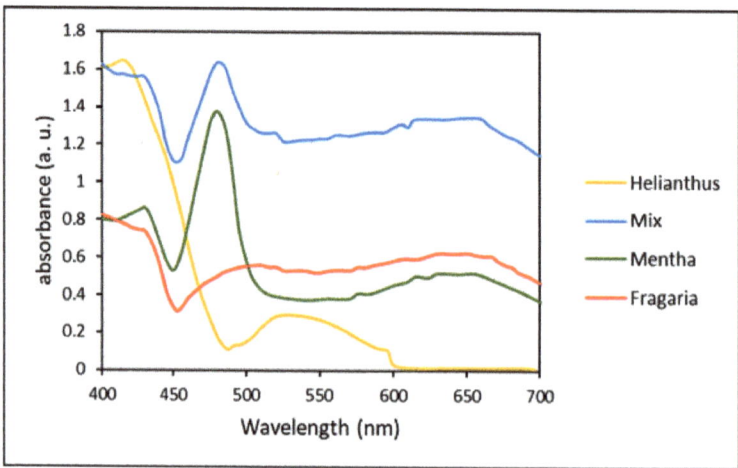

Figure 12. UV-Vis absorption spectra of Mentha leaves, Helianthus annuus, Fragaria, and a mixture of the three showing various peak locations.

In the Helianthus annuus case, the absorption spectrum of Helianthus annuus showed a maximum absorption plausible region from the 350–450 nm to the 500–600 nm regions, where its absorption peaks appeared at 415 and 515 nm, as shown in Figure 12.

In the Fragaria case, the absorbance spectrum for Fragaria showed a maximum absorption peak in the visible regions at 512 and 650 nm wavelengths. It had a wide absorption range in the region of 450–690 nm, as shown in Figure 12.

In the dye mixture case, the optical absorption spectrum of the dye mixture showed absorption in the visible region at wavelengths of 380–445, 575–700 and 510–450 nm, where its absorption peaks appeared at 433, 484, and 652 nm, as shown in Figure 12.

It has been observed that the three pigments can absorb the light in the visible area when melting, which makes them suitable as catalysts in DSSCs. Thus, it can be concluded that the nature of the solvent used to extract the dyes can affect the concentration of the dye due to the presence of different chemical compounds in plants. Accordingly, they have different solubility in different solvents. Additionally, it can be concluded that the optimal combination of dyes showed better cumulative absorption properties, as its absorption spectrum was broader and higher. Taking light in the visible range makes it more likely that higher levels of solar energy can be turned into an electrochemical form.

Photovoltaics Performance of DSSCs

Under white light irradiation (100 mW/cm^2) from a high-pressure mercury arc lamp, photovoltaic studies of DSSCs manufactured using natural dyes as catalysts were performed by measuring the J–V curve of each cell. The short circuit current (J_{sc}), fill factor (FF), open circuit voltage (V_{oc}), and power conversion efficiency (η) were used to evaluate the performance of natural dyes as catalysts in DSSCs. Figure 13 shows J–V curves of DSSCs utilizing sensitizers taken from Mentha leaves, Helianthus annuus, Fragaria, and a mixture of these dyes. The outcomes reveal that the sensitizing dye has a significant impact on the performance of the DSSC, as dyes absorb sunlight and convert it into electrical energy.

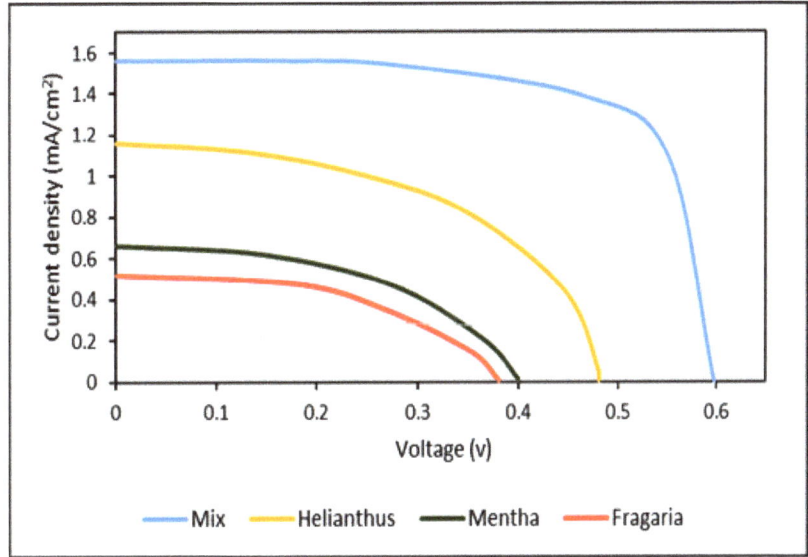

Figure 13. J–V curve for the dyed DSSCs.

The short-circuit current had the highest value in the DSSC based on the mixture of dyes, while it showed the highest open circuit voltage with the DSSC based on Mentha leaves. The power output power of the DSSC was computed using IV data as $P = IV$. Figure 14 shows the power estimated as a function of V for DSSC sensitized by a dye combination as an example. The photoelectron chemical properties of the DSSCs sensitized with natural dyes are listed in Table 1. As displayed in Table 1, the fill factor of the fabricated DSSCs ranged between 46.44% and 73.55%. The V_{oc} changed from 0.38 to 0.59 V and the J_{sc} varied from 0.51 to 1.59 mA/cm^2. The DSSC sensitization based on the mixture of dyes yielded the best results, with a cell efficiency of 0.69%. This dye showed the highest and most absorbance peaks in the visible region, due to carbonyl and hydroxyl groups in its chemistry, which allow it to attach to the TiO$_2$'s surface, enhancing the energy conversion efficiency.

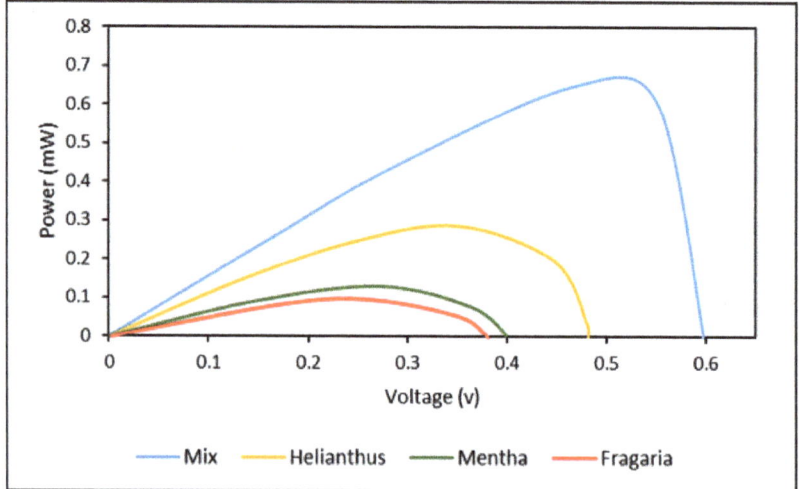

Figure 14. P–V curves for DSSCs for the four types of dyes adopted in this study.

Table 1. Performance of DSSCs.

Dye	V_{oc} (v)	J_{sc} (mA)	η (%)	F (%)
Fragaria	0.38	0.51	0.09	46.44
Mentha leaves	0.41	0.64	0.15	57.16
Helianthus annuus leaves	0.48	1.19	0.29	50.77
Mix	0.59	1.59	0.69	73.55

5. Conclusions

In this study, four different natural dyes extracted from mint, helianthus annuus, Fragaria, and a mixture of the three dyes were investigated, and their performances were calculated. The dye sensitizer is a key factor in DSSCs, as it acts as an electronic pump to transfer light energy from the sun to an electrical power generation device. These dyes were selected, as Fragaria extract and mint leaves contain anthocyanin and chlorophyll pigments, and Helianthus annuus leaves contain auxin. The adopted methodology in this study has extracted the dyes successfully, and then the absorption conditions were examined and the photocurrent activity was tested. The outcomes have shown that a mixture of dyes (anthocyanins, auxin, and chlorophyll) has a better efficiency among the other dyes, reaching approximately 0.69%. This is due to the carbonyl and hydroxyl groups in its chemistry, which allow it to attach to the TiO$_2$'s surface, enhancing energy conversion efficiency. By contrast, the efficiency magnitudes for Mentha leaves, Fragaria, and Helianthus annuus leaves were 0.15%, 0.09%, and 0.29%, respectively. The results obtained lack the detailed research needed to increase the efficiency and stability of DSSCs based on natural dyes. Accordingly, the study concludes that natural dyes have good potential to form photosensitizers in DSSCs, and they are cheap, safe, environmentally friendly, and easy to extract. Therefore, it can be concluded that DSSCs that are based on natural dyes will shortly become one of the photovoltaic sources that overcome traditional electrical energy sources.

Author Contributions: Conceptualization, Z.H.A. and D.M.H.; methodology, Z.H.A.; software, F.K. and D.M.H.; formal analysis, D.M.H.; investigation, Z.H.A.; resources, A.S.N.A.-m.; writing—original draft preparation, Z.H.A.; writing—review and editing, A.A.-M.; supervision T.Y. All authors have read and agreed to the published version of the manuscript.

Funding: This research received no external funding.

Institutional Review Board Statement: Not applicable.

Informed Consent Statement: Not applicable.

Data Availability Statement: Not applicable.

Conflicts of Interest: The authors declare no conflict of interest.

References

1. Ullah, A.; Khan, J.; Sohail, M.; Hayat, A.; Zhao, T.K.; Ullah, B.; Khan, M.; Uddin, I.; Ullah, S.; Ullah, R.; et al. Fabrication of polymer carbon nitride with organic monomer for effective photocatalytic hydrogen evolution. *J. Photochem. Photobiol. A Chem.* **2020**, *401*, 112764. [CrossRef]
2. Medina-Santana, A.A.; Hewamalage, H.; Cárdenas-Barrón, L.E. Deep Learning Approaches for Long-Term Global Horizontal Irradiance Forecasting for Microgrids Planning. *Designs* **2022**, *6*, 83. [CrossRef]
3. Al-Saeed, Y.W.; Ahmed, A. Evaluating Design Strategies for Nearly Zero Energy Buildings in the Middle East and North Africa Regions. *Designs* **2018**, *2*, 35. [CrossRef]
4. Vandewetering, N.; Hayibo, K.S.; Pearce, J.M. Open-Source Design and Economics of Manual Variable-Tilt Angle DIY Wood-Based Solar Photovoltaic Racking System. *Designs* **2022**, *6*, 54. [CrossRef]
5. Bhagavathy, S.M.; Pillai, G. PV Microgrid Design for Rural Electrification. *Designs* **2018**, *2*, 33. [CrossRef]
6. Aykapadathu, M.; Nazarinia, M.; Sellami, N. Design and Fabrication of Absorptive/Reflective Crossed CPC PV/T System. *Designs* **2018**, *2*, 29. [CrossRef]
7. Rehman, A.U.; Shah, M.Z.; Rasheed, S.; Afzal, W.; Arsalan, M.; Rahman, H.U.; Ullah, M.; Zhao, T.; Ullah, I.; Din, A.U.; et al. Inorganic salt hydrates and zeolites composites studies for thermochemical heat storage. *Z. Für Phys. Chem.* **2021**, *235*, 1481–1497. [CrossRef]
8. Maldon, B.; Thamwattana, N. A Fractional Diffusion Model for Dye-Sensitized Solar Cells. *Molecules* **2020**, *25*, 2966. [CrossRef]
9. Su, R.; Lyu, L.; Elmorsy, M.; El-Shafei, A. Novel metal-free organic dyes constructed with the D-D∣A-π-A motif: Sensitization and co-sensitization study. *Sol. Energy* **2019**, *194*, 400–414. [CrossRef]
10. Ramamoorthy, R.; Karthika, K.; Dayana, A.M.; Maheswari, G.; Eswaramoorthi, V.; Pavithra, N.; Anandan, S.; Williams, R.V. Reduced graphene oxide embedded titanium dioxide nanocomposite as novel photoanode material in natural dye-sensitized solar cells. *J. Mater. Sci. Mater. Electron.* **2017**, *28*, 13678–13689. [CrossRef]
11. Cavallo, C.; Di Pascasio, F.; Latini, A.; Bonomo, M.; Dini, D. Nanostructured Semiconductor Materials for Dye-Sensitized Solar Cells. *J. Nanomater.* **2017**, *2017*, 1–31. [CrossRef]
12. Kumar, D.K.; Kříž, J.; Bennett, N.; Chen, B.; Upadhayaya, H.; Reddy, K.R.; Sadhu, V. Functionalized metal oxide nanoparticles for efficient dye-sensitized solar cells (DSSCs): A review. *Mater. Sci. Energy Technol.* **2020**, *3*, 472–481. [CrossRef]
13. Yao, Z.; Wu, H.; Li, Y.; Wang, J.; Zhang, J.; Zhang, M.; Guo, Y.; Wang, P. Dithienopicenocarbazole as the kernel module of low-energy-gap organic dyes for efficient conversion of sunlight to electricity. *Energy Environ. Sci.* **2015**, *8*, 3192–3197. [CrossRef]
14. Bisquert, J.; Cahen, D.; Hodes, G.; Rühle, S.; Zaban, A. Physical Chemical Principles of Photovoltaic Conversion with Nanoparticulate, Mesoporous Dye-Sensitized Solar Cells. *J. Phys. Chem. B* **2004**, *108*, 8106–8118. [CrossRef]
15. Sugathan, V.; John, E.; Sudhakar, K. Recent improvements in dye sensitized solar cells: A review. *Renew. Sustain. Energy Rev.* **2015**, *52*, 54–64. [CrossRef]
16. Akinoglu, B.G.; Tuncel, B.; Badescu, V. Beyond 3rd generation solar cells and the full spectrum project. Recent advances and new emerging solar cells. *Sustain. Energy Technol. Assess.* **2021**, *46*, 101287. [CrossRef]
17. Hug, H.; Bader, M.; Mair, P.; Glatzel, T. Biophotovoltaics: Natural pigments in dye-sensitized solar cells. *Appl. Energy* **2014**, *115*, 216–225. [CrossRef]
18. Kushwaha, R.; Srivastava, P.; Bahadur, L. Natural Pigments from Plants Used as Sensitizers for TiO_2 Based Dye-Sensitized Solar Cells. *J. Energy* **2013**, *2013*, 1–8. [CrossRef]
19. Amogne, N.Y.; Ayele, D.W.; Tsigie, Y.A. Recent advances in anthocyanin dyes extracted from plants for dye sensitized solar cell. *Mater. Renew. Sustain. Energy* **2020**, *9*, 1–16. [CrossRef]

20. Nien, Y.-H.; Chen, H.-H.; Hsu, H.-H.; Rangasamy, M.; Hu, G.-M.; Yong, Z.-R.; Kuo, P.-Y.; Chou, J.-C.; Lai, C.-H.; Ko, C.-C.; et al. Study of How Photoelectrodes Modified by TiO$_2$/Ag Nanofibers in Various Structures Enhance the Efficiency of Dye-Sensitized Solar Cells under Low Illumination. *Energies* **2020**, *13*, 2248. [CrossRef]
21. Yu, Z. Liquid Redox Electrolytes for Dye-Sensitized Solar Cells Ze Yu. Ph.D. Thesis, KTH Royal Institute of Technology, Stockholm, Sweden, 2012. Available online: https://www.diva-portal.org/smash/get/diva2:483008/FULLTEXT01.pdf (accessed on 15 September 2022).
22. Zeng, K.; Chen, Y.; Zhu, W.-H.; Tian, H.; Xie, Y. Efficient Solar Cells Based on Concerted Companion Dyes Containing Two Complementary Components: An Alternative Approach for Cosensitization. *J. Am. Chem. Soc.* **2020**, *142*, 5154–5161. [CrossRef] [PubMed]
23. Zou, J.; Yan, Q.; Li, C.; Lu, Y.; Tong, Z.; Xie, Y. Light-Absorbing Pyridine Derivative as a New Electrolyte Additive for Developing Efficient Porphyrin Dye-Sensitized Solar Cells. *ACS Appl. Mater. Interfaces* **2020**, *12*, 57017–57024. [CrossRef] [PubMed]
24. Jadhav, S.A.; Garud, H.B.; Patil, A.H.; Patil, G.D.; Patil, C.R.; Dongale, T.D.; Patil, P.S. Recent advancements in silica nanoparticles based technologies for removal of dyes from water. *Colloids Interface Sci. Commun.* **2019**, *30*, 100181. [CrossRef]
25. Singh, K.; Arora, S. Removal of Synthetic Textile Dyes From Wastewaters: A Critical Review on Present Treatment Technologies. *Crit. Rev. Environ. Sci. Technol.* **2011**, *41*, 807–878. [CrossRef]
26. Nidheesh, P.; Zhou, M.; Oturan, M.A. An overview on the removal of synthetic dyes from water by electrochemical advanced oxidation processes. *Chemosphere* **2018**, *197*, 210–227. [CrossRef] [PubMed]
27. Diaz-Uribe, C.; Vallejo, W.; Camargo, G.; Muñoz-Acevedo, A.; Quiñones, C.; Schott, E.; Zarate, X. Potential use of an anthocyanin-rich extract from berries of Vaccinium meridionale Swartz as sensitizer for TiO$_2$ thin films—An experimental and theoretical study. *J. Photochem. Photobiol. A Chem.* **2019**, *384*, 112050. [CrossRef]
28. Dong, Y.; Gu, J.; Wang, P.; Wen, H. Developed functionalization of wool fabric with extracts of Lycium ruthenicum Murray and potential application in healthy care textiles. *Dye. Pigment.* **2018**, *163*, 308–317. [CrossRef]
29. Shahid, M.; Shahid-ul-Islam; Mohammad, F. Recent advancements in natural dye applications: A review. *J. Clean. Prod.* **2013**, *53*, 310–331. [CrossRef]
30. Jalali, T.; Arkian, P.; Golshan, M.; Jalali, M.; Osfouri, S. Performance evaluation of natural native dyes as photosensitizer in dye-sensitized solar cells. *Opt. Mater.* **2020**, *110*, 110441. [CrossRef]
31. Golshan, M.; Osfouri, S.; Azin, R.; Jalali, T. Fabrication of optimized eco-friendly dye-sensitized solar cells by extracting pigments from low-cost native wild plants. *J. Photochem. Photobiol. A Chem.* **2019**, *388*, 112191. [CrossRef]
32. Iqbal, M.Z.; Ali, S.R.; Khan, S. Progress in dye sensitized solar cell by incorporating natural photosensitizers. *Sol. Energy* **2019**, *181*, 490–509. [CrossRef]
33. Ammar, A.M.; Mohamed, H.; Yousef, M.M.K.; Abdel-Hafez, G.M.; Hassanien, A.S.; Khalil, A.S.G. Dye-Sensitized Solar Cells (DSSCs) Based on Extracted Natural Dyes. *J. Nanomater.* **2019**, *2019*, 1–10. [CrossRef]
34. Wolf, S.E.; Lieberwirth, I.; Natalio, F.; Bardeau, J.-F.; Delorme, N.; Emmerling, F.; Barrea, R.; Kappl, M.; Marin, F. Merging Models of Biomineralisation with Concepts of Nonclassical Crystallisation: Is a Liquid Amorphous Precursor Involved in the Formation of the Prismatic Layer of the Mediterranean Fan Mussel Pinna nobilis? *Faraday Discuss.* **2012**, *159*, 433–448. [CrossRef]
35. Jia, H.-L.; Chen, Y.-C.; Ji, L.; Lin, L.-X.; Guan, M.-Y.; Yang, Y. Cosensitization of porphyrin dyes with new X type organic dyes for efficient dye-sensitized solar cells. *Dye. Pigment.* **2018**, *163*, 589–593. [CrossRef]
36. Chandra, R.; Iqbal, H.M.N.; Vishal, G.; Lee, H.-S.; Nagra, S. Algal biorefinery: A sustainable approach to valorize algal-based biomass towards multiple product recovery. *Bioresour. Technol.* **2019**, *278*, 346–359. [CrossRef]
37. Park, N.-G.; van de Lagemaat, J.; Frank, A.J. Comparison of Dye-Sensitized Rutile- and Anatase-Based TiO$_2$ Solar Cells. *J. Phys. Chem. B* **2000**, *104*, 8989–8994. [CrossRef]
38. Zatirostami, A. Increasing the efficiency of TiO$_2$-based DSSC by means of a double layer RF-sputtered thin film blocking layer. *Optik* **2020**, *207*, 164419. [CrossRef]
39. Dhas, V.; Muduli, S.; Agarkar, S.; Rana, A.; Hannoyer, B.; Banerjee, R.; Ogale, S. Enhanced DSSC performance with high surface area thin anatase TiO$_2$ nanoleaves. *Sol. Energy* **2011**, *85*, 1213–1219. [CrossRef]
40. Saadaoui, S.; Ben Youssef, M.A.; Ben Karoui, M.; Gharbi, R.; Smecca, E.; Strano, V.; Mirabella, S.; Alberti, A.; Puglisi, R.A. Performance of natural-dye-sensitized solar cells by ZnO nanorod and nanowall enhanced photoelectrodes. *Beilstein J. Nanotechnol.* **2017**, *8*, 287–295. [CrossRef]
41. Shi, Y.; Tang, Y.; Yang, K.; Qin, M.; Wang, Y.; Sun, H.; Su, M.; Lu, X.; Zhou, M.; Guo, X. Thiazolothienyl imide-based wide bandgap copolymers for efficient polymer solar cells. *J. Mater. Chem. C* **2019**, *7*, 11142–11151. [CrossRef]
42. Sclafani, A.; Herrmann, J.M. Comparison of the Photoelectronic and Photocatalytic Activities of Various Anatase and Rutile Forms of Titania in Pure Liquid Organic Phases and in Aqueous Solutions. *J. Phys. Chem.* **1996**, *100*, 13655–13661. [CrossRef]
43. Ezike, S.C. Effect of Concentration Variation on Optical and Structural Properties of TiO$_2$ Thin Films. *J. Mod. Mater.* **2020**, *7*, 1–6. [CrossRef]

Article

Analysis of Forest Biomass Wood Briquette Structure According to Different Tests of Density

Kamil Roman [1,*], Witold Rzodkiewicz [2] and Marek Hryniewicz [3]

[1] Institute of Wood Sciences and Furniture, Warsaw University of Life Sciences, 166 Nowoursynowska St., 02-787 Warsaw, Poland
[2] Central Office of Measures, Electricity and Radiation Department, 2 Elektoralna St., 00-139 Warsaw, Poland
[3] Institute of Technology and Life Sciences—National Research Institute, Falenty, 3 Al. Hrabska, 05-090 Raszyn, Poland
* Correspondence: kamil_roman@sggw.edu.pl

Abstract: X-ray technology is capable of non-destructively testing solid wood samples. The prepared wood briquette samples were identified by X-ray technology. The studies assessed the effect of biomass briquette structure by observing wood chip fractions under an X-ray. Study results show that X-ray technology is an effective tool for analyzing biomass wood-based materials, e.g., density, improving wood products quality and performance. The measurements are consistent with the true density, and chemical properties were measured from chosen material. In the article, a coefficient representing the picture density and true density of briquettes was proposed, and the comparison of both densities was based on the empirical measurements. Probably through an application of the conversion factor, the process of determining material densities could be simplified, cheaper, and quicker. Due to the conducted research, X-ray technology is an effective tool for improving wood products' quality and performance. Combining X-ray technology with laboratory test results can provide quick and easy analysis. For example, the density comparison of shredded forest residues was defined. Based on the results, the mean value of the conversion factor was about 0.6. In addition, the observed results were compared with the doctoral research. Higher durability was found in briquettes with lower fractions of f_1 (31–68%) compared to the higher f_2, f_3, and f_4 (6–37%), which was approved in the spectrum picture.

Keywords: forest residues; briquettes; wooden chips; X-ray radiation

1. Introduction

Plant materials intended for energy purposes have become significantly more popular in recent years. In particular, the biomass of forest origin is increasingly used as an energy carrier in Poland. The described substance is a biodegradable raw material produced in the form of waste. Wood-based biomass is produced as a residue in the processes of wood production, processing, or sanitary cutting [1]. It can be obtained also from forest residues or as a result of care treatments carried out in the so-called general cargo. Huge heating possibilities concentrated in forest areas require supervision, consisting of the regulation of forest resources and ensuring the supply of renewable energy sources [2]. Adaptation of this fuel requires increasing the use of forest residues and improving the quality of the product by using the optimal technology for the production of briquettes from general logging. The idea of adaptation and dissemination of non-destructive tests of wood products may potentially enable a cheaper, faster, and simpler way to measure the specific density of bulk materials. Defining and determining the value of the density conversion factor is an additional value of work.

Preliminary studies indicate significant differences in the composition of biomass from logging waste. It contains more ingredients such as bark and needles; moreover, it is a biomass usually derived from various plant species, namely from coniferous and deciduous wood. For example, after conducting laboratory tests, Ekielski proved that the effectiveness of the production of briquettes from forest biomass is encouraging in terms of calorific value [3]. Forest biomass is similar to other plants intended for energy purposes, i.e., Virginia mallow, sugar miscanthus, or Jerusalem artichoke, which have similar processing schemes. Therefore, all technological operations can be analogous to the processing of straw plants [4]. Co-combustion of forest biomass may be another opportunity to replace fossil fuels, and, thus, there is a possibility of further reduction of harmful gas emissions into the atmosphere.

One of the non-destructive tests of the solid structure is X-ray waves. X-ray radiation is an electromagnetic field that is generated from the emission of electrons. The wave latitude measures 10 pm to 10 nm and is placed between ultraviolet (UV) and gamma radiation. Techniques involving X-ray radiation are mostly popular in medicine to diagnose the internal structure of objects. However, they are also used in defectoscopy or studies of structural and spectral analysis. Wood science uses X-ray technology for various applications, including the study of wood anatomy and the analysis of wood-based materials. An example of X-ray's usage in wood science could be the studies of the internal structure of wood and the distribution of different cell types. This can be studied by using X-ray computed tomography (CT) scans of wood samples. Other wood-based materials analysis can be treated by X-ray diffraction (XRD), and X-ray fluorescence (XRF) can be used to determine the crystal structure and chemical composition.

X-ray technology can be used to non-destructively test wood samples, which means that no physical damage is caused. The prepared wood briquette samples were identified by X-ray technology. The performed studies aim to measure the effect of biomass briquettes' structure by visualization of wood chip fractions under the X-ray. According to the studies, X-ray technology is a valuable tool in wood science for analyzing wood-based materials and improving the quality and performance of wood products.

The shredded Scots pine wood (*Pinus sylvestris* L.) as bases of investigated materials were used. Scots pine wood [5,6] is widely used in a variety of applications, including construction, furniture, flooring, and paper production. The wood is light yellow to reddish brown in color, with a straight grain and a fine, even texture. It is also used as a source of wood pulp for paper production and as a raw material for the production of wood-based panels. Due to its low density and relatively low cost, it is also commonly used as a source of wood chips for the production of energy. Scots pine wood is a versatile and widely used species of tree, known for its light, soft wood that is easy to work with and has a variety of uses.

2. Materials and Methods

2.1. Material for Tests

The research assumptions require a comprehensive analysis of the material by subjecting the raw material to fragmentation and then compaction in specific proportions and parameters. The tests were carried out in a prepared logical sequence, analyzing the properties of the raw material and the product. It was noticed that the forest residues contained a certain amount of fine admixtures in the form of soil, bark, moss, and needles. The material analysis began with the examination of the basic physical characteristics and granulometric composition; then, the physical and chemical parameters were determined.

2.2. Physical Preparation of Raw Material

Shredded forest residues come from the processing of biomass generated during tree felling, cleaning, or sanitary cutting. They contain a certain amount of fine admixtures in the form of soil, bark, moss, and needles. The characterization of the analyzed material began with the examination of the basic physical characteristics and granulometric composition. The study required pre-treatment of collected wood. The material used in the study was shredded by shredder type RTB13, with a fraction on the sieves not exceeding 16 mm. The tests were conducted according to PN-EN 15149-1:2011 [7] and PN-EN 15149-

2:2011 [8] standards. The test takes 120 s, where the fractions of 0.1–1.0, 1.1–4.0, 4.1–8.0, and 8.1–16 mm, corresponding to the groups of $f1$ ($0 \div 1$), $f2$ ($1 \div 4$), $f3$ ($4 \div 8$), and $f4$ ($8 \div 16$), were separated from the test, and the briquettes were prepared as a base for further research.

The other significant parameter that affects the briquetting process in a significant way is the measurement of moisture content; that target moisture was 12%. The material was dried at $103 \pm 2\,°C$, taking into account the constant moisture content of the wood material. The sorted particles before the preparation of the briquettes were dried to a moisture of 12%. The moisture content of the shredded wood material was tested according to PN-ISO 589:2006 [9]. The material was additionally controlled before compaction. The tests were carried out with the use of the Radwag WS30 type weighing dryer.

2.3. Chemical Properties of the Tested Material

The determination of chemical parameters was aimed to measure the C, H, O, N, S content. In order to determine the ash content, the sample was tested. The analysis of carbon, hydrogen, nitrogen, and sulfur was carried out in the specialist laboratory apparatus Vario macro elementar. After determining the elemental content of carbon, hydrogen, and nitrogen, the sample of the raw material was incinerated, determining the ash and oxygen content. The oxygen share was determined as an indication and resulted from the content of other elements and ash in the analyzed material. Prepared material after thermally treating produces ash.

Due to its composition, ash determines mineral saturation, consisting of silicon, iron, aluminum, calcium, magnesium, sodium, and potassium [10]. A muffle oven was used to measure the ash content [11] according to the norm PN-ISO 2144:1996 [12]. The measured raw material was brought to a dry state before ashing. During the experiment, the mass of the weighed sample was recorded in a crucible. In this study, three weights were prepared (with an accuracy of 0.0001 g) with a mass of approximately 3 g per sample. Samples were heated in a muffle oven that had to be warmed to $805 \pm 2\,°C$ before testing. The samples were tested for ash content over a period of two hours. In order to cool the sample, it was placed in a desiccator after being ashed in the crucible. In the following step, the weight of the crucible with the ash inside was determined.

2.4. Helium Pycnometry

In wood science, helium pyrometry is used to determine the density of wood and wood-based materials. It relies on the principle of gas displacement to determine the volume of a sample by determining how much gas (typically helium) is required to fill a known volume. Material porosity can be detected using helium pycnometry. Compared with traditional methods, helium pycnometry captures accurate, specific results that describe absolute volume changes. There is a high degree of correspondence between helium pyrometry and traditional methods [13].

The displacement medium is usually an inert gas, such as helium. The sample is placed in a sealed cup with a known volume. The cup was placed in the sample chamber. It involves gas injection into a sample chamber, expanding it into a known volume chamber, and then expanding it back into the first one. After filling the sample cell and releasing the pressure in the expansion chamber, the volume was calculated. The true density of the sample can be calculated by dividing the weight of the sample by the volume measured. As a result of helium pycnometry true density, measurements results are often referred to as helium density, implying that open pores are excluded from the calculation of true density.

2.5. Briquetting Process in Special Compaction Head

The briquetting process was conducted on the prepared samples. The measurement stand connected the testing machine to the computer. In order to carry out the planned tests, the compaction head was used. Specimens were prepared for testing on an appropriate measuring stand. The study involved determining the compressive strength of test specimens [14]. It was determined that the prepared raw material sample would be compacted inside the chamber at a preferred temperature level. During preparation, the controller was set up to maintain a heater's constant temperature according to the instructions. The temperature requirement, in this case, was 120 °C. The technical drawing of the compaction head and piston is presented in Figure 1.

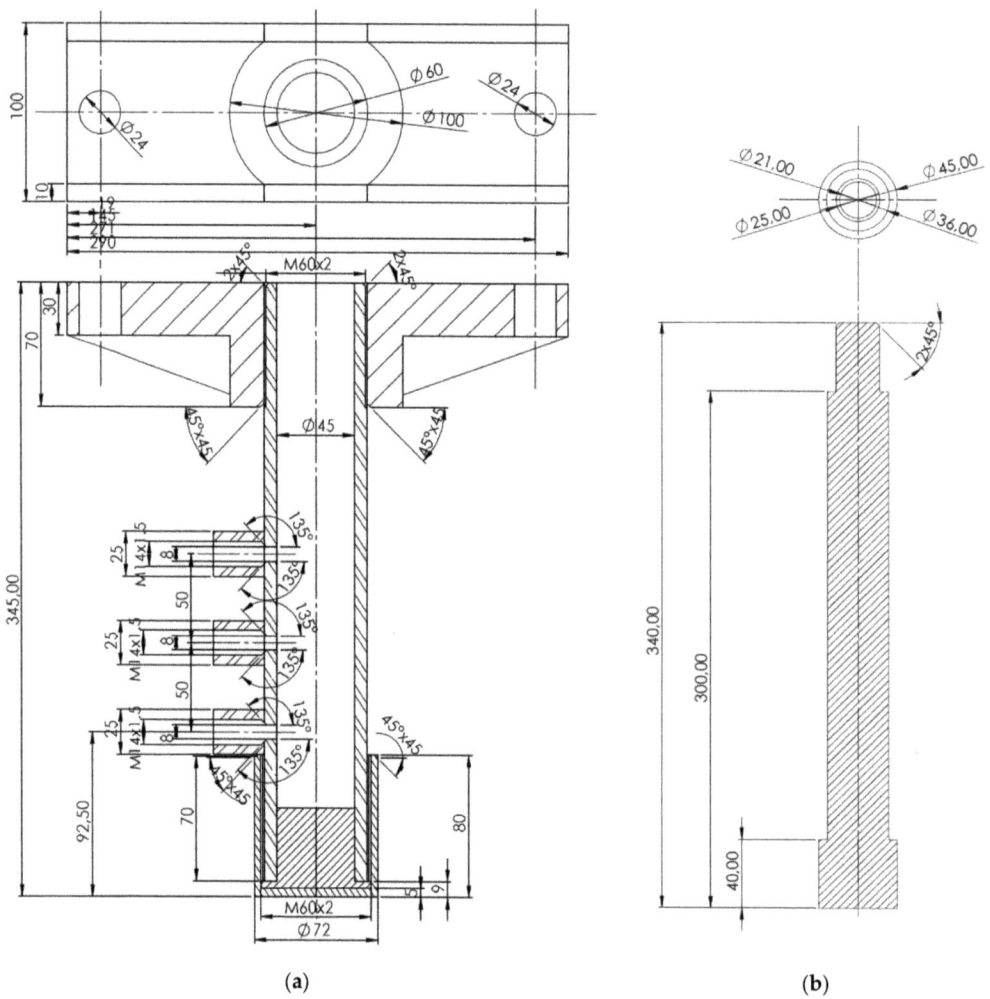

Figure 1. The technical drawing of (**a**) the compaction head and (**b**) the piston.

The compaction process involves compression of force no greater than 100 kN. The force values will be changed if the quality of the produced sample cannot be satisfied. Compression was performed on samples measuring about 100 cm^3. Constant parameters were the inner diameter of the compaction chamber (45 mm) and its length (300 mm). The parameters of the compaction head and the maximum compaction pressure reach 63 MPa. The sample strength is determined by the amount of resistance created when compressive stresses are applied in a given direction. Different materials can have different compressive strengths due to their anisotropic nature. The actual image of the compaction head and displacement simulation is presented in Figure 2.

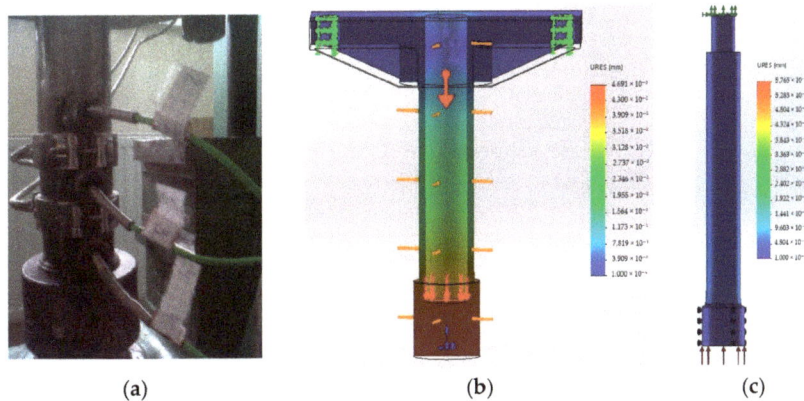

Figure 2. The actual image of (**a**) the compaction head and displacement simulation (**b**) the compaction head and (**c**) the piston.

2.6. X-ray Photography Analysis

The study aims to measure the effect of biomass briquette structure by visualization of wood chip fractions under X-ray. The method of X-ray diagnostic is based on the absorption of wave radiation by various particles of matter. In the softer matter, X-ray waves with higher intensity of radiation are passing through structures more easily, making blackout spots on photographic paper. That process is caused by two electrodes (cathode and anode) sunk in a vacuum or compressed air glass bulb. Cathode reaches the voltage of incandescence Uh causing the electrons' emission, which is accelerated by anode voltage U_a. Electrons reaching the anode are detained with the matter [15]. Tests involve a wood material known as Scots pine (*Pinus Sylvestris* L.) in the fractions of ranges 0.1–1.0, 1.1–4.0, 4.1–8.0, and 8.1–16 mm. Examination of X-ray photography can describe the density coefficient of the briquette divided into fraction ranges, after the briquetting process. The X-ray photography of analyzed briquettes is presented in Figure 3.

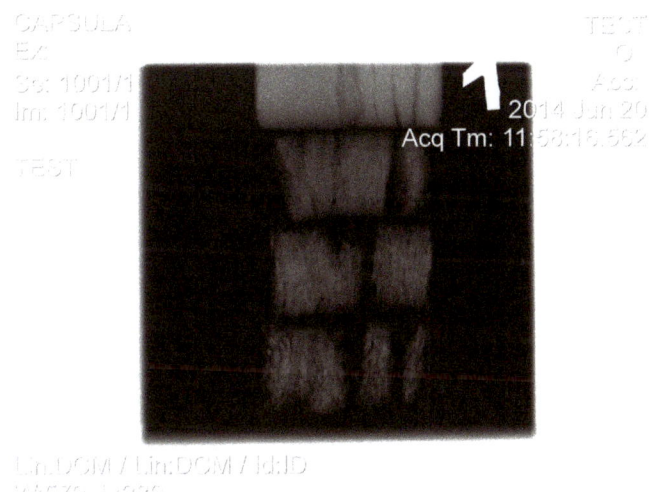

Figure 3. X-ray photography of analyzed briquettes (scale 1:2).

In the upper right corner of the picture (Figure 3), the steel test material that presents the inversion of number one can be seen. The spectrum of steel is white, which means that

X-ray waves do not pass through that material. The X-ray spectrum shows the differentiated briquette density coefficient depending on the fraction from which they were made. The surface of the compression wood of 0–1 fraction appears distinctly lighter-colored, with means it is a denser wood area on the surface. Wood may change its chemical composition and structure after compaction. This image of the X-ray spectrum was the final form needed for the density coefficient analysis of every briquette type. A specially prepared software program was used to examine the images. The desaturation of the picture is analyzed pixel-by-pixel in the loop by the software. In a specially developed program, pixel colors are analyzed based on grayscale values. The computer program interface was designed and programmed in Delphi. The functioning of the algorithm is presented in Figure 4.

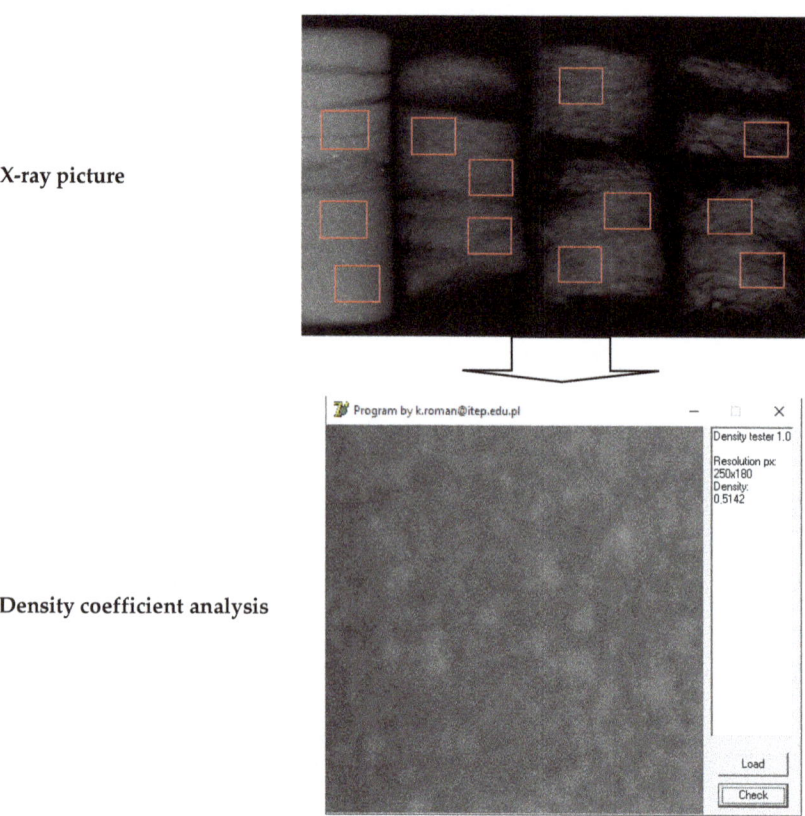

Figure 4. Schematic of screen desaturation.

Figure 4 presents a schematic of screen desaturation, which is required for X-ray image analysis to determine the X-ray spectrum density. According to the schematic, the X-ray spectrum raster picture was loaded into the specially prepared computer program. The computer program analyzed only selected pictures that separately were loaded into the program as individual areas. Those areas were pointed on the schematic (Figure 4) in the red frame and were corresponding with every fraction of the briquette X-ray image spectrum in three repeats. The kernel of a specially prepared program was an algorithm that analyzed the selected area of the X-ray raster picture, reading the color in the loop pixel-by-pixel. The algorithm recognized the color of the pixel, assigning the fractional number value from 0 to 1 in shades of grey. The value 1 corresponded to the white pixel and the value 0 to the black. This was consistent with the white pixels that do not pass the

X-rays waves, which was opposed to the black pixels. The algorithm as a result gave the mean percent of X-ray spectrum density.

Desaturation mode significantly reduces the amount of color saturation information in an image. By reducing the color to grayscale, the hue and saturation of the pixels were removed. In order to save information about a specific image, only luminance data is stored. Using a specially designed program, gray-scale pixels are counted in order to perform analysis. An algorithm is used to count all the pixels and then separate them based on a grayscale. Combining the obtained data enables the estimation of the black-to-white ratio in an image. The study results were analyzed using an analysis of variance (ANOVA) [16]. It was possible to determine the influence of parameters of shredded wood material (pine) on the coefficient density results of the software.

3. Results

3.1. Chemical Analysis

The measurement of mineral compounds was conducted to support the explanation of X-ray spectrum analysis. Expectedly, the higher ash content represented a brighter X-ray spectrum. This means there could be a wide range of ash content in briquettes depending on the fractured structure. The ash measurement process burns the organic macronutrients carbon (C), hydrogen (H), oxygen (O), nitrogen (N), and sulfur (S). In order to determine the ash content, the material left after burning is examined. The ash content determines the mineral saturation of the tested raw material. The percent content of mineral compounds and moisture content from measured raw material are presented in Table 1.

Table 1. Mineral compounds and moisture content.

Fractions	Moisture, %	C	H	O	N	S	Ash Content, %
f_1 (0 ÷ 1)	12.4	50.744	5.441	41.308	0.671	0.377	1.457
	12.4	50.770	5.467	41.334	0.697	0.403	1.327
	12.4	50.737	5.434	41.301	0.664	0.370	1.493
f_2 (1 ÷ 4)	11.9	50.796	5.729	41.423	0.678	0.241	1.133
	12.0	50.793	5.726	41.420	0.675	0.238	1.150
	12.0	50.792	5.725	41.419	0.674	0.237	1.155
f_3 (4 ÷ 8)	12.5	50.808	5.876	41.516	0.614	0.169	1.017
	12.6	50.789	5.857	41.497	0.595	0.150	1.114
	12.5	50.794	5.862	41.502	0.600	0.155	1.088
f_4 (8 ÷ 16)	12.1	50.932	5.802	41.519	0.616	0.163	0.968
	12.1	50.919	5.789	41.506	0.603	0.150	1.031
	12.2	50.928	5.798	41.515	0.612	0.159	0.987

The basic elemental composition of selected wood materials was determined, as part of the research. The results were exposed to statistical analysis to determine the effects of respective parameters on the type of briquettes. The ANOVA indicated some differences between the types of briquettes, and the Duncan test was used to assign the materials to the homogeneous groups. In every chemical property test, the individual material formed an individual or shared homogeneous group. The most confirmed considerable differences were recorded for the ash content, where individual materials create only two homogeneous groups. Basically, the analyzed raw material was characterized by slight differences in chemical composition between the fractions. The difference in the elemental composition can be explained by the content of mineral values. The results of the ash statistical analysis were presented in another summary chart below. The aim of this part of the research was to compare the ash share trend to the density measured in the helium pycnometer and X-ray analysis.

3.2. Computer Program Analyzes

Digitalization of the structure results was accomplished by preparing a 250 × 180 px image. Assuming that grayscale shades are measured horizontally and vertically in a loop, a specially composed program was used for image characterization. The program produced a percentage characterization of grayscale share, which was defined as the density coefficient of the image. A program was used to calculate the density coefficient in X-ray images by using white as a value of "1" and black as a value of "0". The proportion of white to black in the analyzed image can be derived from these values. The parameter breakdown for the respective briquettes is presented in Table 2.

Table 2. Breakdown of parameters for respective briquettes.

Fraction	Density Coefficient	Mean
f_1 (0 ÷ 1)	0.4921	
	0.5142	0.5133
	0.5335	
f_2 (1 ÷ 4)	0.3483	
	0.3432	0.3525
	0.3660	
f_3 (4 ÷ 8)	0.3725	
	0.3358	0.3571
	0.3630	
f_4 (8 ÷ 16)	0.2987	
	0.3383	0.2954
	0.2492	

The results, in accordance with the methodology, were subjected to statistical analysis. The density coefficient statistical analysis was presented in another summary chart below. The results of the analysis allow for observing the relationship between the measurements. It is expected that the relationships between the measured parameters of the density coefficient for the X-ray spectrum and true density will be relative.

3.3. Helium Pycnometry True Density Characteristics

Material density provides a lot of information about the strength and stiffness of wood species and mechanical properties as well. The density of these materials can be determined by hydrogen pycnometry. The measurement of true density is one of the most important elements of the raw material characteristics to specify the density test. The porosity of a material affects its mechanical properties, such as its hardness and elastic modulus. To accurately characterize mechanical properties, the correct porosity must be determined, which depends on the accuracy of the true density measurement. For that purpose, the most common technique for measuring true density is via a helium pycnometer [15]. The measurement of conducted studies was made using a helium pycnometer. The summary of shredded forest residues true density test results is presented in Table 3.

The average true density for all measurements obtained using a helium pycnometer was 1253.7 kg·m^{-3}. It was noticed that the differences in the results of specific density measurements of crushed pine logging residues for each of the measured fractions have a similar value. The use of helium pyrometry is an effective technique for determining the true density of wood and wood-based materials. The information can be utilized to gain valuable insight into their properties, species, and applications.

Table 3. The shredded forest residues true density tests results.

Material	Sample Number	Measurement Number	True Density, kg·m^{-3}	Mean, kg·m^{-3}
f_1 (0 ÷ 1)	1	1	1344.	1344
		2	1345	
		3	1344	
	2	1	1355	1354
		2	1353	
		3	1353	
	3	1	1355	1355
		2	1355	
		3	1354	
f_2 (1 ÷ 4)	1	1	1287	1288
		2	1289	
		3	1290	
	2	1	1282	1287
		2	1290	
		3	1290	
		4	1287	
	3	1	1283	1287
		2	1289	
		3	1290	
f_3 (4 ÷ 8)	1	1	1105	1116
		2	1110	
		3	1133	
	2	1	1080	1105
		2	1116	
		3	1119	
	3	1	1090	1115
		2	1123	
		3	1131	
f_4 (8 ÷ 16)	1	1	1282	1285
		2	1287	
		3	1284	
	2	1	1283	1286
		2	1286	
		3	1290	
	3	1	1278	1278
		2	1274	
		3	1280	

3.4. The Density Comparison of Shredded Forest Residues

The tests were designed to determine factors that have an impact on briquettes' properties. To determine the significant effects of the factors on the parameters, a Student's T-test was used. The empirical measurements provided information on the picture density, and true density of respective briquettes. The ash content was measured additionally to compare with density values. The comparison between the measurements can produce interesting results.

During the statistical analysis, the agreed factors were the results of the empirical measurements, though the briquette type and its kind were determined as a quality predictor. The results of the studies were statistically analyzed to verify the performance of the program in image characterization color percentage. The analysis assumption was to identify the structure percentage, based on the desaturated image in correlation with the percent of ash content and density measured by helium pycnometry. It indicates that there may be significant differences between the measured parameters between cases as shown in the diagram. The sums for the briquettes made from various biofuel materials are presented in Figure 5.

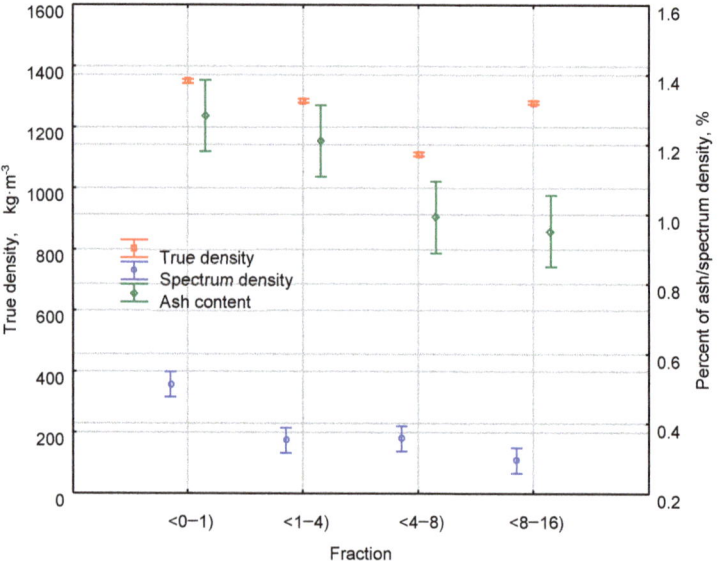

Figure 5. The expected marginal means for the briquettes are made from various fractions.

The figure shows the measurement of true density, spectral density, and ash content results, which were divided by the fraction range. Observing the layout of the measurement results, it was important to conduct statistical analysis in order to compare and determine the existing influences. The Duncan test for X-ray spectrum density (%) showed a partial dependence resulting in three groups that were homogenous. In this study, the separator alpha between the individual homogenous groups was $p = 0.00006$. The statistical test applied to the true density measurement by helium pycnometry (kg·m^{-3}) gives similar results as a test of the X-ray spectrum. Three homogeneous groups were also created based on the faction ranges. In this case, the separator alpha was also less than 0.05. Additional measurement applied to the ash content gives a wide picture of the relationship between elemental composition and material density. Similar results of statistical tests were found for the ash content (%), with creates two homogeneous groups. In the individual homogenous groups, the separator alpha was $p = 0.00181$. The characteristics of the homogeneous samples and the mean characteristic of the effects of the parameters measured are presented in Table 4.

Table 4. Characteristics of the measured parameters to the type of briquettes.

Briquette Name	Density Helium Pycnometry (kg·m^{-3})	X-ray Spectrum Density (%)	Ash Content (%)
		Mean	
f_1 (0 ÷ 1)	1350.8 [a]	0.513 [d]	1.283 [g]
f_2 (1 ÷ 4)	1287.6 [b]	0.353 [e]	1.211 [g]
f_3 (4 ÷ 8)	1111.9 [c]	0.357 [e]	0.992 [h]
f_4 (8 ÷ 16)	1282.8 [b]	0.295 [f]	0.953 [h]

[a, b, c, d, e, f, g, h]—homogenous group.

The statistical analysis allows the characterization of the results of measurements in the context of true density and X-ray spectrum density. The statistical tests of ash content measurements were not mandatory. Based on statistical analysis, all measured parameters showed similarities. The conversion factor between densities can be calculated using the statistical test between measured parameters. The conversion factor varies from 0.55 for the f4 fraction (8 ÷ 16 mm) to 0.67 for the f1 fraction (0 ÷ 1 mm). Considering the X-ray spectrum and true density results, the conversion factor may be valuable for preliminary or indicative measurements. The conversion factor can be used to identify imaging errors. Based on the results, the mean value of the conversion factor was 0.62. The proposed conversion factor can make the process of measuring materials' densities simpler, cheaper, and faster. An additional benefit of the work was the definition of the conversion factor and comparison of ash content with density measurement in order to obtain two different research methods.

4. Discussion

4.1. Chemical Analysis

The samples were properly prepared for laboratory analysis to determine the chemical properties, such as total hydrogen, total carbon, total nitrogen, total sulfur, and oxygen (the calculation method). The carbon content of the tested material (about 50%) classified it into a group of materials with energy purpose potential. The literature does not provide any information about the chemical composition of forest residues. A similar level of carbon content was found in spruce wood and energy chips [17,18]. The hydrogen content of the raw materials analyzed was close to 6%, a much lower percentage than that found in the analyzed chemicals. The obtained results did not correspond to those in the literature. The literature reports that nitrogen and sulfur are not affected by metamorphism processes [19]. The number of environmental parameters could have influenced how these elements were distributed in the analyzed material. For example, cultivation technology, soil conditions, and soil valuation classes could all have had a significant impact. The analyzed energy chips contained comparable amounts of nitrogen to sawmill chips [17]. The sulfur percentage of energy chips matched the values of post-consumer woods [20]. The low levels of sulfur and nitrogen in the raw material confirm the pro-ecological issue of analyzed fuel characteristics. Statistics were used to analyze the results of the experiment.

The muffle furnace was used to determine the ash share of the analyzed raw materials. In order to incinerate the raw material sample, the material was heated to 805 °C. Ash contains mineral substances, based on information from the literature. The literature on the subject reveals that the ash content of forest residues in Scots pine ranges from 0.6 to 1.6%. The results of this research have confirmed this. According to the measurements, the average ash content was 1.07%. There is probably a large share of non-woody green parts in this biomass type, which explains its low ash content. It is possible that this is due to high levels of contamination during the accumulation process in forest areas, during transport and grinding.

A similar share of ash content can be found in the examined logging residues as in the branches of Scots pine (*Pinus sylvestris* L.) and Norway spruce (*Picea abies*) [21]. There was no correlation between the results of the research and any literature values, despite the fact that sawmill chips and energy chips are derivatives of logging residues. It is important to note that the raw material in the study came from different locations and from different stands, so the differences cannot be explained easily. Based on the microelements and ash content percentage in the tested material, the mean value of oxygen content was calculated. Based on the types of raw materials used in the analysis, the oxygen content of each type corresponded to literature values for spruce wood with a high correlation [17,18].

4.2. The X-ray Spectrum Analysis

The most influential factor affecting the X-ray spectrum is the elemental composition of the material. Depending on the data gathered from the chemical composition study, a comparison is made between the results of the study and other measurements taken. The computer program enabled the analysis of the X-ray spectrum and the differences between the structure of the briquette fraction in samples obtained as a result of analyzing the X-ray spectrum. Using the program, it is possible to analyze the structures, their differences depending on fraction content, and how the density changes as mineral compounds are correlated. In briquettes with low fracture dimensions, the structure was more structured, and organic compounds were less concentrated.

Desaturation data can be analyzed to estimate black-to-white shares and identify mineral compounds. The X-ray spectra were analyzed to estimate gray tone density factors according to the fractional ranges. The results of the measurements show significant differences between the highest and lowest fractional ranges. For the lowest fraction range $f1$ $(0 \div 1)$, the density factor was 0.513 and was higher than all measured samples. There is a dependency between fraction content and the density coefficient value from the X-ray spectrum that increases with fraction content.

4.3. The Helium Pycnometry Analysis

Helium pycnometry is a method for determining the true density of wood materials. This method involves immersing a sample in helium gas and measuring the change in the volume of the gas. The mass of the wood sample is then divided by its volume to calculate its density. In the case of helium pycnometric analysis of a wood-based sample, the results usually include the density of the wood-based sample as well as variations in density within the sample, if any are present. The true density measurements were conducted for different materials fractions of $f1, f2, f3,$ and $f4$. The density of each material group was measured multiple times for each material group.

The true density in this study ranged from 1105 kg/m^3 to 1355 kg·m^{-3}, with an average value of 1258 kg·m^{-3} being the highest. The highest value of true density was 1355 kg·m^{-3} and represents the state of being the densest. The lowest value of true density is 1105 kg·m^{-3}. Each of these values falls within a range of 68 kg·m^{-3} that is relatively close to one another. The data on the true density of the material and the mean density are generally consistent with the true density, with a small margin of error. This indicates that the samples have similar densities or that the density of the samples, in general, is consistent throughout the sample set. Based on the experimental results, it is possible to characterize the deep material structure and deduce the conversion factor as well. A description of the observations is provided in the article.

5. Conclusions

To conduct a qualitative analysis of the briquettes, it is necessary to determine the basic elemental composition of the material. It was found that the material included contamination, which was assigned as a percentage of mineral compounds contained in ash. Different types of briquettes contained different compounds that were found to be present at different levels. According to the results, the ash content of briquettes decreases as the fractions in briquettes are increased. Prepared materials have a decreasing ash value as wood fractions increase.

Based on the micro elementary results, the material analyzed contained carbon contents around 51%, hydrogen contents around 6%, nitrogen contents around 0.6%, sulfur contents ranging between 0.2 to 0.4%, oxygen contents around 42%, and ash contents around 1%. In this case, the raw materials appeared to have been diverse, which is likely due to the location of the cutting. Biomass samples were collected from branches, preventing mineral contamination in some ways.

X-ray spectrum analysis shows that the low fraction dimension $f1(0 \div 1)$ helps the composite structure adhere. In the X-ray spectrum, it is difficult to see how compact the material is. Wood fractions with low dimensions are used for briquette manufacturing and have a positive effect on their strength. Based on the above sentence, a conversion factor determining the strength of briquette can be proposed. The results were compared with a doctoral thesis [22]. It was noticed that briquettes with higher durability were characterized by a lighter shade: respectively, f_1 was in the range of 31–68%; f_2 was 6–27%; f_3 was 6–15%; and f_4 was about 8–37%. In the doctoral study, the strength tests were carried out for prepared mixtures; therefore, a direct comparison for strength research was not corresponding and was of an illustrative nature. The combination of high-milled fractions and briquettes can be considered suitable. Research in this field will be developed in the future. In addition, it is also possible to inspect wood-based materials for defects using X-ray technology, such as knots, cracks, and rot, which can affect their strength and durability. The conducted research seems to have a lot of potential and should be further developed.

Author Contributions: Conceptualization, K.R. and M.H.; methodology, K.R.; software, K.R.; validation, K.R., W.R. and M.H.; formal analysis, K.R.; investigation, K.R., W.R. and M.H.; resources, K.R. and W.R.; data curation, K.R., W.R. and M.H; writing—original draft preparation, K.R., W.R. and M.H.; writing—review and editing, K.R.; visualization, K.R.; supervision, K.R.; project administration, K.R.; funding acquisition, K.R. All authors have read and agreed to the published version of the manuscript.

Funding: This research received no external funding.

Data Availability Statement: Not applicable.

Conflicts of Interest: The authors declare no conflict of interest.

References

1. Niedziółka, I.; Zuchniarz, A. Analiza energetyczna wybranych rodzajów biomasy pochodzenia roślinnego. *Motrol* **2006**, *8A*, 232–237.
2. Wasiak, A. *Raport o Stanie Lasów w Polsce*; Lasy Państwowe: Warsaw, Poland, 2013; pp. 1–103.
3. Ekielski, S. *Wytyczne Konstrukcyjne Zespołu Roboczego do Granulatora Trocin Drzewnych*; IBMER: Warsaw, Poland, 1986.
4. Gradziuk, P. *Analiza Możliwości Wykorzystania Słomy na Cele Energetyczne, Expert Opinion Commissioned by the Ekofundusz Foundation*; Agricultural University in Lublin, Institute of Agricultural Sciences in Zamość: Lublin, Poland, 2002; 62p.
5. Krzysik, F. *Nauka o Drewnie*; Wydawnictwo Naukowe: Warsaw, Poland, 1975; 645p.
6. Tomczak, A.; Jelonek, T. Parametry techniczne młodocianego i dojrzałego drewna sosny zwyczajnej (*Pinus sylvestris* L.). *Sylwan* **2012**, *156*, 695–702.
7. PN-EN 15149-1; Wersja Angielska—Biopaliwa Stałe—Oznaczanie Rozkładu Wielkości Ziaren—Część 1: Metoda Przesiewania Oscylacyjnego Przy Użyciu Sit o Szczelinie 1 mm Lub Większej. Polish Committee for Standardization: Warsaw, Poland, 2011; pp. 1–13.
8. PN-EN 15149-2; Wersja Angielska—Biopaliwa Stałe—Oznaczanie Rozkładu Wielkości Ziaren—Część 2: Metoda Przesiewania Wibracyjnego Przy Użyciu Sit o Szczelinie 3,15 mm Lub Mniejszej. Polish Committee for Standardization: Warsaw, Poland, 2011; pp. 1–13.
9. PN-ISO 589; Wersja Polska—Węgiel Kamienny—Oznaczanie Wilgoci Całkowitej. Polish Committee for Standardization: Warsaw, Poland, 2006; pp. 1–12.
10. Babiarz, M.; Bednarczuk, Ł. *Popiół ze Spalania Biomasy i Jego Wykorzystanie*; WPIA AGH: Cracow, Poland, 2013; p. 3.
11. Suchorab, B.; Roman, K. The PLA content influence selected properties of wood-based composites. *Ann. Wars. Univ. Life Sci. SGGW For. Wood Technol.* **2022**, *120*, 57–67. [CrossRef]
12. PN-ISO 2144:1996; Paper and Cardboard—Determination of Ash Content. Polish Committee for Standardization: Warsaw, Poland, 1996.

13. Yang, X.; Sun, Z.; Shui, L.; Ji, Y. Characterization of the absolute volume change of cement pastes in early-age hydration process based on helium pycnometry. *Constr. Build. Mater.* **2017**, *142*, 490–498. [CrossRef]
14. Roman, K.; Barwicki, J.; Rzodkiewicz, W.; Dawidowski, M. Evaluation of Mechanical and Energetic Properties of the Forest Residues Shredded Chips during Briquetting Process. *Energies* **2021**, *14*, 3270. [CrossRef]
15. Czyzewski, A.; Krawiec, F.; Brzezinski, D.; Porebski, P.; Minor, W. Detecting anomalies in X-ray diffraction images using convolutional neural networks. *Expert Syst. Appl.* **2021**, *174*, 114740. [CrossRef] [PubMed]
16. Chang, S.-Y.; Wang, C.; Sun, C.C. Relationship between hydrate stability and accuracy of true density measured by helium pycnometry. *Int. J. Pharm.* **2019**, *567*, 118444. [CrossRef] [PubMed]
17. Bach, Q.; Chen, W.; Chu, Y.; Skreiberg, Ø. Predictions of biochar yield and elemental composition during torrefaction of forest residues. *Bioresour. Technol.* **2016**, *215*, 239–246. [CrossRef] [PubMed]
18. Kajda-Szcześniak, M. Evaluation of the basic properties of the wood waste and woodbasedwastes. *Arch. Waste Manag. Environ. Prot.* **2013**, *15*, 1–10.
19. Kraszkiewicz, A. Analiza wybranych właściwości chemicznych drewna i kory robinii akacjowej (*Robinia pseudoacacia* L.). *Inż. Rol.* **2009**, *8*, 69–75.
20. Wandrasz, J.W.; Wandrasz, A.J. *Paliwa formowane: Biopaliwa i Paliwa z Odpadów w Procesach Termicznych*; Wydawnictwo Seidel-Przywecki Sp. z o.o.: Warsaw, Poland, 2006; pp. 1–56.
21. Moriana, R.; Vilaplana, F.; Ek, M. Cellulose Nanocrystals from Forest Residues as Reinforcing Agents for Composites: A Study from Macro- to Nano-Dimensions. *Carbohydr. Polym.* **2016**, *139*, 139–149. [CrossRef] [PubMed]
22. Roman, K. Dobór Parametrów Technicznych Procesu Brykietowania Biomasy Leśnej. Ph.D. Thesis, Warsaw University of Life Sciences, Warsaw, Poland, 2017.

Disclaimer/Publisher's Note: The statements, opinions and data contained in all publications are solely those of the individual author(s) and contributor(s) and not of MDPI and/or the editor(s). MDPI and/or the editor(s) disclaim responsibility for any injury to people or property resulting from any ideas, methods, instructions or products referred to in the content.

Article

Assessment of the Suitability of Coke Material for Proppants in the Hydraulic Fracturing of Coals

Tomasz Suponik *, Krzysztof Labus and Rafał Morga

Faculty of Mining, Safety Engineering and Industrial Automation, Silesian University of Technology, Akademicka 2, 44-100 Gliwice, Poland
* Correspondence: tomasz.suponik@polsl.pl

Abstract: To enhance the extraction of methane gas from coal beds, hydraulic fracturing technology is used. However, stimulation operations in soft rocks, such as coal beds, are associated with technical problems related mainly to the embedment phenomenon. Therefore, the concept of a novel coke-based proppant was introduced. The purpose of the study was to identify the source coke material for further processing to obtain a proppant. Twenty coke materials differing in type, grain size, and production method from five coking plants were tested. The values of the following parameters were determined for the initial coke: micum index 40; micum index 10; coke reactivity index; coke strength after reaction; and ash content. The coke was modified by crushing and mechanical classification, and the 3–1 mm class was obtained. This was enriched in heavy liquid with a density of 1.35 g/cm^3. The crush resistance index and Roga index, which were selected as key strength parameters, and the ash content were determined for the lighter fraction. The most promising modified coke materials with the best strength properties were obtained from the coarse-grained (fraction 25–80 mm and greater) blast furnace and foundry coke. They had crush resistance index and Roga index values of at least 44% and at least 96%, respectively, and contained less than 9% ash. After assessing the suitability of coke material for proppants in the hydraulic fracturing of coal, further research will be needed to develop a technology to produce proppants with parameters compliant with the PN-EN ISO 13503-2:2010 standard.

Keywords: hydraulic fracturing; proppant; coke; coal bed methane

Citation: Suponik, T.; Labus, K.; Morga, R. Assessment of the Suitability of Coke Material for Proppants in the Hydraulic Fracturing of Coals. *Materials* **2023**, *16*, 4083. https://doi.org/10.3390/ma16114083

Academic Editors: Saeed Chehreh Chelgani and Farooq Sher

Received: 31 January 2023
Revised: 6 March 2023
Accepted: 4 April 2023
Published: 30 May 2023

Copyright: © 2023 by the authors. Licensee MDPI, Basel, Switzerland. This article is an open access article distributed under the terms and conditions of the Creative Commons Attribution (CC BY) license (https://creativecommons.org/licenses/by/4.0/).

1. Introduction

Coal bed methane is natural gas (NG) found in coal deposits. During the formation of coal, large amounts of this gas are generated, and their part is trapped by the coal matter. Coal has a huge internal surface area; therefore, it can store considerable volumes of methane, more than conventional porous rocks.

As the coal bed methane is held in place by water pressure, to extract it, the water pressure should be reduced by pumping out the fluid through wells, which enables the release of natural gas from the coal. The gas should then be separated from the water and transported to compression plants.

To improve methane gas extraction, hydraulic fracturing technology (HF) has been used since the 1970s [1]. HF consists of injecting a fracturing fluid (FF), under high pressure, into a coal seam to propagate existing fractures and induce a new artificial fracture network. This ensures hydraulic conductivity in the created fractures to achieve a continuous high gas production rate. When the fracture network is induced, in most cases, a propping agent—proppant, commonly sieved sand or ceramic spheres [2], added to the injected fluid—prevents fracture closure after the release of pressure from the injected fracturing fluid once the HF operation is completed. The remaining proppant grains in the fracture prevent it from fully closing and losing hydraulic conductivity. As a result, the propped fractures constitute highly permeable arteries that enable communication between the well and the reservoir [3]. This facilitates the removal of water and extraction of the gas.

HF technology in the oil and gas industry has long reached the application phase [4–11]. However, hydraulic fracturing in soft rocks, such as coal beds, is associated with technical problems related mainly to the embedment process, which reduces the fracture conductivity (Figure 1). On the other hand, the fracturing operations will take place at depths generally not exceeding 1500 m, which does not require the use of high-strength proppants. Therefore, it is necessary to develop innovative proppant materials that will be suitable for these specific conditions.

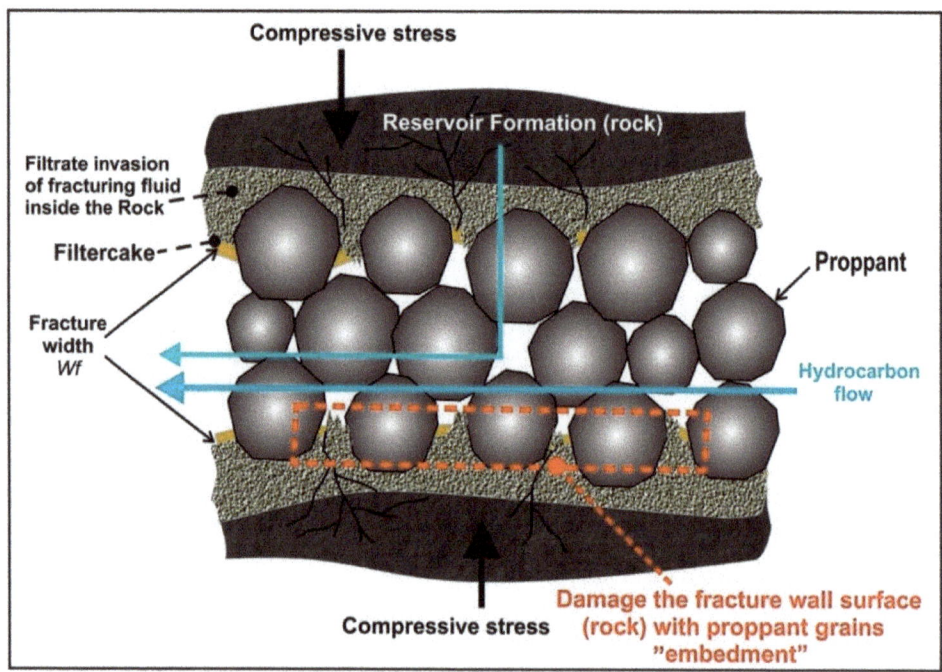

Figure 1. Embedment and other phenomena that affect the surface of the fracture face during hydraulic fracturing [12].

Some solutions have already been proposed, an example of which is an ultra-lightweight proppant (ULWP) for the hydraulic fracturing of hydrocarbon deposits and non-hydrocarbon formations, in particular, coal seams, used to increase methane production [13]. This ultra-light backfill, as described, consists of a base material partially covered with a protective coating or hardened. The basic material can be ground nut shells, quartz, glass, sand, or silicates. The particle size of this material is from 12/20 mesh (1.680/0.841 mm) to 40/70 mesh (0.400/0.210 mm). Such small particles are better suspended in the low-viscosity fracturing fluid and easily move in narrow natural fissures.

Although proppant use has been common in deep reservoir fracturing, its use in soft rocks, such as coal, should be carefully investigated [14]. Typical proppants are prone to the embedment phenomenon, and, therefore, are not suitable for soft coal seams. Furthermore, their composition does not correspond to the chemical composition of coal seams, and the proppant used wears the surface of the fracture and reduces its width and conductivity.

The main challenges related to the properties and parameters of proppants for fracking operations in coals are the following [15]:

1. Compressive strength of 13.8 MPa, at which over 90% of grains remain undestroyed;
2. Proppant pack conductivity for 2% KCl solution is higher than 13×10^{-14} m^2·m;
3. Dimensionless fracture conductivity is higher than 40 [-];
4. The optimal transport and settling of proppant in the created fracture;

5. Possibility to control the quality of filling the fracture with proppant;
6. Compatibility of the proppant material with the formation rock and fracking fluid;
7. Maintaining a fracture system of desired conductivity inside the coal rock;
8. Limitation of the embedment phenomenon, consisting of (A) minimizing the depth of indentation of the proppant grains into the fracture wall (carbon rock) to a value not greater than 20% of the average diameter of the grains and (B) maintaining the surface of the fracture wall with less than 35% damage;
9. Reduction in the fracture permeability damage during coal seam demethanization;
10. Reduction in water and material consumption.

Given the desired proppant properties stated above, we developed the concept of a novel, innovative, coke-based proppant:

1. The chemical affinity of coke proppant to coal facilitates underground coal gasification (UCG), or the extraction of demethanized coals as energy carriers or for use in the raw material chemical industry;
2. The considerable porosity and permeability of the proppant allow for more intensive gas migration. The total porosity of the coke proppant can be as high as 50%, with effective porosity reaching 15%. This facilitates gas migration in the propped fracture as well as through the coke proppant itself, unlike most traditional proppants. Gas migration through coke proppant grains may be maintained even when they are partially embedded in the coal rock surface;
3. The porous nature of the coke proppant allows it to be saturated with chemicals designed to control the viscosity of the fracking fluid after stimulation;
4. The structure of the coke proppant limits its embedment. Due to its roughness, it supports the fracture wall at multiple points and reduces embedment compared to proppants of a smoother grain surface, such as treated sand, ceramics, and resin-coated proppants;
5. The low density makes it easy to suspend the proppant in the fracturing fluid and pump it easily into the fractures. The bulk density of the coke material (e.g., coke breeze) can reach 0.57 g/cm^3, which is lower than for typical frac sand $-1.49\ g/cm^3$ (for 40/70 mesh) and even ultra-lightweight proppants (ULWP), e.g., $-0.66\ g/cm^3$ (for 30/80 mesh). The low bulk and specific densities of coke proppants allow them to be used more efficiently with low-viscous foams and energized fracking fluids. It also facilitates the injection of large volumes of proppant from a specific well into the distant parts of the fractures [16].

These characteristics indicate that coke may be a useful material for the production of a proppant for the hydraulic fracturing of coals.

Coke is a material obtained from the pyrolysis of hard coal carried out at a temperature of approximately 1000 °C. Its parameters depend mainly on the petrographic composition and the coking properties of the coal mixture [17–22]. The quality of coke is estimated by determining the content of ash, moisture, and volatile matter, as well as the elemental composition [17,23]. Furthermore, when coke is used in the classic blast furnace process, its mechanical properties, such as strength, abrasiveness, and coke reactivity to CO_2, are assessed, comprising the coke reactivity index (CRI) and coke strength after reaction (CSR) [23,24]. Coke that achieves a low CRI value and a high CSR value is highly valued, especially because it has higher mechanical strength and better gas permeability in metal production processes.

The aim of this preliminary study was to identify the most promising source of coke material for further processing in order to obtain a suitable coke proppant. Due to pending patent proceedings, the authors cannot provide any details of the processing procedures or the final properties of the proppants.

2. Materials and Methods

2.1. Materials

Twenty coke samples that differed in type, grain size, origin, and method of production were the initial material (Table 1). They were obtained from five coking plants located in southern Poland: "Częstochowa Nowa" in Częstochowa; "Zdzieszowice" in Zdzieszowice; "Przyjaźń" in Dąbrowa Górnicza; "Jadwiga" in Zabrze; and "Radlin" in Radlin.

Table 1. Type, grain size, and production method of the tested coke materials.

Sample	Producer	Type	Method of Filling the Chambers and Quenching the Coke; Coking Time	Grain Size, mm
I	Coking plant 1	Blast furnace coke	Hopper system; wet quenching; 16:30 h	25–80
II	Coking plant 1	Coke dust (coke breeze) after production of blast furnace coke nr I	Hopper system; wet quenching; 16:30 h	0–3
III	Coking plant 1	Blast furnace coke	Hopper system; dry quenching; 16:30 h	25–80
IV	Coking plant 1	Coke dust (coke breeze) after production of dry-quenched blast furnace coke nr III	Hopper system; dry quenching; 16:30 h	0–10
V	Coking plant 2	Foundry coke	Stamper system; wet quenching; 29:00 h	>80
VI	Coking plant 2	Coke dust (coke breeze) after production of blast furnace coke nr V	Stamper system; wet quenching; 29:00 h	0–10
VII	Coking plant 2	Blast furnace coke	Stamper system; wet quenching; 26:00 h	25–80
VIII	Coking plant 2	Coke dust (coke breeze) after production of blast furnace coke nr VII	Stamper system; wet quenching; 26:00 h	0–10
IX	Coking plant 2	Fuel coke	Stamper system; wet quenching; 26:00 h	10–25
X	Coking plant 3	Blast furnace coke stabilized	Stamper system; wet quenching; 21:00 h	25–80
XI	Coking plant 3	Fuel coke	Stamper system; wet quenching; 21:00 h	10–30
XII	Coking plant 4	Blast furnace coke	Stamper system; wet quenching; 41:09 h	25–80
XIII	Coking plant 4	Coke dust (coke breeze) after production of blast furnace coke nr XII	Stamper system; wet quenching; 41:09 h	0–10
XIV	Coking plant 4	Fuel coke	Stamper system; wet quenching; 41:09 h	10–25
XV	Coking plant 5	Blast furnace coke	Stamper system; wet quenching; 20:35 h	40–80
XVI	Coking plant 5	Foundry coke	Stamper system; wet quenching; 25:39 h	60–100
XVII	Coking plant 5	Coke dust (coke breeze) after production of foundry coke nr XVI	Stamper system; wet quenching; 25:39 h	0–10
XVIII	Coking plant 1	Coke dust (coke breeze) after production of blast furnace coke nr I	Hopper system; wet quenching; 24:30 h	0–10
XIX	Coking plant 5	Coke dust (coke breeze) after production of blast furnace coke nr XV	Stamper system; wet quenching; 20:35 h	0–10
XX	Coking plant 5	Fuel coke	Stamper system; wet quenching; 26:24 h	10–20

2.2. Methods

2.2.1. Analytical procedures performed on the initial coke materials

The following parameters of the initial cokes were analyzed (Figure 2; Table 2), which characterize their properties and may determine their suitability for use as proppants.

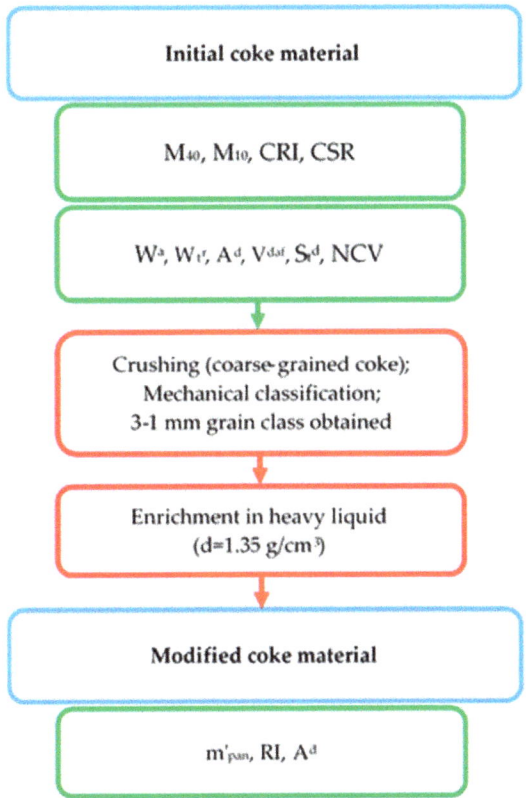

Figure 2. Processing and analytical procedures performed in the study.

For micum index 40 (M_{40}) and micum index 10 (M_{10}), the drum method according to [25] was applied to assess the mechanical properties of the coke material, including its strength and abrasiveness (Table 2).

Micum index 40 (M_{40}), also known as mechanical strength, is the percentage of the coke residue above the 40 mm sieve size (Equation (1)):

$$M_{40} = \frac{m_2}{m_1} 100\% \qquad (1)$$

where

m_1—coke mass before tumbling with grain size > 40 mm, g;
m_2—coke mass after tumbling with grain size > 40 mm (residue in a sieve with a mesh size of 40 mm), g;

Micum index 10 (M_{10}), referred to as abrasiveness, is the percentage of the coke residue below the 10 mm sieve size (Equation (2)):

$$M_{10} = \frac{m_2'}{m_1'} 100\% \qquad (2)$$

where

m_1'—coke mass before tumbling, g;
m_2'—the loss after tumbling through a 10 mm mesh sieve, g;

CRI and CSR indices were determined via a widely used method developed by Nippon Steel Corporation [26] (Table 2).

Coke reactivity index (CRI), being the measure of coke reactivity, is the percentage mass loss of the coke sample treated with CO_2 (Equation (3):

$$CRI = \frac{m_1'' - m_2''}{m_1''} 100\% \tag{3}$$

where

m_1''—coke mass before determination of reactivity, g
m_2''—coke mass after determination of reactivity, g;

Coke strength after reaction with CO_2 (CSR) is the mass of coke that remains on a 10 mm square mesh sieve after mechanical treatment in a standardized rotary drum related to the mass of the sample remaining after the determination of reactivity (Equation (4)):

$$CSR = \frac{m_3'''}{m_2'''} 100\% \tag{4}$$

where

m_2'''—coke mass after determination of reactivity before tumbling, g;
m_3'''—coke mass after determination of reactivity and with a grain size greater than 10 mm after tumbling, g;

Ash content (A^d) was determined according to [27] (Table 2). The value of this parameter is important as an increase in the ash content reduces the mechanical strength of coke [23].

Furthermore, the following parameters were determined for a better understanding of the basic properties of the initial coke materials (Table 2): moisture content (total, W^a, and in the analytical sample, W_t^r) [28,29], volatile matter content (V^{daf}) [30], total sulfur content (S_t^d)—measured using a LECO SC 132 analyzer—and net calorific value (NCV) [31].

Table 2. Parameters of the coke materials determined in the study.

Parameter	Standard/Method
Micum index 40 (M_{40})	PN-C-04305:1998 [25]
Micum index 10 (M_{10})	PN-C-04305:1998 [25]
Coke reactivity index (CRI)	ISO 18894:2006 [26]
Coke strength after reaction with CO_2 (CSR)	ISO 18894:2006 [26]
Ash content (A^d)	PN-ISO 1171:2002 [27]
Total moisture content (W^a)	PN-ISO 579:2002 [28]
Moisture content in the analytical sample (W_t^r)	PN-ISO 687:2005 [29]
Volatile matter content (V^{daf})	PN-G-04516:1998 [30]
Total sulfur content (S_t^d)	Combustion of samples at 1350 °C using the LECO SC 132 analyzer
Net calorific value (NCV)	PN-ISO 1928:2020-05 [31]
Crush resistance index (m'_{pan})	PN-EN ISO 13503-2:2010 [32]
Roga index (RI)	PN-ISO 15585:2009 [33]

2.2.2. Modification of the initial coke materials

To make an in-depth preliminary assessment of their suitability for use as proppants, the initial coke materials were subjected to modification (Figure 2). The method of modification depended on the grain size of the cokes. For coarse-grained coke (samples I, III, V, VII, IX, X, XI, XII, XIV, XV, XVI, and XX), it consisted of successive crushing in two jaw crushers, the first with a gap of 20 mm (the average grain size after crushing varied from 20 mm to 8 mm), the second with a gap of 6 mm (the average grain size after crushing varied from 6 to 0.5 mm), and mechanical classification using sieves with a mesh size of 1 and 3 mm. For coke breeze (samples II, IV, VI, VIII, XIII, XVII, XVIII, XIX), it involved mechanical classification using sieves of 1 and 3 mm mesh size only.

The resulting materials in class 3–1 mm were subjected to a simplified densiometric analysis [34] in a heavy liquid with a density of 1.35 g/cm^3 (calcium chloride solution was used). The lighter density fraction was dried in an oven at 105 °C to a constant sample weight.

2.2.3. Analytical procedures performed on the modified coke materials

The lighter density fraction was further tested to determine technological parameters (Table 2).

A crush resistance test of the modified coke material (m'_{pan}) was performed [32]. The test is useful for determining and comparing the strength of proppants. It was carried out on samples that were sieved so that all of the particles tested were within the specified size range. The amount of material crushed under a stress of 13.8 MPa was measured.

Mechanical strength according to the modified Roga method (RI_{mod}) [33,35] was also analyzed. The coke material was weighed and subjected to three five-minute tumblings at 50 rpm in a Roga drum filled with 100 g of coke and 1000 g of steel balls of 12 mm diameter. The operation of sieving through a 1 mm square mesh sieve and weighing was repeated after each cycle (Equation (5)). The tests were carried out twice. The arithmetic mean was taken as a result.

$$RI = \frac{100}{3Q}\left[\frac{a+d}{2} + b + c\right] \qquad (5)$$

where

Q—weight of the sample after coking (before the first sieving), g;
a—weight of the sample on the sieve before the first tumbling, g; $Q = a = 100$ g;
b—weight of the sample on the sieve after the first tumbling, g;
c—weight of the sample on the sieve after the second tumbling, g;
d—mass of the sample on the sieve after the third tumbling, g.

The ash content (A^d) in the modified coke materials was measured again [27].

3. Results and Discussion

3.1. Properties of the initial and modified coke materials

Micum index 40 (M_{40}) for the initial coke material was at least 70 (mostly exceeding 75), while micum index 10 (M_{10}) remained below 10 (Figure 3). This shows the relatively high strength and resistance to breakage and abrasion of the studied cokes, adequately meeting the European requirements [17,19]. The coke reactivity index (CRI) ranged between 27.3 and 39.4, with most values being below 30, while coke strength after the reaction (CSR) varied from 42.2 to 69.0, usually exceeding 60 (Figure 3). CRI<30 and CSR>60 are the required values for high-quality blast furnace cokes [17–19]. Moreover, CSR>60 indicates the high mechanical strength of the studied material [17]. Foundry coke (sample V) was found to have an exceptionally low CSR value (42.2). The ash content (A^d) ranged from 7.7% to 13.1%, at an average value of 10.5% (Figure 4), being typical for different kinds of cokes, including the European blast furnace cokes [17,19].

The other parameters of the initial coke material, such as total moisture content (W^a), the moisture content in the analytical sample (W_t^r), volatile matter content (V^{daf}), total sulfur content (S_t^d), and net calorific value (NCV), are given in Figures 3 and 4.

The feasibility of enriching coke materials to improve their strength depends not only on their initial ash content but also on their homogeneity. If the material is isotropic, enrichment is difficult or impossible. Only the coke materials that are inhomogeneous in terms of their mineral content are enrichable. In this study, it was assumed that coke materials of the highest strength parameters and with an ash content (A^d) <9% would be selected for further detailed work covering technological (processing) operations to obtain the best proppant for the hydraulic fracturing of coals. Therefore, the grain class of 3–1 mm obtained by the mechanical classification of each coke material was enriched in heavy liquid with a density of 1.35 g/cm^3, and a set of parameters (m'$_{pan}$, RI and A^d) was determined for the lighter fraction.

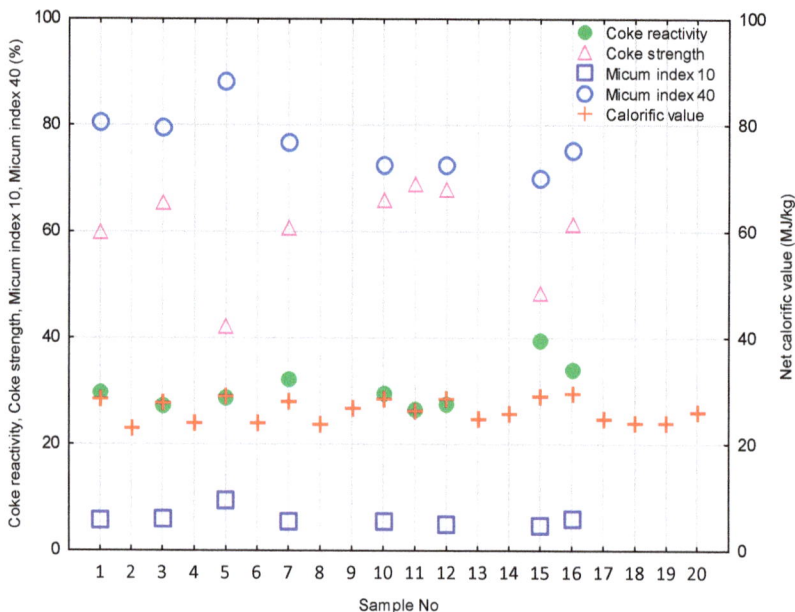

Figure 3. Strength parameters and net calorific value of the coke materials tested.

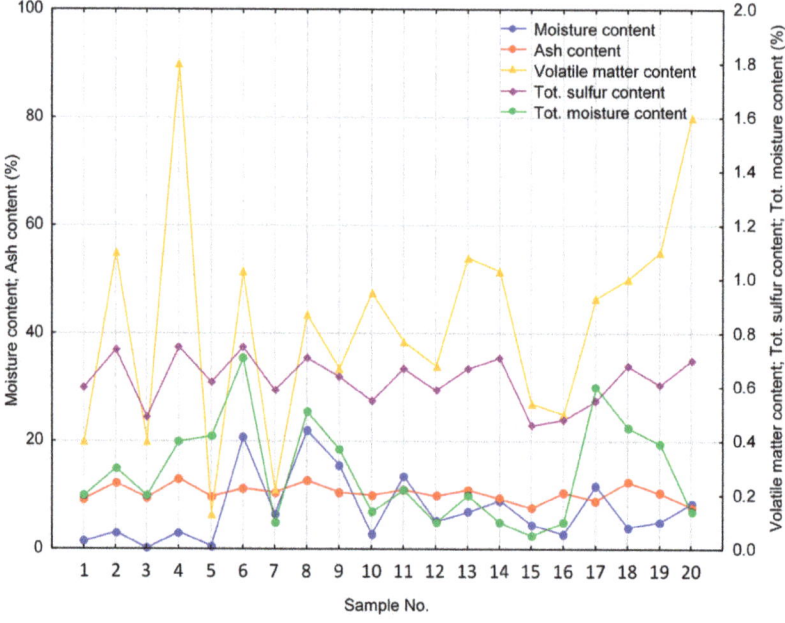

Figure 4. Results of the technical analysis of the coke materials tested.

The modified coke materials were characterized by a varied crush resistance index (m'_{pan}) (40.25–63.28%) and Roga index (85.30–97.55), reflecting their different strengths (Figure 5).

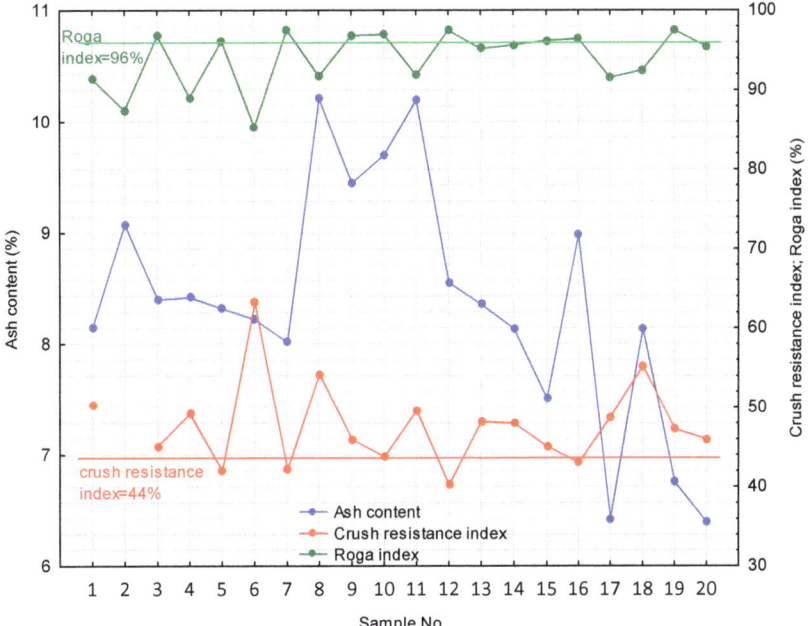

Figure 5. Technological parameters of the modified coke materials.

The ash contents were significantly lower than in the initial coke materials, with most samples meeting the $A^d<9\%$ criterion (Figure 5). The average ash contents in the light fraction for coarse-grain coke and coke breeze were 8.49 and 8.08%, which means that the difference between them was small, and, in the heavy fraction, they were 16.70 and 19.85%, respectively, which proves that in the coke breeze, there was 3.15 p.p. (percentage point) more mineral matter than in coarse coke. The average ash contents in the combined density fractions of coarse-grained coke and coke breeze are 11.11% and 12.85%. The results indicate a significant loss of mineral substance in the light fraction of the coke materials, which contributed to the improvement in the strength parameters of the modified cokes. Samples II and VIII-XI were difficult to enrich. They represented different types of coke, but they were usually fine-grained and were obtained from different coking plants (Table 1). Although the ash content was reduced with reference to the initial material, it remained above the desired level (Figure 5).

3.2. Selection of the Most Promising Material for the Final Processing to Obtain a Coke Proppant

As mentioned earlier, parameters that have a strong influence on the decision on the suitability of coke materials for use as proppants are coke strength indicators such as M_{40}, M_{10}, CRI, CSR, m'_{pan}, and RI. The parameters m'_{pan} and RI were chosen as key parameters because the tests were carried out for modified samples, and this is the kind of coke material that is to be evaluated as a potential raw material for proppant production.

From this point of view, the studied coke materials were divided into three groups: cokes of the highest, medium, and lowest strength. The following values of the technological parameters were used as a criterion for the division:

1. <44%/44–49%/>49% (highest/average/lowest strength)—for the crush resistance of the coke material, m'_{pan};

2. >98% ±2%/96–92% ±2%/<90% ±2% (highest/average/lowest strength) for RI, that is, for mechanical strength according to the modified Roga method.

Consequently, on the basis of the criteria adopted, the coke materials were divided into three groups:
1. Cokes of the highest strength—samples V, VII, X, XII, XVI;
2. Cokes of medium strength—samples III, IX, XIII, XIV, XV, XVII, XIX, XX;
3. Cokes of the lowest strength—samples I, IV, VI, VIII, XI, XVIII.

The values of the micum index 40 (M_{40}) for the coke materials in the highest strength group ranged from 72.4% (sample XII) to 88.2% (sample V), while the values of the micum index 10 (M_{10}) ranged from 5.0% (sample XII) to 9.6% (sample V) (Figure 3). All materials in this group had similar coke reactivity to CO_2 (CRI). Sample V was characterized by the lowest strength after reaction (CSR), 42.2%, compared to an average of 64.1% for the others in this group (Figure 3).

High values of M_{40} and CSR and low values of M_{10} and CRI were also characteristic of samples I and III (Figure 3). This predisposed them to the group of cokes with the highest strength. However, the values of the m'_{pan} and RI determined after the modification of the coke materials were relatively low. This caused them to be placed in the group of samples with the lowest and medium strength, respectively.

The parameters M_{40}, M_{10}, CRI, and CSR were not investigated for fuel coke and coke breeze, as they are only determined for large grain classes. Coke breeze, as a waste material from the production of foundry coke, blast furnace coke, or fuel coke, could, by definition, be an interesting alternative for the production of proppants. As a waste material, it is inexpensive, and its grain size range from 10 to 0 mm means that it does not require grinding as part of the coke material modification process. By omitting this energy-consuming operation, production costs could be reduced. However, all of the breeze cokes, due to their low values of the m'_{pan} and RI parameters, were classified into the group of medium- and lowest-strength cokes.

It should be added at this point that sample II, i.e., the coke breeze from the production of blast furnace coke in the 3–0 mm class, contained only 1.76% of the 3–1 mm grain class and 98.24% of the 1–0 mm grain class. The negligible amount of grain class 3–1 mm and the potential for its use made it impossible (owing to the unprofitability of future production) to process it. Therefore, this sample was rejected. The analyses carried out on the RI and ash content of the enrichment product (Figure 5) further indicate that this material has a low potential for use as a proppant.

In summary, it can be concluded that on the basis of parameters M_{40} and M_{10}, it is not possible to clearly determine whether coke will be suitable for the preparation of the proppant. It can only be said that their values should be at least 70% and at most 10%, respectively. To initially select the best cokes for proppant preparation, further tests should be carried out to determine the strength parameters of crushed coke to the 3–1 mm class after its initial modification consisting of the removal of heavy coke fractions containing an increased content of mineral substance expressed as ash content. These parameters include crush resistance (m'_{pan}) and mechanical strength, according to the modified Roga method (RI_{mod}). These parameters should be at least 44% and at least 96%, respectively.

As this is the first time that the raw material for the coke-based proppant has been sought, it is not possible to compare the properties of the best materials found in this study with those of any other cokes investigated for a similar purpose.

3.3. The role of fracture conductivity

It should also be noted here that one of the key factors that affects the productivity of coal bed gas wells is fracture conductivity [36]. CBM reservoirs, due to their low values of permeability, Young's modulus, relatively low pressure, and high Poisson ratio are significantly different from those of conventional hydrocarbon deposits. It has been found that fracture conductivity is related to reservoir properties, closure pressure, proppant properties, and concentration, as well as fracturing fluid composition and rheology. The conductivity increases with the diameter of the proppant particles and the number of proppant layers, whereas it decreases with the decreasing closing pressure. At the same closing pressure, the conductivity of a multi-layer proppant pack is greater than the conductivity of a single-layer pack. In the early stages of fracturing, with fine-grained proppants, the extent of the fracture may be longer. For the final stage of fracturing, injected coarser proppants are able to improve near-wellbore conductivity [37]. The complicated issues described above mean that for each type of coal and the proppant and fracturing fluid used, hydraulic fracture conductivity tests should be performed.

4. Conclusions

This is the first time that the source material for coke-based proppants has been identified. On the basis of the tests carried out, the following conclusions can be drawn:

1. The preliminary assessment of the suitability of the source coke material for the final processing is based on a set of tests performed on the initial coke and the 3–1 mm grain fraction enriched in a heavy liquid with a density of 1.35 g/cm^3. For the initial coke, these include the strength parameters M_{40}, M_{10}, CRI, CSR, and ash content (A^d). For the modified cokes, they comprise m'_{pan} and RI, which were selected as key strength parameters, and A^d.
2. The most promising modified coke materials with the best strength properties had m'_{pan} and RI values of at least 44% and at least 96%, respectively, and contained less than 9% ash. They were obtained from a coarse-grained (fraction 25–80 mm or larger) blast furnace and foundry coke. For these materials, the values of the individual parameters are as follows: M_{40} > 72.4%, M_{10} < 9.6%, CRI < 34.1%, and CSR > 42.2%.
3. Further work on the highest strength coke materials described in this article will consist of developing a technology for the production of proppants in order to obtain a product with parameters compliant with the PN-EN ISO 13503-2:2010 standard. They will consist, in particular, of the selection of effective and environmentally friendly methods of crushing, shaping, and enriching coke materials.

Author Contributions: Conceptualization, K.L., T.S. and R.M.; methodology, T.S.; investigation, T.S.; writing—original draft preparation, K.L., T.S. and R.M.; writing—review and editing, R.M. and T.S.; visualization, K.L. All authors have read and agreed to the published version of the manuscript.

Funding: This study was performed within the INGA Project entitled "Coke-Based Proppants for Coal Bed Fracturing", funded by the National Centre for Research and Development of The Republic of Poland and by the Polish Oil and Gas Company (PGNiG) under grant agreement no. POIR.04.01.01-00-0016/18.

Institutional Review Board Statement: Not applicable.

Informed Consent Statement: Not applicable.

Data Availability Statement: The data presented in this study are available on request from the corresponding authors.

Conflicts of Interest: The authors declare that they have no conflicts of interest.

References

1. Elder, C.H. *Effects of Hydraulic Stimulation on Coalbeds and Associated Strata*; USBM report RI 8260; Department of the Interior, Bureau of Mines: Washington, DC, USA, 1977.
2. Jeffrey, R.G.; Boucher, C. Sand propped hydraulic fracture stimulation of horizontal in-seam gas drainage holes at Dartbrook Coal Mine. In Proceedings of the COAL 2004 Symposium, University of Wollongong, Wollongong, New Zealand, 4–6 February 2004; pp. 169–179.
3. Moska, R.; Labus, K.; Kasza, P. Hydraulic Fracturing in Enhanced Geothermal Systems—Field, Tectonic and Rock Mechanics Conditions—A Review. *Energies* **2021**, *14*, 5725. [CrossRef]
4. Su, S.; Beath, A.; Guo, H.; Mallett, C. An assessment of mine methane mitigation and utilisation technologies. *Prog. Energ. Combust.* **2005**, *31*, 123–170. [CrossRef]
5. Mazzotti, M.; Pini, R.; Storti, G. Enhanced coalbed methane recovery. *J. Supercrit. Fluid* **2009**, *47*, 619–627. [CrossRef]

6. Karacan, C.Ö.; Ruiz, F.A.; Cotè, M.; Phipps, S. Coal mine methane: A review of capture and utilization practices with benefits to mining safety and to greenhouse gas reduction. *Int. J. Coal Geol.* **2011**, *86*, 121–156. [CrossRef]
7. Moore, T.A. Coalbed methane—A review. *Int. J. Coal Geol.* **2012**, *101*, 36–80. [CrossRef]
8. Cheng, L.; Ge, Z.; Xia, B.; Li, Q.; Tang, J.; Cheng, Y.; Zhuo, S. Research on Hydraulic Technology for Seam Permeability Enhancement in Underground Coal Mines in China. *Energies* **2018**, *11*, 427. [CrossRef]
9. Fan, C.; Elsworth, D.; Li, S.; Chen, Z.; Luo, M.; Song, Y.; Zhang, H. Modelling and optimization of enhanced coalbed methane recovery using CO_2/N_2 mixtures. *Fuel* **2019**, *253*, 1114–1129. [CrossRef]
10. Lakirouhani, A.; Detournay, E.; Bunger, A.P. A reassessment of in situ stress determination by hydraulic fracturing. *Geophys. J. Int.* **2016**, *205*, 1859–1873. [CrossRef]
11. LeBlanc, D.; Martel, T.; Graves, D.; Tudor, E.; Lestz, R. Application of Propane (LPG) Based Hydraulic Fracturing in The McCully Gas Field, New Brunswick, Canada. In Proceedings of the SPE North American Unconventional Gas Conference and Exhibition, The Woodlands, TX, USA, 14–16 June 2011; Paper 144093. OnePetro: Richardson, TX, USA, 2011. [CrossRef]
12. Masłowski, M.; Labus, M. Preliminary Studies on the Proppant Embedment in Baltic Basin Shale Rock. *Rock Mech. Rock Eng.* **2021**, *54*, 2233–2248. [CrossRef]
13. Brannon, H.D.; Wood, W.D.; Rickards, A.R.; Stephenson, C.J. Method of Enhancing Hydraulic Fracturing Using Ultra Lightweight Proppants. U.S. Patent 7,726,399, 1 June 2010.
14. Ahamed, M.A.A.; Perera, M.S.A.; Elsworth, D.; Ranjith, P.G.; Matthai, S.K.M.; Dong-yin, L. Effective application of proppants during the hydraulic fracturing of coal seam gas reservoirs: Implications from laboratory testings of propped and unpropped coal fractures. *Fuel* **2021**, *304*, 121394. [CrossRef]
15. Labus, K.; Morga, R.; Suponik, T.; Masłowski, M.; Wilk, K.; Kasza, P. The concept of coke based proppants for coal bed fracturing. *IOP C Ser. Earth Env.* **2019**, *261*, 012026. [CrossRef]
16. Labus, K.; Kasza, P.; Masłowski, M.; Czupski, M.; Wilk, K. Proppant for Use in Hydraulic Fracturing of Coals. Polish Patent 418,516, 2016 08 31, 2016.
17. Díez, M.A.; Álvarez, R.; Barriocanal, C. Coal for metallurgical coke production: Predictions of coke quality and future requirements for cokemaking. *Int. J. Coal Geol.* **2002**, *50*, 389–412. [CrossRef]
18. Álvarez, R.; Díez, M.A.; Barriocanal, C. An approach to blast furnace coke quality prediction. *Fuel* **2007**, *86*, 2159–2166. [CrossRef]
19. Koszorek, A.; Krzesińska, M.; Pusz, S.; Pilawa, B.; Kwiecińska, B. Relationship between the technical parameters of cokes produced from blends of three Polish coals of different coking ability. *Int. J. Coal Geol.* **2009**, *77*, 363–371. [CrossRef]
20. Rejdak, M.; Strugała, A.; Sobolewski, A. Stamp-Charged Coke-Making Technology—The Effect of Charge Density and the Addition of Semi-Soft Coals on the Structural, Textural and Quality Parameters of Coke. *Energies* **2021**, *14*, 3401. [CrossRef]
21. Pusz, S.; Buszko, R. Reflectance parameters of cokes in relation to their reactivity index (CRI) and the strength after reaction (CSR), from coals of the Upper Silesian Coal Basin, Poland. *Int. J. Coal Geol.* **2012**, *90–91*, 43–49. [CrossRef]
22. North, L.; Blackmore, K.; Nesbitt, K.; Mahoney, M. Models of coke quality prediction and the relationships to input variables: A review. *Fuel* **2018**, *219*, 446–466. [CrossRef]
23. Rutkowski, P.; Chomiak, K. *Determination of Reactivity and Mechanical Strength of Coke*; Politechnika Wrocławska: Wrocław, Poland, 2020. Available online: https://iptm.pwr.edu.pl/fcp/VGBUKOQtTKlQhbx08SlkAVgBeUTgtCgg9ACFDCwgCFiFPFRYqCl5 tDXdAGHpEQVgQaxMDOCAEDgMdLA5fRE0OPxZSBw/154/public/gmw_dydaktyka/am_w1_reakcyjnosc_koksu.pdf (accessed on 14 November 2022). (In Polish)
24. Mianowski, A.; Radko, T.; Koszorek, A. Assessment of the high-quality blast furnace coke by using the reactivity and strength integrated Nippon Steel Corporation test. *Przem. Chem* **2009**, *88*, 692–698, (In Polish with English abstract).
25. PN-C-04305:1998; Hard Coal Coke. Determination of Mechanical Strength. Polish Committee for Standardization: Warsaw, Poland, 1998. (In Polish)
26. ISO 18894:2006; Coke—Determination of Coke Reactivity Index (CRI) and Coke Strength after Reaction (CSR). ISO: Geneva, Switzerland, 2010.
27. PN-ISO 1171:2002; Solid Fuels. Ash Determination. ISO: Geneva, Switzerland, 2002. (In Polish)
28. PN-ISO 579:2002; Hard Coal Coke. Determination of Total Moisture. ISO: Geneva, Switzerland, 2002. (In Polish)
29. PN-ISO 687:2005; Hard Coal Coke. Determination of Moisture in the Analytical Sample. ISO: Geneva, Switzerland, 2005. (In Polish)
30. PN-G-04516:1998; Solid Fuels. Determination of Volatile Matter Content by Weight Method. Polish Committee for Standardization: Warsaw, Poland, 1998. (In Polish)
31. PN-ISO 1928:2020-05; Solid Fuels. Determination of Gross Calorific Value by the Calorimetric Bomb Method and Calculation of Calorific Value. ISO: Geneva, Switzerland, 2020. (In Polish)
32. PN-EN ISO 13503-2:2010; Petroleum and Natural Gas Industries—Completion Fluids and Materials—Part 2: Measurement of Properties of Proppants Used in Hydraulic Fracturing and Gravel-Packing Operations. ISO: Geneva, Switzerland, 2010. (In Polish)
33. PN-ISO 15585:2009; Hard Coal. Determination of Caking Index. ISO: Geneva, Switzerland, 2009. (In Polish)
34. PN-ISO 7936: 1999P; Hard Coal. Determination and Presentation of Gravity Enrichment Characteristics—General Guidelines for Apparatus and Procedures. ISO: Geneva, Switzerland, 1999. (In Polish)
35. Roga, B.; Wnękowska, L. *Coal and Coke Analysis*; Wydawnictwo Naukowo-Techniczne: Warsaw, Poland, 1966.

36. Li, Y.; Meng, W.; Rui, R.; Wang, J.; Jia, D.; Chen, G.; Patil, S.; Dandekar, A. The calculation of coal rock fracture conductivity with Different Arrangements of proppants. *Geofluids* **2018**, *2018*, 4938294. [CrossRef]
37. Meng, W.; Li, Z.; Guo, Z. Calculation model of fracture conductivity in coal reservoir and its application. *J. China Coal Soc.* **2014**, *39*, 1852–1856. [CrossRef]

Disclaimer/Publisher's Note: The statements, opinions and data contained in all publications are solely those of the individual author(s) and contributor(s) and not of MDPI and/or the editor(s). MDPI and/or the editor(s) disclaim responsibility for any injury to people or property resulting from any ideas, methods, instructions or products referred to in the content.

Article

Partial Oxidation Synthesis of Prussian Blue Analogues for Thermo-Rechargeable Battery

Yutaka Moritomo [1,2,3,*], Masato Sarukura [1], Hiroki Iwaizumi [1,†] and Ichiro Nagai [2,‡]

1. Graduate School of Pure & Applied Science, University of Tsukuba, Tennodai 1-1-1, Tsukuba 305-8571, Ibaraki, Japan
2. Faculty of Pure & Applied Science, University of Tsukuba, Tennodai 1-1-1, Tsukuba 305-8571, Ibaraki, Japan
3. Research Center for Energy Materials Science (TREMS), University of Tsukuba, Tennodai 1-1-1, Tsukuba 305-8571, Ibaraki, Japan
* Correspondence: moritomo.yutaka.gf@u.tsukuba.ac.jp
† Current address: Research Institute of Material and Chemical Measurement, AIST, Umezono 1-1-1, Tsukuba 305-8569, Ibaraki, Japan.
‡ Current address: Faculty of Marine Technology, Tokyo University of Marine Science and Technology, Etchujima 2-1-6, Koto-ku, Tokyo 135-8533, Japan.

Abstract: A thermo-rechargeable battery or tertiary battery converts thermal energy into electric energy via an electrochemical Seebeck coefficient. The manufacturing of the tertiary batteries requires a pre-oxidation step to align and optimize the cathode and anode potentials. The pre-oxidation step, which is not part of the secondary battery manufacturing process, makes the manufacturing of tertiary batteries complex and costly. To omit the pre-oxidation step, we used partially oxidized Prussian blue analogs, i.e., $Na_xCo[Fe(CN)_6]_y zH_2O$ (Co-PBA) and $Na_xNi[Fe(CN)_6]_y zH_2O$ (Ni-PBA), as cathode and anode materials. The modified tertiary battery without the pre-oxidation step shows good thermal cyclability between 10 °C and 50 °C without detectable deterioration of the thermal voltage (V_{cell}) and discharge capacity (Q_{cell}).

Keywords: energy harvesting; thermo-rechargeable battery; tertiary battery; Prussian blue analogue; partial oxidization synthesis

1. Introduction

Energy harvesting from environmental heat is an important technology for our future society. Among the energy harvesting devices, thermo-rechargeable batteries [1–8] that can be charged by the warming/cooling of the device are promising. In thermo-rechargeable batteries, the temperature coefficient ($\alpha \equiv dV/dT$) of redox potential (V) is different between the cathode and anode. The thermo-rechargeable batteries convert thermal energy into electric energy between the low (T_L) and high (T_H) heat sources, analogously to a heat engine. Hereafter, we call the thermo-rechargeable battery composed of solid active materials a tertiary battery, because it has the same device configuration as an ion secondary battery. With an increase in the device temperature from T_L to T_H, cell voltage (v_{cell}) increases from 0 V to V_{cell} (≥ 0). The battery produces electric energy through the discharge processes at T_H. With a decrease in the device temperature to T_L, v_{cell} decreases from 0 V to $-V_{cell}$. The battery produces electric energy through the discharge processes at T_L. A tertiary battery can generate electrical energy from daily temperature cycles such as day and night or sunlight and shade. Unlike a thermo-electric conversion device that converts the temperature difference within the device into electric energy, the tertiary battery converts the temperature change of the device into electric energy. Like batteries, tertiary batteries are mobile because the device does not need to be in contact with a heat source. Thus, a tertiary battery is a device that has both mobility like a dry battery and independence from a power grid like a thermoelectric conversion device.

The thermal voltage (V_{cell}) and discharge capacity (Q_{cell}) are important performance parameters of a battery. The thermal voltage is expressed by electrode parameters as $V_{cell}^{cal} = (\alpha^+ - \alpha^-)(T_H - T_L)$, where α^+ (α^-) is α of cathode (anode) material. We emphasize that cathode and anode potentials should be the same when they are assembled into a tertiary battery. We called this potential adjustment step the pre-oxidization step. Generally, the voltage of a battery decreases as the extracted charge increases. The discharge capacity of a tertiary battery is defined as the capacity at which the voltage drop is equal to V_{cell}. Since V_{cell} of the tertiary battery is relatively small (several tens mV), the charge coefficient ($\beta \equiv -dV/dq$, where q is capacity per unit weight of active material) of V can be regarded as constant. Considering the weight ratio ($r = \frac{m^+}{m^+ + m^-}$) of the cathode ($m^+$) and anode ($m^-$) active materials, the discharge capacity per unit weight of total active material is expressed by electrode parameters as $Q_{cell}^{cal} = \frac{V_{cell}^{cal}}{\frac{\beta_+}{r} + \frac{\beta_-}{1-r}}$, where β^+ (β^-) and is β of the cathode (anode) material [7,8]. The formula tells us that Q_{cell}^{cal} is maximized when the cathode and anode potentials are optimized so that $\frac{\beta_+}{r} + \frac{\beta_-}{1-r}$ is minimized. The adjustment of the cathode and anode potential is also performed at the pre-oxidization step. Thus, the pre-oxidization step is indispensable for the fabrication of a tertiary battery.

Prussian blue analogs, $Na_xM[Fe(CN)_6]_y zH_2O$ (M-PBAs; M is divalent metal element), are promising electrode materials for lithium-ion secondary battery (LIBs) [9–14], sodium-ion secondary batteries (SIBs) [15–34], and tertiary batteries [4–6]. M-PBAs show a jungle-gym-type framework structure; the intersection points and edges are metal elements and cyano groups, respectively. The framework can accommodate Na^+ and H_2O molecules. Importantly, the accommodated Na^+ can be electrochemically moved in and out accompanying reduction and oxidation of the framework. In the reduced state of M-PBA, Fe and M takes a divalent state and each nanopore accommodates one Na^+. In the oxidization process, the valence of Fe and/or M increases to trivalent with removing Na^+ from the nanopore. Most of M-PBA in reduced state show face-centered cubic (fcc) ($Fm\bar{3}m$; $Z = 4$) or trigonal ($R\bar{3}m$; $Z = 3$) structures [35–37]. As active materials for tertiary batteries, M-PBAs have three advantages. First, they can be synthesized as precipitates from aqueous solutions. In addition, they can be synthesized at a low cost since they do not contain rare elements. Second, M-PBA and its oxidants are stable in air and in aqueous solutions. In particular, a reversible redox reaction is possible in aqueous solution. Third, M-PBA shows positive and negative α. Therefore, a tertiary battery can be composed of M-PBA with positive α and M-PBA with negative α. The α values are negative in Mn-, Ni-, Cu-, and Zn-PBAs and are positive in Fe- and Co-PBAs [38]. Among M-PBAs, Co-PBA (α = +0.57 mV/K [38]) and Ni-PBA (-0.42 mV/K [38]) are prototypical combinations as cathode and anode materials for tertiary batteries [5], because their cycle stability is good as compared with the other combination.

The conventional tertiary battery is fabricated by four steps, i.e., (i) material synthesis, (ii) electrode fabrication, (iii) pre-oxidization of electrodes, and (iv) battery assembly (Figure 1a). Unlike LIBs/SIBs (Figure 1c), the manufacturing of tertiary batteries requires the (iii) pre-oxidation step for the following two reasons. First, the cathode and anode potentials should be the same before they are assembled into a tertiary battery, Second, the potentials should be optimized so that $\frac{\beta_+}{r} + \frac{\beta_-}{1-r}$ is minimized and Q_{cell} is maximized. In the as-grown reduced state, the potential gradient (β) against charge is significantly large because the charge curve steeply increases with q. The pre-oxidation step flattens the potential gradient. The pre-oxidization step, which is performed by electrochemical oxidization of the electrodes, makes the manufacture of tertiary batteries complex and costly. If we directly synthesize partially oxidized M-PBA, we can omit the pre-oxidation step from the manufacturing process (Figure 1b). We will call such a battery a modified tertiary battery.

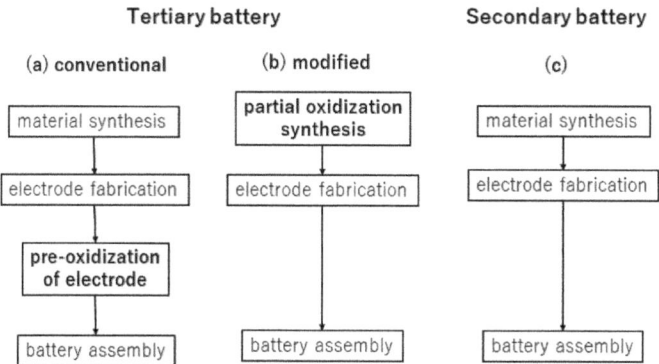

Figure 1. Manufacturing processes of (**a**) conventional, (**b**) modified tertiary batteries, and (**c**) secondary battery.

In this study, we performed partial oxidation synthesis of Co-PBA and Ni-PBA to omit the pre-oxidation step in the manufacturing of the tertiary batteries. The modified tertiary battery without the pre-oxidation step shows good thermal cyclability between T_L (=10 °C) and T_H (=50 °C) without detectable deterioration of V_{cell} and Q_{cell}. The V_{cell} and Q_{cell} values are close to the ideal V_{cell}^{cal} and Q_{cell}^{cal} values. By incorporating the partial oxidation synthesis, the tertiary battery can be prepared by the same processes as LIBs/SIBs and will possess an industrial advantage.

2. Materials and Methods

2.1. Partial Oxidation Synthesis

Partial oxidation synthesis of Co-PBAs and Ni-PBAs, i.e., $Na_xCo[Fe(CN)_6]_y zH_2O$ and $Na_xNi[Fe(CN)_6]_y zH_2O$, were performed by dipping solution A into solution B at a speed of 100 mL/h at 35 °C Solution A was an aqueous solution containing 10 mM MCl_2 and 4 M NaCl. Solution B was an aqueous solution containing $10(1 - n^0_{Fe3+})$ mM $Na_4[Fe(CN)_6]$, $10 n^0_{Fe3+}$ mM $K_3[Fe(CN)_6]$, and 4 M NaCl. The degree of the oxidation was controlled by the initial Fe^{3+} ratio (n^0_{Fe3+}). In preparation of solution B ($n^0_{Fe3+} > 0.0$), K^+ in the solution was removed by an ion-exchange method. The aqueous solution containing $10(1 - n^0_{Fe3+})$ mM $Na_4[Fe(CN)_6]$ and $10 n^0_{Fe3+}$ mM $K_3[Fe(CN)_6]$ was stirred for 3 h at 50 °C with $50 n^0_{Fe3+}$ mM Ni-PBA ($n^0_{Fe3+} = 0.0$) powder. In this process, K^+ in solution exchanges with Na^+ in Ni-PBA. The ion-exchanged solutions were filtered and added 4 M NaCl.

The molar ratios of Fe and K to Co/Ni (Table 1) were determined using an SEM-EXS machine (JST-IT2000; JEOL, Ltd., Akishima, Japan). The molar ratios of C, N, and H (Table 2) were evaluated by a CHN analyzer (elementar; UNICUBE). The water content (z) was evaluated by $z = 3y \frac{m_H}{m_{CN}}$, where m_{CN} and m_H are the molar ratio of CN and H. Thus, the obtained chemical compositions of Co-PBAs and Ni-PBAs are listed in Table 3. The Fe concentration (y) ranged from 0.83 to 1.00. The molar ratio of K is less than 0.05. The Na concentration (x) was calculated from y and the degree of oxidation (p; *vide infra*) assuming charge neutrality. In the charge process of Co-PBA, Co^{2+} is oxidized first and then Fe^{3+} is oxidized [18]. Then, we formally obtained the relation $x = (1-p)(4y-2)$. In the charging process of Ni-PBA, only Fe^{2+} is oxidized [14]. Then, we obtained the relation $x = (1-p)y + (3y-2)$.

Table 1. Molar ratios of Fe and K to Co/Ni of partially-oxidized Co-PBAs and Ni-PBAs. n^0_{Fe3+} are the initial Fe^{3+} ratio. Quantification of Na moles was difficult due to the weak intensity of the EDX signal.

	n^0_{Fe3+}	Co/Ni	Fe	K
Co-PBA	0.0	1	0.88	below the limits of detection
Co-PBA	0.6	1	0.84	0.01
Co-PBA	0.8	1	0.85	0.01
Co-PBA	1.0	1	0.88	0.02
Ni-PBA	0.0	1	0.83	below the limits of detection
Ni-PBA	0.5	1	0.86	below the limits of detection
Ni-PBA	0.9	1	0.91	below the limits of detection
Ni-PBA	1.0	1	1.00	0.05

Table 2. Weight percent of C, N, and H of partially-oxidized Co-PBAs and Ni-PBAs. The experimental error was less than 0.1 wt%. n^0_{Fe3+} are the initial Fe^{3+} ratio.

	n^0_{Fe3+}	C (wt%)	N (wt%)	H (wt%)
Co-PBA	0.0	19.07	22.07	1.78
Co-PBA	0.6	19.08	21.75	1.96
Co-PBA	0.8	19.07	21.85	1.90
Co-PBA	1.0	19.43	22.17	1.77
Ni-PBA	0.0	17.88	20.65	2.25
Ni-PBA	0.5	18.23	20.57	2.11
Ni-PBA	0.9	18.42	20.74	2.26
Ni-PBA	1.0	18.49	20.86	2.23

Table 3. Chemical composition and lattice constant of partially-oxidized Co-PBAs and Ni-PBAs. n^0_{Fe3+} and p are the initial Fe^{3+} ratio and degree of the oxidization, respectively. Fcc means face-centered-cubic cell. Subscript H means hexagonal setting.

	n^0_{Fe3+}	p	Composition	Structure	Lattice Constant (Å)
Co-PBA	0.0	0.00	$Na_{1.52}Co[Fe(CN)_6]_{0.88}2.96H_2O$	trigonal	$a_H = 7.428(3), c_H = 17.57(1)$
Co-PBA	0.6	0.06	$Na_{1.28}Co[Fe(CN)_6]_{0.84}3.14H_2O$	fcc	$a = 10.247(4)$
Co-PBA	0.8	0.07	$Na_{1.31}Co[Fe(CN)_6]_{0.85}3.07H_2O$	fcc	$a = 10.206(3)$
Co-PBA	1.0	0.10	$Na_{1.37}Co[Fe(CN)_6]_{0.88}2.91H_2O$	fcc	$a = 10.181(3)$
Ni-PBA	0.0	0.00	$Na_{1.32}Ni[Fe(CN)_6]_{0.83}3.77H_2O$	fcc	$a = 10.290(6)$
Ni-PBA	0.5	0.10	$Na_{1.35}Ni[Fe(CN)_6]_{0.86}3.65H_2O$	fcc	$a = 10.297(4)$
Ni-PBA	0.9	0.16	$Na_{1.57}Ni[Fe(CN)_6]_{0.93}4.17H_2O$	fcc	$a = 10.291(3)$
Ni-PBA	1.0	0.15	$Na_{1.85}Ni[Fe(CN)_6]_{1.00}4.41H_2O$	fcc	$a = 10.272(4)$

Figure 2 shows the X-ray diffraction (XRD) patterns of (a) Co-PBA and (b) Ni-PBA against n^0_{Fe3+}. The XRD patterns were obtained using an XRD machine (MultiFlex; Rigaku) at room temperature. The X-ray source was the CuKα line. Except for Co-PBA (n^0_{Fe3+} = 0.0) (Figure 2a), all the diffraction peaks can be indexed in the fcc cell. For Co-PBA (n^0_{Fe3+} = 0.0), all the diffraction peaks can be indexed in the trigonal cell (hexagonal setting). With the use of the Rietan-FP program [39], we refined the lattice constants. The obtained lattice constants are listed in Table 3.

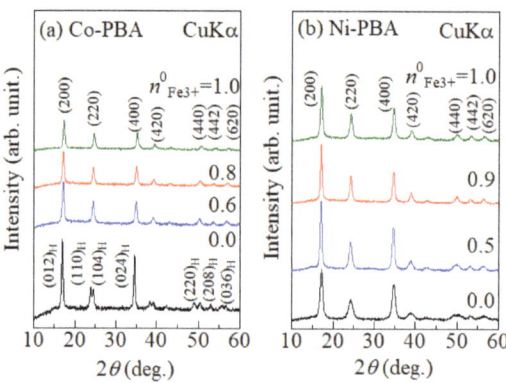

Figure 2. XRD patterns of (**a**) Co-PBA and (**b**) Ni-PBA against initial Fe^{3+} ratio (n^0_{Fe3+}). Parentheses without subscripts represent indexes in the fcc cell. Subscript H means the hexagonal setting in the trigonal cell.

2.2. Electrode Fabrication

Paste-type electrodes of Co-PBAs and Ni-PBAs were prepared as follows. First, M-PBA, acetylene black, and PVDF were well mixed with DMF at a ratio of 7:2:1. Then, the slurry was applied on an indium tin oxide (ITO) electrode and dried in a vacuum at 60 °C. The slurry was carefully applied by hand so that the thickness was uniform. The area of the electrode was 1.0 cm². To reduce the weight error, the weights of the electrodes with and without slurry were measured for each electrode. The weight of the active material, which is 70% of the slurry weight, was evaluated from the difference in the electrode weights.

We fabricated a three-pole beaker-type cell and measured charge and discharge curves with a potentiostat (HJ1001SD8; HokutoDENKO) at a rate of 0.3 C. The working, counter, and referential electrodes of the cell were the sample electrode, Pt, and Ag/AgCl standard electrodes, respectively. The electrolyte was an aqueous solution containing 17 mol/kg $NaClO_4$. We used theoretical capacity in the fully discharged state to define the C rate. The ideal capacities of Co-PBA in the fully discharged state are 122, 113, 116, and 123 mAh/g at n^0_{Fe3+} = 0.0, 0.6, 0.8, and 1.0, respectively. The currents at 1 C for a 1 mg electrode are 1.32, 1.26, 1.28, and 1.35 mA at n^0_{Fe3+} = 0.0, 0.6, 0.8, and 1.0, respectively. The ideal capacities of Ni-PBA in the fully discharged state are 67, 68, 68, and 68 mAh/g at n^0_{Fe3+} = 0.0, 0.5, 0.9, and 1.0, respectively. The currents at 1 C for a 1 mg electrode are 0.72, 0.73, 0.67, and 0.63 mA at n^0_{Fe3+} = 0.0, 0.5, 0.9, and 1.0, respectively. The potential range was from 1.1 and 0.3 V (vs. Ag/AgCl).

2.3. Battery Assembly

The Co-PBA/Ni-PBA tertiary batteries were two-pole beaker-type cell. The cathode, anode, and electrolyte were the Co-PBA electrode, Ni-PBA electrode, and aqueous solution containing 17 mol/kg $NaClO_4$, respectively. We note that Q^{cal}_{cell} (= $\frac{V^{cal}_{cell}}{\frac{\beta_+}{r} + \frac{\beta_-}{1-r}}$) is maximized when the β values of the cathode and anode are minimized. Figure 3 shows β of (a) Ni-PBA and (b) Co-PBA electrodes against V. The β values of Co-PBA and Ni-PBA are minimized at $V \approx 510$ mV (Figure 3). As will be described in the next section, Co-PBA at n^0_{Fe3+} = 1.0 and Ni-PBA at 1.0 satisfy this condition.

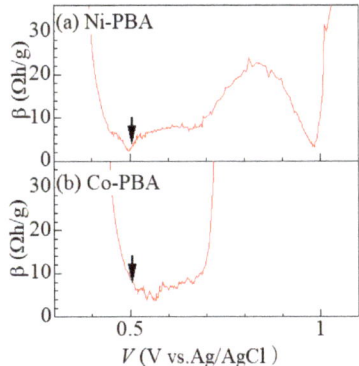

Figure 3. Charge coefficients ($\beta = -dV/dq$) of redox potential (V) of (**a**) Ni-PBA and (**b**) Co-PBA electrodes against V. Downward arrows indicate the initial potential for the tertiary battery.

We assembled a Co-PBA(n^0_{Fe3+} = 1.0)/Ni-PBA(1.0) tertiary battery (modified cell) without the pre-oxidation step (Figure 1b). For comparison, we assembled a conventional Co-PBA(n^0_{Fe3+} = 0.0)/Ni-PBA(0.0) tertiary battery with the pre-oxidation step (Figure 1a). The pre-oxidization was performed with the use of the three-pole beaker-type cell. In Table 4, we summarized the electrode parameters for the modified and conventional cells.

Table 4. Parameters of cathodes and anodes for the modified and conventional cells. β^+ (β^-), m^+ (m^-), R^+ (R^-), and V^+ (V^-) are the charge coefficient ($\beta \equiv -dV/dq$) of V, the weight of active material, internal resistivity, and initial potential (vs. Ag/AgCl) of cathode (anode), respectively. The electrolyte was 17 kg/L NaClO$_4$.

Cathode	Material	β^+ (Ωg/h)	m^+ (mg)	V^+ (mV)	α^+ (mV/K)	R^+ (kΩ)
modified cell	Co-PBA	4.1	0.70	510	0.57 [38]	0.40
conventional cell	Co-PBA	5.0	0.26	520	0.57 [38]	0.71
Anode	Material	β^- (Ωg/h)	m^- (mg)	V^- (mV)	α^+ (mV/K)	R^- (kΩ)
modified cell	Ni-PBA	7.6	0.27	510	−0.42 [38]	0.74
conventional cell	Ni-PBA	5.3	0.13	520	−0.42 [38]	0.80

2.4. Thermal Cycle Measurement

The thermal cycle properties of the tertiary batteries were investigated between T_L (=10 °C) and T_H (=50 °C). The temperature (T) of the battery was controlled in a thermobath. To stabilize the potential of the electrodes, the cell was connected to an external resistor and held for several hours at T_L (=10 °C) before the measurement. The thermal cycle consists of four processes, (i) heating, (ii) discharge at T_H, (iii) cooling, and (iv) discharge at T_L. In the (i) heating process, T was slowly increased from T_L to T_H under open circuit conditions. Then, the cell was kept at T_H for 15 min to stabilize the potential. At (ii) T_H, the discharge process was investigated at 0.03 C until the v_{cell} reached 0 mV. The C rate was defined by the theoretical capacity of the Co-PBA electrode in the fully discharged state. The currents at 1 C for a 1 mg Co-PBA electrode are 1.35 and 1.32 mA for the modified and conventional cells, respectively. In the (iii) cooling process, T was slowly decreased to T_L. Then, the cell was kept at T_L for 15 min. At (iv) T_L, the discharge process was investigated at 0.03 C until v_{cell} reaches 0 mV. The thermal cycle measurements were repeated five times.

3. Results and Discussion

3.1. Degree of Oxidization

Figure 4a shows first (red) and second (blue) charge curves of partially oxidized Co-PBA at 0.3 C and 25 °C against n^0_{Fe3+}. At n^0_{Fe3+} = 0.0, the first and second charge curves coincide with each other. The charge curves show a two-plateau structure at ~1.0 V and ~0.6 V. The upper and lower plateaus are ascribed to the redox process of $[Fe(CN)_6]^{4-}/[Fe(CN)_6]^{3-}$ and Co^{3+}/Co^{2+}, respectively [18]. The observed charge capacity (=113 mAh/g) is close to the ideal value (=122 mAh/g). At $n^0_{Fe3+} = 1.0$, the first charge capacity (=98 mAh/g) is smaller than second charge capacity (=109 mAh/g), reflecting the the partial oxidization of Co-PBA. The first charge curve is overlapped with the second charge curve if the curve is shifted by 11 mAh/g. The partially oxidized Co-PBA ($n^0_{Fe3+} = 1.0$) shows initial potential of 0.50 V and is suitable for the cathode material for the Co-PBA/Ni-PBA tertiary battery. Similar behaviors are observed at $n^0_{Fe3+} = 0.6$ and 0.8. Figure 5a shows the first (red) and second (blue) charge curves of partially oxidized Co-PBA ($n^0_{Fe3+} = 0.8$) at 0.3 C and 25 °C. The first and second discharge curves coincide with each other even though the first and second charge curves (Figure 4a) differ significantly. Thus, the partially oxidized Co-PBA shows stable charge properties.

Figure 4b shows the first and second charge curves of partially oxidized Ni-PBA at 0.3 C and 25 °C against n^0_{Fe3+}. At $n^0_{Fe3+} = 0.0$, the first and second charge curves coincide with each other. The charge curves show a single plateau structure. The plateau is ascribed to the redox process of $[Fe(CN)_6]^{4-}/[Fe(CN)_6]^{3-}$ [14]. The observed charge capacity (=54 mAh/g) is close to the ideal value (=67 mAh/g). At $n^0_{Fe3+} = 1.0$, the first charge capacity (=62 mAh/g) is smaller than the second charge capacity (=73 mAh/g), reflecting the partial oxidization of Ni-PBA. The first charge curve is overlapped with the second charge curve if the curve is shifted by 11 mAh/g. The partially oxidized Ni-PBA ($n^0_{Fe3+} = 1.0$) shows an initial potential of 0.53 V and is suitable for the anode material for the Co-PBA/Ni-PBA tertiary battery. Similar behaviors are observed at $n^0_{Fe3+} = 0.5$ and 0.9. Figure 5b shows the first (red) and second (blue) charge curves of partially oxidized Ni-PBA ($n^0_{Fe3+} = 0.9$) at 0.3 C and 25 °C. The first and second discharge curves coincide with each other even though the first and second charge curves (Figure 4b) differ significantly. Thus, the partially oxidized Co-PBA shows stable charge properties.

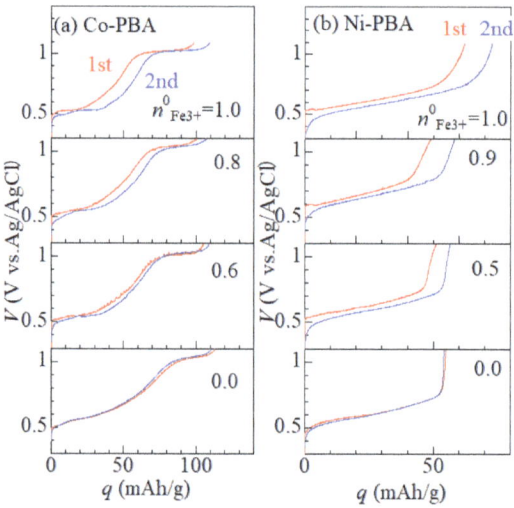

Figure 4. First (red) and second (blue) charge curves of partially oxidized (**a**) Co-PBA and (**b**) Ni-PBA at 0.3 C and at 25 °C against n^0_{Fe3+}.

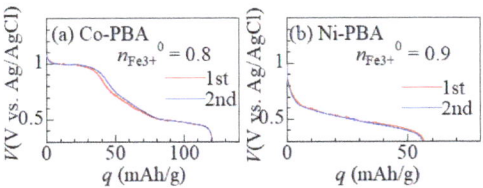

Figure 5. First (red) and second (blue) discharge curves of (**a**) Co-PBA (n^0_{Fe3+} = 0.8) and (**b**) Ni-PBA (n^0_{Fe3+} = 0.9) at 0.3 C and at 25 °C.

Here, we define the degree of oxidization (p) as the ratio between the oxidization charge and the observed capacity of the fully reduced compound. When the active material is partially oxidized, the initial capacity is the difference between the observed capacity in the fully reduced compound and the oxidation charge. The second cycle capacity corresponds to the observed capacity in the fully reduced compound. Then, p is expressed as 1- (initial capacity)/(second cycle capacity). Figure 6 shows the p values of (a) Co-PBA and (b) Ni-PBA against n^0_{Fe3+}. In both the compounds, the p value linearly increases with an increase in n^0_{Fe3+}, as indicated by straight lines. In Ni-PBA, p is proportional to the concentration of Fe^{3+} in the as-grown sample. Then, the observed proportional relationship indicates that the as-grown Ni-PBA contains Fe^{3+} in proportion to n^0_{Fe3+}. In Co-PBA, p is not directly proportional to the concentration of Fe^{3+} in the as-grown sample, since charge transfer from Co^{2+} to Fe^{3+} occurs within the sample. The observed proportional relationship is probably because the amount of charge transfer is proportional to the concentration of Fe^{3+} incorporated during crystal growth.

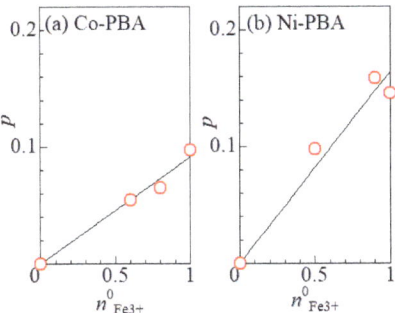

Figure 6. Degree of oxidization (p; open circles) of (**a**) Co-PBA and (**b**) Ni-PBA against n^0_{Fe3+}. Straight lines are the results of least-squares fitting.

3.2. Thermal Cycle Properties

Figure 7 shows the first thermal cycle of the modified Co-PBA/Ni-PBA tertiary battery between T_L (=10 °C) and T_H (=50 °C). The modified cell was fabricated without the pre-oxidization step (Figure 1b). In the (i) heating process, v_{cell} increases from 0 mV to 40 mV (=V_{cell}). In the (ii) discharge at T_H, v_{cell} linearly decreases from V_{cell} to 0 mV. In the (iii) cooling process, v_{cell} decreases from 0 mV to −40 mV (=−V_{cell}). In the (iv) discharge at T_L, v_{cell} linearly increases from −V_{cell} to 0 mV. Figure 7b shows the thermal cycle of the modified cell between T_L (=10 °C) and T_H (=50 °C). We observed no serious deterioration of V_{cell} and Q_{cell} up to five cycles. For comparison, we show in Figure 7c the thermal cyclability of the conventional cell.

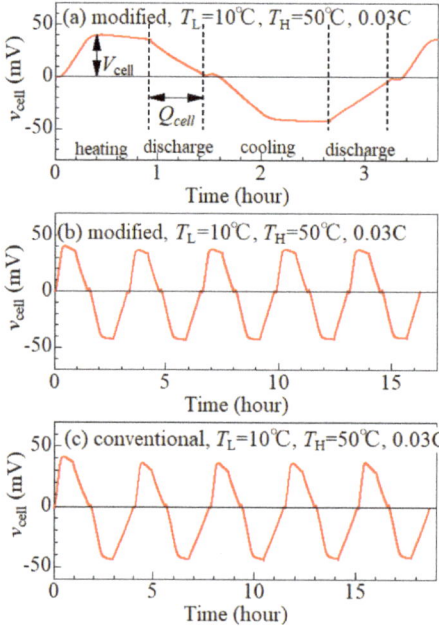

Figure 7. (a) The first thermal cycle of modified Co-PBA/Ni-PBA tertiary battery between T_L (= 10 °C) and T_H (= 50 °C). V_{cell} and Q_{cell} represent thermal voltage and discharge capacity, respectively. (b) Thermal cycle of modified cell. (c) Thermal cycle of conventional cell.

Figure 8a shows V_{cell} and Q_{cell} of the modified cell against cycle number. The Q_{cell} value was evaluated from the discharge time. V_{cell} and Q_{cell} are nearly constant against cycle number. We show in Figure 8b V_{cell} and Q_{cell} of the conventional cell against cycle number. V_{cell} are nearly constant against cycle number. Q_{cell}, however, gradually decreases with an increase in cycle number. Thus, the capacity cyclability of the modified cell (Figure 8a) is improved as compared with that of the conventional cell (Figure 8b).

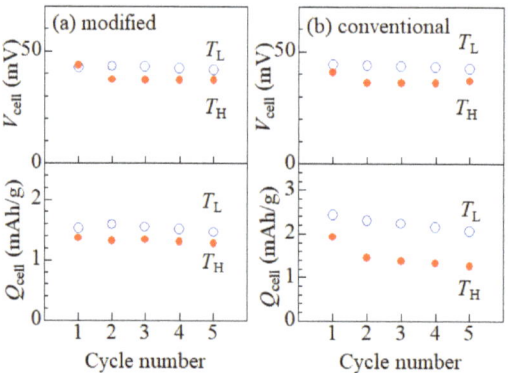

Figure 8. V_{cell} and Q_{cell} against cycle number; (a) modified and (b) conventional cells. Open and closed circles represent the data obtained at T_L (= 10 °C) and T_H (= 50 °C).

Takahara et al. [5] fabricated two tertiary batteries made of thin films of M-PBA, i.e., $Na_xCo[Fe(CN)_6]_{0.90}/Na_xCo[Fe(CN)_6]_{0.71}$ and $Na_xCo[Fe(CN)_6]_{0.90}/Na_xNi[Fe(CN)_6]_{0.68}$ cells, and investigated their thermal cyclabilities between 10 °C and 50 °C. They found that the capacity cyclability is much improved in the latter cell and ascribed the improvement to better capacity cyclability of the $Na_xNi[Fe(CN)_6]_{0.68}$ anode at 50 °C. Then, the better capacity cyclability observed in the modified cell (Figure 8a) is considered to be ascribed to better capacity cyclability of the active materials used in the modified cell. With use of the Scherrer's equation, we evaluated the grain sizes of Co-PBAs and Ni-PBAs used in the modified and conventional cells. The grain sizes of the active materials in the modified cell were 20 nm (Co-PBA, n^0_{Fe3+} = 1.0) and 10 nm (Ni-PBA, n^0_{Fe3+} = 1.0) whereas those in the conventional cell were 100 nm (Co-PBA, n^0_{Fe3+} = 0.0) and 10 (Ni-PBA, n^0_{Fe3+} = 0.0). The smaller grain size in the modified cell is advantageous for the stable electrochemical process, and hence, the capacity cyclability.

3.3. Comparison between Observed and Calculated V_{cell} Anmd Q_{Cell}

Finally, let us compare the observed V_{cell} and Q_{cell} values with the calculated values, i.e., V^{cal}_{cell} [= $(\alpha^+ - \alpha^-)(T_H - T_L)$] and Q^{cal}_{cell} [= $\frac{\alpha^+ - \alpha^-}{\beta^+ + \beta^-}(T_H - T_L)$] [7,8]. The electrode parameters, α^+ (α^-), β^+ (β^-), and r ($\frac{m^+}{m^+ + m^-}$), are listed in Table 4. The temperature difference (=$T_H - T_L$) is 40 K. In the modified cell, V^{cal}_{cell} and Q^{cal}_{cell} are 39.6 mV and 1.2 mAh/g, respectively. The observed V_{cell} and Q_{cell} (Figure 8a) are close to the calculated values. In the conventional cell, V^{cal}_{cell} and Q^{cal}_{cell} are 39.6 mV and 1.7 mAh/g, respectively. The observed V_{cell} values (Figure 8b) are close to the calculated value (=39.6 mV). In the conventional cell (Figure 8b), Q_{cell} at T_H is much smaller than Q_{cell} at T_L even though their average is around the calculated value (=1.7 mAh/g). We note that the difference in Q_{cell} between T_H and T_L has a serious effect on the long-term thermal stability. The cathode (anode) material in a tertiary battery is reduced (oxidized) at T_H and oxidized (reduced) at T_L. Therefore, if Q_{cell} differs significantly, the Na^+ concentration (x) of the electrode materials will change as the cycles number increases.

4. Conclusions

We proposed a modified manufacturing method for the Co-PBA/Ni-PBA tertiary battery free from the pre-oxidation step. Especially, the capacity cyclability of the modified cell between T_L (=10 °C) and T_H (=50 °C) is better than that of the conventional one. The observed V_{cell} and Q_{cell} are reproduced by the calculation based on the electrode parameters. Our method is applicable to other electrode materials and enables the manufacturing of tertiary batteries in the same steps as the commercialized LIBs. From the viewpoint of the manufacturing process and manufacturing costs, the tertiary battery is considered to be one of the promising energy harvesting devices.

Author Contributions: Conceptualization, supervision, and writing, Y.M., investigation, M.S., data curation, H.I. and I.N. All authors have read and agreed to the published version of the manuscript.

Funding: This research was funded by the Yazaki Memorial Foundation for Science and Technology.

Data Availability Statement: The data presented in this study are available on request from the corresponding author.

Acknowledgments: The CNH analyses were outsourced to the Chemical Analysis Division, Research Facility Center for Science and Engineering, University of Tsukuba.

Conflicts of Interest: The authors declare no conflict of interest.

References

1. Lee, S.W.; Yang, Y.; Lee, H.-W.; Ghasemi, H.; Kraemer, D.; Chen, G.; Cui, Y. An electrochemical system for efficiently harvesting low-grade heat energy. *Nat. Commun.* **2014**, *5*, 3942. [CrossRef] [PubMed]
2. Yang, Y.; Lee, S.W.; Ghasemi, H.; Loomis, J.; Li, X.; Kraemer, D.; Zheng, G.; Cui, Y.; Chen, G. Charging-free electrochemical system for harvesting low-grade thermal energy. *Proc. Natl. Acad. Sci. USA* **2014**, *111*, 17011–17016. [CrossRef] [PubMed]
3. Wang, J.; Feng, S.-P.; Yang, Y.; Hau, N.Y.; Munro, M.; Ferreira-Yang, E.; Chen, G. Thermal charging phenomenon in electrical double layer capacitors. *Nano Lett.* **2015**, *15*, 5784–5790. [CrossRef]

4. Shibata, T.; Fukuzumi, Y.; Kobayashi, W.; Moritomo, Y. Thermal power generation during heat cycle near room temperature. *Appl. Phys. Express* **2018**, *11*, 017101. [CrossRef]
5. Takahara, I.; Shibata, T.; Fukuzumi, Y.; Moritomo, Y. Improved thermal cyclability of tertiary battery made of Prussian blue analogues. *ChemistrySelect* **2018**, *4*, 8558–8563. [CrossRef]
6. Nagai, I.; Shimaura, Y.; Shibata, T.; Moritomo, Y. Performance of tertiary battery made of Prussian blue analogues. *Appl. Phys. Express* **2021**, *14*, 094004. [CrossRef]
7. Shimaura, Y.; Shibata, T.; Moritomo, Y. Interrelation between discharge capacity and charge coefficient of redox potential in tertiary batteries made of transition metal hexacyanoferrate. *Jpn. J. Appl. Phys.* **2022**, *61*, 044004. [CrossRef]
8. Shibata, T.; Nakamura, K.; Nozaki, S.; Iwaizumi, H.; Ohnuki, H.; Moritomo, Y. Optimization of electrode parameters of $Na_xCo[Fe(CN)_6]_{0.88}/Na_xCd[Fe(CN)_6]_{0.99}$ tertiary battery. *Sustain. Mater. Technol.* **2022**, *33*, e00483.
9. Imanishi, N.; Morikawa, T.; Kondo, J.; Takeda, Y.; Yamamoto, O.; Kinugasa, N.; Yamagishi, T. Lithium intercalation behavior into iron cyanide complex as positive electrode of lithium secondary battery. *J. Power Sources* **1999**, *79*, 215–219. [CrossRef]
10. Imanishi, N.; Morikawa, T.; Kondo, J.; Yamane, R.; Takeda, Y.; Yamamoto, O.; Sakaebe, H.; Tabuchi, M. Lithium intercalation behavior of iron cyanometallates. *J. Power Sources* **1999**, *81–82*, 530–534. [CrossRef]
11. Okubo, M.; Asakura, D.; Mizuno, Y.; Kim, J.-D.; Mizokawa, T.; Kudo, T.; Honma, I. Switching redox-active sites by valence tautomerism in prussian blue analogues $A_xMn_y[Fe(CN)_6]nH_2O$ (A = K, Rb): Robust frameworks for reversible Li storage. *J. Phys. Chem. Lett.* **2010**, *1*, 2063–2071. [CrossRef]
12. Matsuda, T.; Moritomo, Y. Thin film electrode of Prussian blue analogue for Li-ion battery. *Appl. Phys. Express* **2010**, *4*, 047101. [CrossRef]
13. Takachi, M.; Matsuda, T.; Moritomo, Y. Structural, electronic, and electrochemical properties of $Li_xCo[Fe(CN)_6]_{0.90}2.9H_2O$. *Jpn. J. Appl. Phys.* **2013**, *52*, 044301. [CrossRef]
14. Moritomo, Y.; Takachi, M.; Kurihara, Y.; Matsuda, T. Thin Fflm electrodes of Prussian blue analogues with rapid Li^+ intercalation. *Appl. Phys. Express* **2012**, *5*, 041801. [CrossRef]
15. Lu, Y.; Wang, L.; Cheng, J.; Goodenough, J.B. Prussian blue: A new framework of electrode materials for sodium batteries. *Chem. Commun.* **2012**, *48*, 6544–6546. [CrossRef]
16. Matsuda, T.; Takachi, M.; Moritomo, Y. A sodium manganese ferrocyanide thin film for Na-ion batteries. *Chem. Commun.* **2013**, *49*, 2750–2752. [CrossRef] [PubMed]
17. Takachi, M.; Matsuda, T.; Moritomo, Y. Cobalt hexacyanoferrate as cathode material for Na^+ secondary battery. *Appl. Phys. Express* **2013**, *6*, 025802. [CrossRef]
18. Takachi, M.; Matsuda, T.; Moritomo, Y. Redox reactions in Prussian blue analogue films with fast Na^+ intercalation. *Jpn. J. Appl. Phys.* **2013**, *52*, 090202. [CrossRef]
19. Yang, D.; Xu, J.; Liao, X.-Z.; He, Y.-S.; Liu, H.; Ma, Z.-F. Structure optimization of Prussian blue analogue cathode materials for advanced sodium ion batteries. *Chem. Commum.* **2014**, *50*, 13377–13380. [CrossRef]
20. Lee, H.W.; Wang, R.Y.; Pasta, M.; Lee, S.W.; Liu, N.; Cui, Y. Manganese hexacyanomanganate open framework as a high-capacity positive electrode material for sodium-ion batteries. *Nat. Commun.* **2014**, *5*, 5280. [CrossRef]
21. Wang, L.; Song, J.; Qiao, R.Q.; Wray, L.A.; Hossain, M.A.; Chuang, Y.-D.; Yang, W.; Lu, Y.; Evans, D.; Lee, J.-J.; et al. Rhombohedral Prussian white as cathode for rechargeable sodium-ion batteries. *J. Am. Chem. Soc.* **2015**, *137*, 2548–2554. [CrossRef] [PubMed]
22. You, Y.; Wu, X.-L.; Yin, Y.-X.; Guo, Y.-G.; You, Y.; Wu, X.-L.; Yin, Y.-X.; Guo, Y.-G. A zero-strain insertion cathode material of nickel ferricyanide for sodium-ion batteries. *J. Mater. Chem. A* **2013**, *1*, 14061–14065. [CrossRef]
23. Yu, S.; Li, Y.; Lu, Y.; Xu, B.; Wang, Q.; Yan, M.; Jiang, Y.A. A promising cathode material of sodium iron–nickel hexacyanoferrate for sodium ion batteries. *J. Power Sources* **2015**, *275*, 45–49. [CrossRef]
24. Xu, L.; Li, H.; Wu, X.; Shao, M.; Liu, S.; Wang, B.; Zhao, G.; Sheng, P.; Chen, X.; Han, Y.; et al. Well-defined $Na_2Zn_3[Fe(CN)_6]_2$ nanocrystals as a low-cost and cycle-stable cathode material for Na-ion batteries. *Electrochem. Commun.* **2019**, *98*, 78–81. [CrossRef]
25. Peng, J.; Zhang, W.; Liu, Q.; Wamg, J.; Chou, S.; Liu, H.; Dou, S. Prussian Blue Analogues for Sodium-Ion Batteries: Past, Present, and Future. *Adv. Mater.* **2022**, *34*, 2108384. [CrossRef]
26. Geng, W.; Zhang, Z.; Yang, Z.; Tang, H.; He, G. Non-aqueous synthesis of high-quality Prussian blue analogues for Na-ion batteries. *Chem. Commun.* **2022**, *58*, 4472–4475. [CrossRef]
27. Liu, X.; Cao, Y.; Sun, J. Defect Engineering in Prussian Blue Analogs for High-Performance Sodium-Ion Batteries. *Adv. Energy Mater.* **2022**, *12*, 2205232. [CrossRef]
28. Zhang, H.; Peng, J.; Li, L.; Zhao, Y.; Guo, Y.; Wang, J.; Cao, Y.; Dou, S.; Chou, S. Low-Cost Zinc Substitution of Iron-Based Prussian Blue Analogs as Long Lifespan Cathode Materials for Fast Charging Sodium-Ion Batteries. *Adv. Funct. Mater.* **2023**, *33*, 2210725. [CrossRef]
29. Wang, W.; Gang, Y.; Peng, J.; Hu, Z.; Yan, Z.; Lai, W.; Ahu, Y.; Appadoo, D.; Ye, M.; Cao, Y.; et al. Effect of Eliminating Water in Prussian Blue Cathode for Sodium-Ion Batteries. *Adv. Funct. Mater.* **2022**, *32*, 2111721 [CrossRef]
30. He, M.; Davis, R.; Chartouni, D.; Johnson, M.; Abplanalp, M.; Troendle, P.; Suetterlin, R.-P. Assessment of the first commercial Prussian blue-based sodium-ion battery. *J. Power Sources* **2023**, *548*, 232036. [CrossRef]
31. Wu, C.; Hu, J.; Chen, H.; Zhang, C.; Xu, M.; Zhuang, L.; Ai, X.; Qian, J. Chemical lithiation methodology enabled Prussian blue as a Li-rich cathode material for secondary Li-ion batteries. *Energy Storage Mater.* **2023**, *60*, 102803. [CrossRef]

32. Wang, P.; Li, Y.; Zhu, D.; Gong, F.; Fang, S.; Zhang, Y.; Sun, S. Treatment dependent sodium-rich Prussian blue as a cathode material for sodium-ion batteries. *Dalton Trans.* **2023**, *51*, 9622–9626. [CrossRef] [PubMed]
33. Naskar, P.; Debnath, S.; Biplab, B.; Laha, S.; Benerjee, A. High-Performance and Scalable Aqueous Na-Ion Batteries Comprising a Co-Prussian Blue Analogue Framework Positive Electrode and Sodium Vanadate Nanorod Negative Electrode for Solar Energy Storage. *ACS Appl. Energy Mater.* **2023**, *6*, 4604–4617. [CrossRef]
34. Liu, X.; Gong, H.; Han, C.; Cao, Y.; Li, Y.; Sun, J. Barium ions act as defenders to prevent water from entering Prussian blue lattice for sodium-ion battery. *Energy Storage Mater.* **2023**, *57*, 118–124. [CrossRef]
35. Buser, H.J.; Schwarzenbach, D.; Petter, W.; Ludi, A. The crystal structure of Prussian Blue: $F_4[Fe(CN)_6]_3 \cdot xH_2O$. *Inorg. Chem.* **1977**, *16*, 2704–2710. [CrossRef]
36. Herren, F.; Fischer, P.; Ludi, A.; Halg, W. Neutron diffraction study of Prussian Blue, $Fe_4[Fe(CN)_6]_3xH_2O$. Location of water molecules and long-range magnetic order. *Inorg. Chem.* **1980**, *19*, 956–1959. [CrossRef]
37. Niwa, H.; Kobayashi, W.; Shibata, T.; Nitani, H.; Moritomo, Y. Invariant nature of substituted element in metal-hexacyanoferrate. *Sci. Rep.* **2017**, *7*, 13225. [CrossRef] [PubMed]
38. Moritomo, Y.; Yoshida, Y.; Inoue, D.; Iwaizumi, H.; Kobayashi, S.; Kawaguchi, S.; Shibata, T. Origin of the material dependence of the temperature coefficient of the redox potential in coordination polymers. *J. Phys. Soc. Jpn.* **2021**, *90*, 063801. [CrossRef]
39. Izumi, F.; Momma, K. Three-dimensional visualization in powder diffraction. *Solid State Phenom.* **2007**, *130*, 15–20. [CrossRef]

Disclaimer/Publisher's Note: The statements, opinions and data contained in all publications are solely those of the individual author(s) and contributor(s) and not of MDPI and/or the editor(s). MDPI and/or the editor(s) disclaim responsibility for any injury to people or property resulting from any ideas, methods, instructions or products referred to in the content.

MDPI
St. Alban-Anlage 66
4052 Basel
Switzerland
www.mdpi.com

MDPI Books Editorial Office
E-mail: books@mdpi.com
www.mdpi.com/books

Disclaimer/Publisher's Note: The statements, opinions and data contained in all publications are solely those of the individual author(s) and contributor(s) and not of MDPI and/or the editor(s). MDPI and/or the editor(s) disclaim responsibility for any injury to people or property resulting from any ideas, methods, instructions or products referred to in the content.

www.ingramcontent.com/pod-product-compliance
Lightning Source LLC
LaVergne TN
LVHW070404100526
838202LV00014B/1389